Studies in Fuzziness and Soft Computing 287

Editor-in-Chief

Prof. Janusz Kacprzyk
Systems Research Institute
Polish Academy of Sciences
ul. Newelska 6
01-447 Warsaw
Poland
E-mail: kacprzyk@ibspan.waw.pl

For further volumes:
http://www.springer.com/series/2941

Anna M. Gil-Lafuente, Jaime Gil-Lafuente,
and José M. Merigó-Lindahl (Eds.)

Soft Computing in Management and Business Economics

Volume 2

Springer

Editors
Anna M. Gil-Lafuente
Department of Business Administration
University of Barcelona
Barcelona
Spain

José M. Merigó-Lindahl
Department of Business Administration
University of Barcelona
Barcelona
Spain

Jaime Gil-Lafuente
Department of Business Administration
University of Barcelona
Barcelona
Spain

ISSN 1434-9922
e-ISSN 1860-0808
ISBN 978-3-642-43718-2
ISBN 978-3-642-30451-4 (eBook)
DOI 10.1007/978-3-642-30451-4
Springer Heidelberg New York Dordrecht London

Printed on acid-free paper

Springer is part of Springer Science+Business Media (www.springer.com)

Preface

The European Academy of Management and Business Economics (AEDEM) and the University of Barcelona are pleased to present the main results of the XXVI Annual Conference, held in Barcelona, 5–7 June, 2012, through this Book of Proceedings divided in two volumes published in the Springer Series "*Studies in Fuzziness and Soft Computing*".

AEDEM 2012 Barcelona is co-organized by the AEDEM Academy and the University of Barcelona, Spain. It is co-supported by the Spanish Royal Academy of Financial and Economic Sciences and the *Obra Social "La Caixa"*. It offers a unique opportunity for researchers, professionals and students to present and exchange ideas concerning management and business economics and see its implications in the real world.

In this edition of the AEDEM Annual Conference, we have presented the slogan "Creating new opportunities in an uncertain environment" being conscious of the importance that new technologies represent for dealing with uncertainty. There are different ways for assessing uncertainty in management but in this book we want to departure from the theories concerning soft computing that are fundamental for understanding our complex world. Thus, the title of this book is "*Soft Computing in Management and Business Economics*".

The increasing importance of Soft Computing in Management and Business Economics and more generally in all the Social Sciences, is giving a lot of new opportunities to researchers in these areas to find better solutions that explain the complexities of our world. By using these techniques researchers and practitioners can develop more efficient models that adapt better to the market and permit to maximize the benefits or minimize the costs in a more appropriate way.

AEDEM 2012 Proceedings is constituted by 56 papers from a total of 200 papers presented at the conference. Volume I contains 30 papers and Volume II, 26 papers. The book is divided in 6 parts: (1) Management, (2) Marketing, (3) Business Statistics, (4) Innovation and Technology, (5) Finance, (6) Sports and Tourism. The first three parts are in Volume I while the other three parts are in Volume II.

We would like to thank all the contributors, referees and the scientific and honorary committees for their kind co-operation with AEDEM 2012 Barcelona; to

Jaime Gil Aluja for his leading role at the AEDEM; to Encarna González, President of AEDEM for her support in the organization of the conference; to the whole team of the organizing committee, including Victor Alfaro, Lluis Amiguet, Sefa Boria, Aras Keropyan, Anna Klimova, Salvador Linares, Pilar López-Jurado, Carolina Luis, Carles Mulet, Josep Pons, Sabria Saidj, Mel Solé, Josep Torres Prunyonosa and Emilio Vizuete; and to Leontina Di Cecco (Editor, Springer) for her kind advise and help to publish this volume. Finally, we would like to express our gratitude to Springer and in particular to Janusz Kacprzyk (editor-in-chief of the book series "*Studies in Fuzziness and Soft Computing*") for his support in the preparation of this book.

March 2012
Barcelona

Anna M. Gil-Lafuente, AEDEM 2012 Barcelona Chair
Jaime Gil-Lafuente, AEDEM 2012 Barcelona Co-chair
José M. Merigó-Lindahl, AEDEM 2012 Barcelona Co-chair

The AEDEM 2012 Annual Conference is supported by:

Honorary Committee

Special thanks to the members of the Honorary Committee for their support in the organization of the AEDEM 2012 Annual Conference.

Isidre Fainé Casas	President of "*La Caixa*" and Member of the *Spanish Royal Academy of Financial and Economic Sciences*
Jaime Gil-Aluja	President of the *Spanish Royal Academy of Financial and Economic Sciences*, President of *AMSE* and President of *SIGEF*
Fynn E. Kydland	*Nobel Prize in Economic Sciences*, *University of California at Santa Barbara*

Scientific Committee

Thanks to all the members of the Scientific Committee for their kind support in the organization of the AEDEM 2012 Annual Conference.

Anna M. Gil-Lafuente
(President)

Janusz Kacprzyk
(President)

Jaime Gil-Lafuente
(Vice-President)

José M. Merigó-Lindahl
(Vice-President)

Glòria Barberà
Jon Barrutia Güenaga
Carmen Barroso Castro
Enrique Bigné Alcañiz
Michel Desbordes
Luis Tomás Díez de Castro
Matilde Fernández Blanco
Joao Ferreira
Joan Carles Ferrer Comalat
Llorenç Gascón
Federico González Santoyo
Encarnación González Vázquez
Montserrat Guillén Estany
Manuel Guisado Tato
Leonid Klimenko
Anna Klimova

Pilar Laguna Sánchez
Jaume Lanaspa Gatnau
Jon Landeta Rodríguez
Antonio Leal Millán
Vicente Liern Carrión
Lionel Maltese
Enrique Martín Armario
Carmelo Mercado Idoeta
José Antonio Redondo López
Arturo Rodríguez Castellanos
Antonio Terceño Gómez
Gary Tribou
Emilio Vizuete Luciano
Volodymir Yemelyanov
Constantin Zopounidis

Scientific Committee

Organizing Committee

Special thanks to all the members of the Organizing Committee for their support during the preparation of the AEDEM 2012 Annual Conference.

Chair of the Organizing Committee

Anna M. Gil-Lafuente	President of AEDEM 2012 Barcelona

Co-chair of the Organizing Committee

Jaime Gil-Lafuente	Vicepresident of AEDEM 2012 Barcelona
José M. Merigó-Lindahl	Vicepresident of AEDEM 2012 Barcelona

Organizing Committee

Victor Alfaro, Mexico
Lluis Amiguet Molina, Spain
Sefa Boria, Spain
Aras Keropyan, Turkey
Anna Klimova, Russia
Salvador Linares, Spain
Pilar López-Jurado, Spain
Carolina Luis Bassa, Spain
Carles Mulet, Spain
Josep Pons, Spain
Sabria Saidj, Algeria
Mel Solé, Spain
Emilio Vizuete Luciano, Spain

Acknowledgement to Reviewers

We thank all the referees for their advice in the revision process of the papers submitted to the AEDEM 2012 Annual Conference at Barcelona:

Ramón Alemany
Luciano Barcelos
Sefa Boria
Francisco J. Callado
Francisco J. Cossío
María Rosa Cruz
Jorge de Andrés
César Castillo
Cristina Estevao
Eugenio M. Fedrianni
Santiago Forgas
Ana Gessa
Anna M. Gil-Lafuente
Jaime Gil-Lafuente
Ana Maria Godeanu
Montserrat Guillén
Milagros Gutiérrez
Rubén Huertas
Txomin Iturralde
Nadia Jiménez
Aras Keropyan
Anna Klimova
Paula Lima Ribeiro
Salvador Linares
Carolina Luis
Natalia Medrano
José M. Merigó
Alvaro F. Moncada
Inmaculada Piédrola
Camilo Prado Román
Berta Quintana
María Ángeles Revilla
Leire San José
María Luisa Solé
David Urbano
Teresa J. de Vargas
Juan J. de la Vega
Estrella Vidal
Emilio Vizuete

Contents

Innovation and Technology

Bibliometric Analysis of Business and Economics in the Web of Science

José M. Merigó-Lindahl

Department of Business Administration, University of Barcelona,
Av. Diagonal 690, 08034 Barcelona, Spain
jmerigo@ub.edu

Abstract. We present a general overview of the most influential results found in the Web of Science in the subject area of Business & Economics that includes the categories of Business, Economics, Business Finance and Management. We analyse the most cited papers in the history and rank the most influential institutions by number of papers published. We analyse the most relevant journals, the temporal evolution and the countries with the highest number of publications. We also develop a similar analysis to the Spanish case studying the most cited papers, the most influential institutions and the temporal evolution. Note that this study is only based on the results found on the Web of Science with the objective of giving a general overview of the research done in Business & Economics especially over the last half century. However, many exceptions and particularities may be found throughout the results.

Keywords: Bibliometrics, Web of Science, Business, Economics, Management.

1 Introduction

The Web of Science is a database system that includes all the papers published in journals and conference proceedings indexed in this database. The journals and conference proceedings included in the Web of Science are those journals that are recognized to have the highest quality and accomplish a series of criteria required in order to be a Top journal. It permits to analyse any material available there including a specific topic, country or journal. It encompasses all the known sciences of the World including mathematics, physics, chemistry, medicine and economics. The general database that contains the Web of Science is the Web of Knowledge recently acquired by the Thomson & Reuters Group. It includes other databases such as the Journal Citation Reports that analyses the quality of the journals, the Essential Science Indicators that analyses the key papers, authors, countries and institutions over the last ten years and many others.

This database permits to detect the research that has been more significant worldwide according to citation count. The evaluation of the citation count is subjective because some areas have more journals or tend to cite more than others.

A.M. Gil-Lafuente et al. (Eds.): Soft Comput. in Manag. and Bus. Econ., STUDFUZZ 287, pp. 3–17.
springerlink.com

Therefore, it is difficult to make final conclusions with this information. However, it clearly permits us to identify key research and some main trends in the scientific community. Thus, it is very useful to provide a general overview of the key research done in a specific area. When analysing research fields, there are different ways for classifying them. One common way is by searching some keywords related to the research field that we are interested in. Another way that can be done consists in following the classification provided by the Web of Science where they classify science in more than 200 categories. Some of these categories are grouped in one subject area that includes all of them. This is very common in key areas such as mathematics where we have several categories that are all grouped in the subject area mathematics. In Social Sciences, we also find this situation. For example, Business & Economics is the global subject area that includes all the research done in this field. It includes business, business finance, economics and management. However, it is worth noting that all these categories are usually interrelated finding key papers in one area published in a journal from another area. This is very common because often the journals have a wide range of topics that includes several categories. The Web of Science also recognizes this issue by classifying a journal in several categories. In the case of Business & Economics this is very common when analysing for example, Mathematics, Statistics, Computer Science, Operations Research, Politics, Sociology and Law.

Assuming all these issues, in this paper we want to give a general overview of the key research done in Business & Economics analysing the most cited papers of the history of the Web of Science, the most cited institutions and the most cited journals. Furthermore, we also develop a similar analysis for the Spanish case analysing the most cited papers and the most cited institutions. We will see that most of the results found are those that we could intuitively imagine as they have been classically recognized as the most relevant ones. However, it will be very useful for giving us a general picture of the key research done in this area.

2 Most Cited Articles in the Web of Science

In this Section, we analyse the most cited papers in the history of the Web of Science in Business & Economics. Note that as mentioned before, this subject area is divided in four categories: Business, Business Finance, Economics and Management. Thus, in further studies it could be interesting to analyse these results by categories instead of the whole subject area. The reason for our methodology is that we want to give a general overview independently on the specific topic that it comes from. Today, 10-02-2012, we find about 1.100.000 papers in the Web of Science in this subject area with 575.000 papers in the category of Economics and the rest in the other more business oriented categories. It is also worth noting that of the 1.100.000 papers, about 600.000 papers are strictly articles while the other ones are book reviews, proceedings and other related material. In Table 1 we present the 30 most cited papers in the history of the Web of Science.

Table 1 Top 30 papers in the Web of Science in Business & Economics

	Author	Journal	Times Cited
1-	Kahneman & Tversky (1979)	Economet.	7424
2-	Jensen & Meckling (1976)	JFE	6490
3-	White (1980)	Economet.	6235
4-	Ajzen (1991)	OBHDP	5865
5-	Barney (1991)	JM	5551
6-	Engle & Granger (1987)	Economet.	4934
7-	Heckmann (1987)	Economet.	4874
8-	Black & Scholes (1973)	JPE	4763
9-	Fornell & Larcker (1981)	JMKR	4712
10-	Chen & Levinthal (1990)	ASQ	4511
11-	Eisenhardt (1989)	AMR	4094
12-	Engle (1982)	Economet.	3693
13-	Coase (1960)	JLE	3615
14-	Charnes, Cooper & Rhodes (1978)	EJOR	3602
15-	Teece, Pisano & Shuen (1997)	SMJ	3228
16-	Davis (1989)	MISQ	3126
17-	Bollerslev (1986)	JEconomet.	2967
18-	Hausman (1978)	Economet.	2964
19-	Johansen (1988)	JEDC	2959
20-	Wernerfelt (1984)	SMJ	2928
21-	Akerlof (1970)	QJE	2925
22-	Newey & West (1987)	Economet.	2828
23-	Tiebout (1956)	JPE	2796
24-	Romer (1986)	JPE	2792
25-	Lucas (1988)	JME	2773
26-	Alchian & Demsetz (1972)	AER	2493
27-	Prahalad & Hamel (1990)	HBR	2462
28-	Hansen (1982)	Economet.	2425
29-	Armstrong & Overton (1977)	JMKR	2409
30-	Jensen (1986)	AER	2403

Abbreviations: AER: American Economic Review; AMR: Academy of Management Review; ASQ: Administrative Science Quarterly; Economet: Econometrica; EJOR: European Journal of Operational Research; HBR: Harvard Business Review; JEconomet: Journal of Econometrics; JEDC: Journal of Economic Dynamics & Control; JFE: Journal of Financial Economics; JLE: Journal of Law & Economics; JM: Journal of Management; JME: Journal of Monetary Economics; JMKR: Journal of Marketing Research; JPE: Journal of Political Economy; MISQ: MIS Quarterly; OBHDP: Organizational Behaviour and Human Decision Processes; QJE: Quarterly Journal of Economics; SMJ: Strategic Management Journal.

As we can, see, the classical paper of Daniel Kahneman and Amos Tversky published in 1979 that gave Prof. Kahneman the Nobel Prize in Economics in 2002, is the most cited paper in Business & Economics. Obviously the rest of Top 30 papers are classical well-known papers by the scientific community. It is worth noting that Econometrica is the journal with the most number of papers in this list with 8 papers. Next, we have the Journal of Political Economy with 3 papers and the Strategic Management Journal, the Journal of Marketing Research and the American Economic Review with 2 papers.

Regarding authors, we see that Michael Jensen and Robert Engle have two papers in the list being the only authors with more than one paper. Note that there are a lot of other very good papers that are not included in this list. By showing the 30 papers most cited in the Web of Science, what we are indicating is those papers that have become more popular according to the publications available in the Web of Science. However, we can always find exceptions because in one topic there are not many journals, it is not popular to cite so much and so on.

A further interesting issue is to analyse the evolution of the number of papers published yearly in the Web of Science. The results are shown in Figure 1.

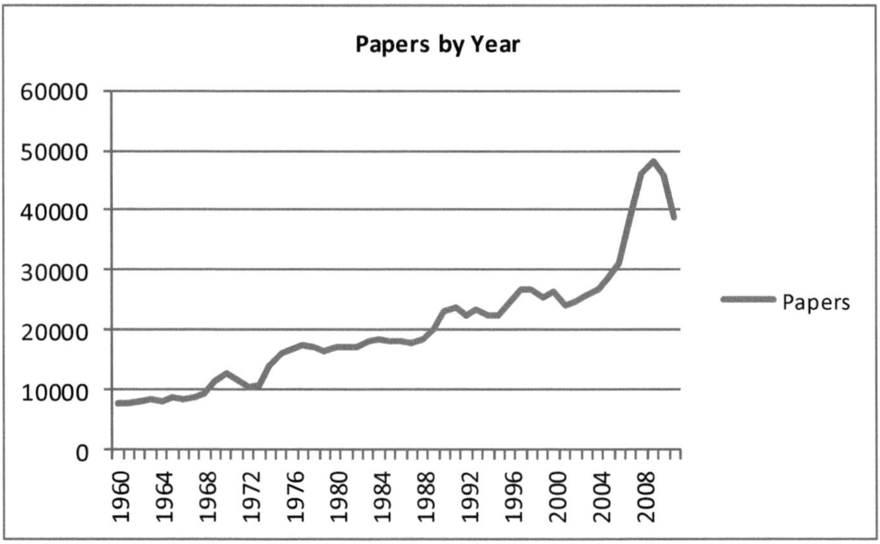

Fig. 1 Number of papers by year

As we can see, there is an increasing trend that has been occurring over the last 50 years. There are several reasons for this including the increase in the number of journals, institutions and researchers. Note that for the year 2011 the results are not complete since some proceedings papers plus some delayed journals are still missing in the publication record. Note that in 1960 the Web of Science was counting about 10.000 papers in Business & Economics while know the number is close to 50.000 papers per year. With this situation, it becomes more and more difficult to control all the research done in a general field since the number of papers published is increasing constantly and at a very high level. Thus, it is clear that the good researchers need to be conscious on this adapting their research profile becoming more specialized in some particular area.

3 Institutions with the Highest Number of Papers

A further interesting issue when making a bibliometric analysis is to analyse the best institutions in the specific field of research considered. In order to do so, we analyse the Top 30 institutions in the World by number of publications in the Web of Science. Note that it is more common to consider the ranking by the number of citations. However, in order to give a different perspective, we have done it in this way. Apart from the number of papers and citations, we also consider the papers published in the last 10 years and the general H-index of the institution. Note that the H-index (Hirsch, 2005) is a measure for evaluating the total quality of a group of papers, either for a researcher, journal, institution or country. It consists in finding the connecting point between the number of papers and number of citations. Thus, for example, if a group of papers have an H-index of 30, it means that in this group of papers there are 30 papers with at least 30 citations each. If the H-index is 31, then, in this group of papers there are 31 papers with at least 31 citations each, and so on. The results are shown in Table 2.

As we can see, most of the universities are from the United States of America (USA) with the exception of the London School of Economics and Political Science, the University of Manchester and the University of Toronto. As expected, Harvard University is the university with the highest number of papers and citations in Business & Economics in the Web of Science. It also gets the first position according to the papers published in the last 10 years and the H-index. Although by number of papers Illinois, Texas and Wisconsin are the in the next positions, we see that by citations and by the H-index Chicago, Massachusetts Institute of Technology (MIT), Stanford and the University of Pennsylvania are the universities that complete the Top 5. In general, according to the number of citations and the H-index, most of the universities in the Top positions are those usually regarded as the best universities in Business & Economics. However, according to the number of papers we can see some differences as mentioned before.

Note that one of the key reasons for having mainly English-speaking universities is that most of the journals in the Web of Science are published in English. Therefore, those countries that do not speak English have a strong disadvantage in this context. However, it is worth noting that in the last years the

Table 2 Ranking of the Top 30 Institutions

	University	Tot. Art.	Tot. Cit.	2002-2011	H-index
1-	Harvard Univ.	11364	342564	3796	241
2-	Univ. Illinois	8331	112004	2276	129
3-	Univ Texas	8299	127003	3289	134
4-	Univ. Wisconsin	8161	126812	1991	146
5-	Univ. Penn	8072	231672	2397	193
6-	MIT	7066	254159	1975	213
7-	Stanford Univ.	6951	234391	2208	204
8-	UC – Berkeley	6774	165051	2191	166
9-	Univ. Michigan	6633	152831	2303	158
10-	Columbia Univ.	6401	144528	1980	157
11-	NYU	6329	126223	2128	148
12-	Cornell Univ.	6307	95736	2041	130
13-	Univ. Chicago	6059	267151	1805	231
14-	London Sch. Econ.	5996	73989	2142	109
15-	Northwestern U.	5577	161865	1734	170
16-	Univ. Maryland	5517	98232	1915	132
17-	Univ. Minnesota	5316	115749	1668	142
18-	Ohio State Univ.	5151	80762	1452	118
19-	Indiana Univ.	5147	72841	1396	113
20-	Michigan State U.	5134	71682	1628	112
21-	Penn State Univ.	5064	76049	1586	116
22-	Univ. Manchester	4869	27273	1539	59
23-	U. North Carolina	4813	69674	1733	108
24-	Yale Univ.	4490	114773	1536	145
25-	UC – Los Angeles	4489	107974	1464	142
26-	Univ. Toronto	4161	49696	1693	90
27-	U. Southern Calif.	4097	70264	1432	114
28-	Duke Univ.	4007	80440	1669	123
29-	Rutgers State U.	3999	46970	1231	88
30-	Purdue Univ.	3970	58392	1131	99

non-English-speaking universities are clearly promoting a policy of improving their knowledge in English and making a strong investment in research. However, there are still huge differences between English and non-English-speaking universities in the area of Business & Economics although the differences are being reduced over the years. Moreover, the Web of Science has developed a new regional policy that will include a wide range of non-English-speaking journals in the Web of Science so all the well-known languages in the World will have at least some journals indexed in the Web of Science.

4 Analysis of the Top Journals in Business and Economics

In this section we develop a bibliometric analysis of the journals commonly regarded as the Top journals in their fields. Note that we have made a subjective selection according to the number of citations. However, it is worth noting that there are other journals that could be included in this list according to the number of citations. Our objective with this analysis is to provide a general overview of the number of citations and papers of most of the top journals in Business & Economics. The results are shown in Table 3.

Table 3 Ranking of the 25 Top Journals in Business, Finance and Management

	Journal	Total Articles	Total Citations	Art. 2010	Cit. 2010	H-index
1-	Management Science	5747	188255	140	18787	175
2-	J. Finance	7080	171896	69	17621	173
3-	Acad. Management J.	2966	156959	63	17239	182
4-	Administ. Science Quart.	3419	144671	12	11539	195
5-	Strategic Management J.	1927	135856	74	15626	172
6-	Acad. Management Rev.	2050	127748	27	15782	184
7-	Operations Research	8067	126494	127	8250	135
8-	J. Financial Economics	1900	118850	100	11815	158
9-	J. Marketing	4286	114864	49	13189	161
10-	J. Marketing Research	3657	106214	90	9586	138
11-	J. Consumer Research	1852	94306	71	8508	138
12-	Harvard Bus. Review	12488	75619	119	9000	121
13-	Organization Science	1109	65690	71	9120	118
14-	Research Policy	2342	55984	112	7539	105
15-	Org.Beh.&Human Dec. Pr.	1524	55569	37	6391	93
16-	J. Management	1323	53937	56	7184	103
17-	MIS Quarterly	913	52695	38	7419	117
18-	Human Relations	3107	51864	79	4234	87
19-	J. Internat. Bus. Studies	1840	45652	81	6307	100
20-	J. Organizational Behavior	1416	35086	49	4747	86
21-	J. Management Studies	2885	34866	60	4457	79
22-	Rev. Financial Studies	1166	32462	119	4958	87
23-	J. Business Research	3060	30347	185	5141	60
24-	J. Accounting Research	1778	29033	34	3400	72
25-	Accounting Review	6105	27980	72	3569	65

Table 4 Ranking of the 25 Top Journals in Economics

	Journal	Tot. Art.	Tot. Cit.	Art. 2010	Cit. 2010	H-index
1-	American Economic Rev.	11971	333740	215	25786	232
2-	Econometrica	6896	282454	65	19858	220
3-	J. Political Economy	4874	219252	25	15535	205
4-	Quarterly J. Economics	2732	143294	44	13983	174
5-	Rev. Economics and Stat.	4562	94479	78	7462	114
6-	Rev. Economic Studies	2417	86847	50	6831	127
7-	J. Econometrics	3024	79111	139	8246	121
8-	J. Economic Theory	3605	78823	109	4884	113
9-	Economic Journal	10081	75543	78	7039	100
10-	American J. Agric. Econ.	13687	60010	93	4256	74
11-	J. Monetary Economics	2204	58903	71	4822	103
12-	J. Economic Literature	7370	56813	19	4846	136
13-	World Development	4824	54750	140	5687	81
14-	J. Economic Perspectives	1435	52488	49	5399	120
15-	J. Public Economics	2896	49608	93	4799	84
16-	J. Law & Economics	1216	44153	16	2911	86
17-	European Economic Rev.	3132	42604	63	3535	80
18-	J. International Economics	2553	40231	66	3666	83
19-	Rand Journal of Econ.	1179	38979	34	3521	94
20-	J. Human Resources	1885	33815	33	2718	75
21-	International Econ. Rev.	1997	33101	48	2985	75
22-	J. Health Economics	1408	30764	73	3381	71
23-	J. Development Economics	2583	28801	72	2847	69
24-	J. Urban Economics	1796	28801	55	2554	65
25-	J. Bus. Economics & Stat.	1439	27723	40	2616	77

As we can see, in this case the results are also clear being the Top 4 journals the American Economic Review, Econometrica, the Journal of Political Economy and the Quarterly Journal of Economics. The rest of well-known economic journals are also in very good positions such as the Review of Economics & Statistics, the Review of Economic Studies, the Journal of Econometrics, the Journal of Economic Theory, the Economic Journal, the Journal of Monetary Economics, the Journal of Economic Literature and the Journal of Economic Perspectives.

Finally, it is also worth noting that there are many other journals with relation to Business & Economics that are not included here because usually their main target is a different one such as the Journal of the American Statistical Association, the Annals of Statistics, the European Journal of Operational Research and the American Journal of Political Science.

5 Bibliometric Analysis of Spanish Institutions

In this section we are going to consider the Spanish results in the Web of Science in the field of Business & Economics. Note that the Spanish case considers only Spanish institutions. That is, any paper where at least one author was affiliated in the paper to a Spanish institution. This is important because many of the best Spanish authors are working outside Spain, especially in the USA and in the United Kingdom (UK). Therefore, all these publications do not count to a Spanish institution although it was written by a Spanish author. Having this restriction in mind, we are going to present the most cited papers in Business & Economics where at least one Spanish institution participated. The results are presented in Table 5.

Table 5 Top 20 papers published by a Spanish institution

	Author	Journal	Times Cited
1-	Arellano & Bover (1990)	JEconomet.	862
2-	Clarida, Galí & Gertler (1999)	JEL	843
3-	Clarida, Galí & Gertler (2000)	QJE	658
4-	Galí & Gertler (1999)	JME	505
5-	Ciccone & Hall (1996)	AER	430
6-	Jarillo (1988)	SMJ	414
7-	Sala-i-Martín (1997)	AER	373
8-	Cooke, Uranga & Etxebarría (1997)	RP	260
9-	Kremers, Ericsson & Dolado (1992)	OBES	259
10-	Harvey, Ruiz, Shephard (1994)	RES	249
11-	Boisot & Child (1996)	ASQ	248
12-	Vives (1990)	JMathE	222
13-	Sala-i-Martín (1996)	EER	218
14-	Esteban & Ray (1994)	Economet.	212
15-	Galí, Gertler & López-Salido (2001)	EER	212
16-	Petrongolo & Pissarides (2001)	JEL	201
17-	Nagel (1995)	AER	197
18-	Sala-i-Martín, Doppelhofer & Miller (2004)	AER	196
19-	Trejos & Wright (1995)	JPE	191
20-	Van Doorslaer, Wagstaff, et al. (1997)	JHE	191

Abbreviations not included in Table 1: EER: European Economic Review; JEL: Journal of Economic Literature; JHE: Journal of Economics; JMathE: Journal of Mathematical Economics; OBES: Oxford Bulletin of Economics & Statistics; RES: Review of Economic Studies; RP: Research Policy.

As we can see, the paper by Arellano and Boyer from the Centro de Estudios Financieros y Monetarios (CEMFI) is the most cited paper and Jordi Galí from the Centre de Recerca en Economia Internacional (CREI) at Universitat Pompeu

Fabra (UPF) has the second, third and fourth most cited papers. In general, the most dominant author is Galí with four papers in the Top 20 and next, we have Sala-i-Martín with three papers.

Regarding journals, in this ranking we see that the American Economic Review has four papers and the Journal of Economic Literature and the European Economic Review two papers. In this case, we see a clear dominance in the area of economics while in Business, Finance and Management we only find 3 papers.

As mentioned before, these are the most cited papers published under the name of a Spanish institution but many other papers published when Spanish authors were in a foreign institution should also be included. For example, the most cited paper by a Spanish author is Arelllano and Bond (1991) with 2223 citations. But at the time of publication of the paper he was at the London School of Economics and Political Science. Therefore, this paper is not included in the list of papers published by a Spanish institution although Arellano is Spanish and now is working at the CEMFI in Madrid. Note that many other similar cases are found in the Web of Science. Furthermore, note that some papers have been published by Spanish institutions but none of the authors was Spanish. For example, the paper by Ciccone and Hall (1996) where Ciccone is Italian but he is working at the UPF.

Next, we are going to analyse the evolution over the years of the number of papers published by Spanish institutions. The results are shown in Figure 2.

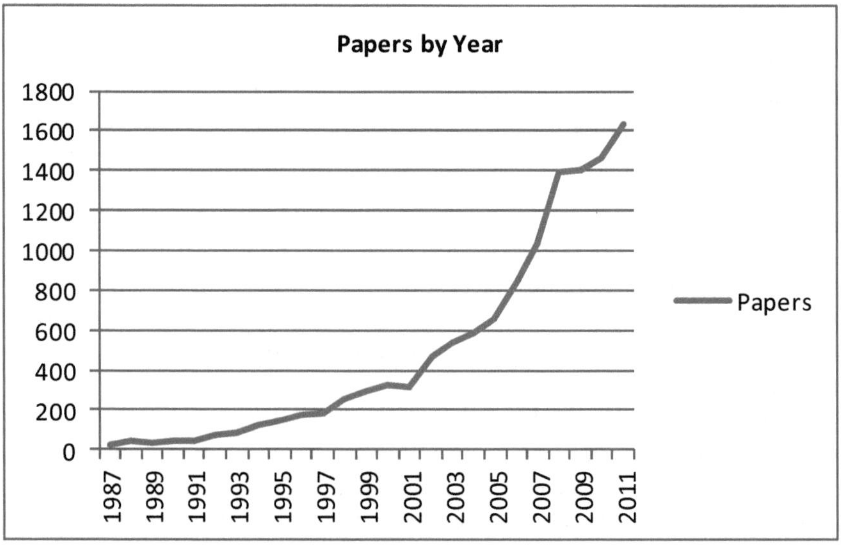

Fig. 2 Number of papers by year by Spanish institutions

As we can see, before the entrance of Spain in the European Union (EU) the number of publications in Business & Economics in the Web of Science was below 50 papers per year. Since the nineties, Spain has been experiencing a strong increase in research in all fields and this is also shown in this area. Since the end of the nineties, the publication record per year was above 200 papers and in 2005 it was already over 600 papers. The number of publications per year is still increasing at a very high level and today it is already over 1500 papers per year.

Note that there are several factors that explain this situation. The main reasons are the number of researchers in Spain that is increasing a lot and the regional expansion of the Web of Science to non-English-speaking journals that have given the opportunity to many Spanish journals to be included in the system. Today, the main Spanish journals in the area of Business & Economics are the SERIE - Journal of the Spanish Economic Association which represents the merger between *Investigaciones Económicas* and the Spanish Economic Review, the *Revista de Economía Aplicada*, *Revista de Historia Económica*, *Hacienda Pública Española*, *Universia Business Review*, *Cuadernos de Economía y Dirección de Empresas* and *Revista Española de Financiación y Contabilidad.* Note that there are other Spanish journals that only publish in English and are not included in this list such as the TOP and the SORT – Statistics and Operations Research Transactions.

Table 6 Top 20 Spanish institutions in Business & Economics

		Tot. Art.	Tot. Cit.	2002-11	H-index
1-	Univ. Carlos III	827	6092	645	33
2-	Univ. Autón. Barcelona	765	6343	529	37
3-	Univ. Valencia	751	3574	615	25
4-	Univ. Pompeu Fabra	727	11806	515	46
5-	Univ. Complut. Madrid	612	2338	442	24
6-	Univ. Barcelona	537	2274	465	21
7-	Univ. Zaragoza	432	1545	363	17
8-	Univ. Alicante	431	2169	315	22
9-	Univ. País Vasco	396	1840	286	19
10-	Univ. Oviedo	332	1545	285	20
11-	Univ. Navarra	321	3506	274	28
12-	Univ. Autónoma Madrid	320	1179	261	17
13-	Univ. Sevilla	296	1366	241	19
14-	Univ. Murcia	257	1089	232	18
15-	Univ. Granada	244	1852	221	19
16-	Univ. Pública Navarra	214	772	172	13
17-	Univ. Vigo	213	762	178	14
18-	Univ. Jaume I	197	943	180	16
19-	Univ. Málaga	194	630	166	13
20-	Univ. Politéc. Valencia	191	1071	178	16

Moreover, there are a lot of other Spanish-speaking journals mainly from Central and South America such as the *Revista INNOVAR* (Colombia), *Academia – Revista Latinoamericana de Administración* (Colombia) and the *Revista de Ciencias Sociales* (Venezuela). Furthermore, there are a lot of other Spanish journals that are in the process of entering the Web of Science such as the *Investigaciones Europeas de Dirección y Economía de la Empresa* and the *Revista Europea de Dirección y Economía de la Empresa* from the European Academy of Management and Business Economics (AEDEM).

Next, we are going to analyse the Spanish institutions with the highest number of papers in the Web of Science. The results are shown in Table 6.

As we can see, the usually regarded as the best institutions in Business & Economics are in the first positions. By the number of papers the University Carlos III of Madrid is in the first position while the University Pompeu Fabra of Barcelona leads the ranking by the number of citations and the H-index.

6 Conclusions

We have studied some of the most relevant research in Business & Economics found in the Web of Science. First we have analysed the most cited papers in the history and we have found some of the classical papers in this area including some of the papers that led to the Nobel Prize in Economics. However, we have seen that only the key papers of the last half century are highly cited while the other older papers have not received much attention especially due to the technological infrastructure available at that time. Moreover, we have seen that over the years, the number of papers published is increasing a lot especially because of the increase in the number of researchers and journals. Next, we have analysed the most relevant institutions and we have found those usually regarded as the most important although by quantity of papers we have seen that some huge universities such as Illinois and Texas are well positioned. We have also studied those journals with the highest number of citations and we have found the classical Top journals. We have separated this analysis between economics and business, finance and management.

We have also developed a similar analysis for Spain studying the most cited papers, the evolution by years and the key institutions. The results have been more or less those expected although we have seen that many Spanish authors work in a foreign institution and therefore are included in the Web of Science as a foreign author. Regarding the evolution by years we have seen that Spain did not published many papers before the nineties. Since then, it has been increasing the number of papers published yearly and this increase seems to continue in the future. Concerning institutions, the results have been clear and those universities usually recognized as the Top institutions in Spain have been found in the first positions.

In future research, we expect to develop further developments to this approach by analysing other particular cases and separating the four key categories that constitute the Business & Economics area. Moreover, we will also consider other countries and journals in the analysis and extend this approach to other scientific categories.

References

Ajzen, I.: The theory of planned behavior. Organizational Behavior and Human Decision Processes 50, 179–211 (1991)

Akerlof, G.A.: Market for lemons – Quality uncertainty and market mechanism. Quarterly Journal of Economics 84, 488–500 (1970)

Alchian, A.A., Demsetz, H.: Production, information costs, and economic organization. American Economic Review 62, 777–795 (1972)

Armstrong, J.S., Overton, T.S.: Estimating nonresponse bias in mail surveys. Journal of Marketing Research 14, 396–402 (1977)

Arellano, M., Bover, O.: Another look at the instrumental variable estimation of error-components models. Journal of Econometrics 68, 29–51 (1995)

Arellano, M., Bond, S.: Some tests of specification for panel data – Monte-Carlo evidence and an application to employment equations. Review of Economic Studies 58, 277–297 (1991)

Barney, J.: Firm resources and sustained competitive advantage. Journal of Management 17, 99–120 (1991)

Black, F., Scholes, M.: Pricing of options and corporate liabilities. Journal of Political Economy 81, 637–654 (1973)

Bollerslev, T.: Generalized autoregressive conditional heteroskedasticity. Journal of Econometrics 31, 307–327 (1986)

Boisot, M., Child, J.: From fiefs to clans and network capitalism: Explaining China's emerging economic order. Administrative Science Quarterly 41, 600–628 (1996)

Charnes, A., Cooper, W.W., Rhodes, E.: Measuring efficiency of decision-making units. European Journal of Operational Research 2, 429–444 (1978)

Ciccone, A., Hall, R.E.: Productivity and the density of economic activity. American Economic Review 86, 54–70 (1996)

Clarida, R., Gali, J., Gertler, M.: The science of monetary policy: A new Keynesian perspective. Journal of Economic Literature 37, 1661–1707 (1999)

Clarida, R., Gali, J., Gertler, M.: Monetary policy rules and macroeconomic stability: Evidence and some theory. Quarterly Journal of Economics 115, 147–180 (2000)

Coase, R.H.: The problem of social cost. Journal of Law & Economics 3, 1–44 (1960)

Cohen, W.M., Levinthal, D.A.: Absorptive-capacity – A new perspective on learning and innovation. Administrative Science Quarterly 35, 128–152 (1990)

Cooke, P., Uranga, M.G., Etxebarria, G.: Regional innovation systems: Institutional and organisational dimensions. Research Policy 26, 475–491 (1997)

Davis, F.D.: Perceived usefulness, perceived ease of use, and user acceptance of information technology. MIS Quarterly 13, 319–340 (1989)

Eisenhardt, K.M.: Building theories from case-study research. Academy of Management Review 14, 532–550 (1989)

Engle, R.F.: Autoregressive conditional heteroskedasticity with estimates of the variance of United Kingdom inflation. Econometrica 50, 987–1007 (1982)

Engle, R.F., Granger, C.W.J.: Cointegration and error correction - representation, estimation, and testing. Econometrica 55, 251–276 (1987)

Esteban, J.M., Ray, D.: On the measurement of polarization. Econometrica 62, 819–851 (1994)

Fornell, C., Larcker, D.F.: Evaluating structural equation models with unobservable variables and measurement error. Journal of Marketing Research 18, 39–50 (1981)

Gali, J., Gertler, M.: Inflation dynamics: A structural econometric analysis. Journal of Monetary Economics 44, 195–222 (1999)
Gali, J., Gertler, M., Lopez-Salido, J.D.: European inflation dynamics. European Economic Review 45, 1237–1270 (2001)
Hansen, L.P.: Large sample properties of generalized method of moments estimators. Econometrica 50, 1029–1054 (1982)
Harvey, A., Ruiz, E., Shephard, N.: Multivariate stochastic variance models. Review of Economic Studies 61, 247–264 (1994)
Hausman, J.A.: Specification tests in econometrics. Econometrica 46, 1251–1271 (1978)
Heckman, J.J.: Sample selection bias as a specification error. Econometrica 47, 153–161 (1979)
Hirsch, J.E.: An index to quantify an individual's scientific research output. Proceedings of the National Academy of Science 102, 16569–16572 (2005)
Jarillo, J.C.: On strategic networks. Strategic Management Journal 9, 31–41 (1988)
Jensen, M.C.: Agency costs of free cash flow, corporate finance, and takeovers. American Economic Review 76, 323–329 (1986)
Jensen, M.C., Meckling, W.H.: Theory of firm – managerial behavior, agency costs and ownership structure. Journal of Financial Economics 3, 305–360 (1976)
Johansen, S.: Statistical analysis of cointegration vectors. Journal of Economic Dynamics & Control 12, 231–254 (1988)
Kahneman, D., Tversky, A.: Prospect theory – Analysis of decision under risk. Econometrica 47, 263–291 (1979)
Kremers, J.J.M., Ericsson, N.R., Dolado, J.J.: The power of cointegration tests. Oxford Bulletin of Economics and Statistics 54, 325–348 (1992)
Lucas, R.E.: On the mechanics of economic development. Journal of Monetary Economics 22, 3–42 (1988)
Nagel, R.: Unraveling in guessing games: An experimental study. American Economic Review 85, 1313–1326 (1995)
Newey, W.K., West, K.D.: A Simple, positive semidefinite, heteroskedasticity and autocorrelation consistent covariance matrix. Econometrica 55, 703–708 (1987)
Petrongolo, B., Pissarides, C.A.: Looking into the black box: A survey of the matching function. Journal of Economic Literature 39, 390–431 (2001)
Prahalad, C.K., Hamel, G.: The core competence of the corporation. Harvard Business Review 68, 79–91 (1990)
Romer, P.M.: Increasing returns and long run growth. Journal of Political Economy 94, 1002–1037 (1986)
Sala-i-Martin, X., Doppelhofer, G., Miller, R.I.: Determinants of long-term growth: A Bayesian averaging of classical estimates (BACE) approach. American Economic Review 94, 813–835 (2004)
Sala-i-Martin, X.X.: Regional cohesion: Evidence and theories of regional growth and convergence. European Economic Review 40, 1325–1352 (1996)
Sala-i-Martin, X.X.: I just ran two million regressions. American Economic Review 87, 178–183 (1997)
Teece, D.J., Pisano, G., Shuen, A.: Dynamic capabilities and strategic management. Strategic Management Journal 18, 509–533 (1997)
Tiebout, C.M.: A pure theory of local expenditures. Journal of Political Economy 64, 416–424 (1956)
Trejos, A., Wright, R.: Search, bargaining, money, and prices. Journal of Political Economy 103, 118–141 (1995)

van Doorslaer, E., Wagstaff, A., Bleichrodt, H., Calonge, S., Gerdtham, U.G., Gerfin, M., Geurts, J., Gross, L., Hakkinen, U., Leu, R.E., Odonnell, O., Propper, C., Puffer, F., Rodriguez, M., Sundberg, G., Winkelhake, O.: Income-related inequalities in health: Some international comparisons. Journal of Health Economics 16, 93–112 (1997)

Vives, X.: Nash equilibrium with strategic complementarities. Journal of Mathematical Economics 19, 305–321 (1990)

Wernerfelt, B.: A resource-based view of the firm. Strategic Management Journal 5, 171–180 (1984)

White, H.: A heteroskedasticity consistent covariance matrix estimator and a direct test for heteroskedasticity. Econometrica 48, 817–838 (1980)

Is Human Capital the Key Factor in Explaining Business Location Differences? Knowledge Intensive Business Services in Portugal

João J.M. Ferreira[1] and Cristina I. Fernandes[2]

[1] Management and Economics Department, University of Beira Interior. Researcher at NECE – Research of Business Science of University of Beira Interior, Pólo IV – Edifício Ernesto Cruz, 6200-209 Covilhã, Portugal
jjmf@ubi.pt

[2] Polithecnic Institute of Bragança, and Instituto Superior de Línguas e Administração de Leiria (ISLA). Researcher at NECE – Research of Business Science of University of Beira Interior, Pólo IV – Edifício Ernesto Cruz, 6200-209 Covilhã, Portugal
kristina.fernandes81@gmail.com

Abstract. The purpose of this paper is to verify if the human capital is the key factor in explaining business location differences on Portuguese knowledge intensive business services (KIBS) sector. A logistic regression model is applied, adopting rural versus urban and human resources as the key dimensions. We combine business locations with graduate employee rates to assess whether these aspects are determinant in the KIBS sector or if they vary from firm to firm. Our results demonstrate that rural professional and technological KIBS employ more graduate human resources than urban KIBS. As regards their urban professional and technological KIBS counterparts, we may statistically state that the employment of higher education qualified professionals is not related to KIBS type. Furthermore, we conclude that the employment of graduate human resources, the age and the academic background of business owners do have a statistically significant impact on the logit probability of a KIBS company locating in a rural environment.

Keywords: Human resources, location, rural and urban KIBS, professional and technological KIBS.

1 Introduction

Interest in research into service based innovation has surged ever since the early 1980s among both academics and decision makers in general (Johne and Storey, 1998; de Jong et al., 2003; Miles, 2000). There has been a corresponding rise in the awareness that service companies are not mere passive recipients of innovations produced by transformation companies but, on the contrary, bring about their own innovation (Gallouj and Weinstein, 1997; Tether, 2003). This perspective contrasts sharply with the "supplier-dominated" view (Pavitt, 1984; Barras, 1986 and 1990),

A.M. Gil-Lafuente et al. (Eds.): Soft Comput. in Manag. and Bus. Econ., STUDFUZZ 287, pp. 19–30.
springerlink.com

within the scope of which the service sector is portrayed as the receiver of waves of innovation driven by transformation companies. In addition, greater attention has always been paid to non-technological aspects of service innovation, such as the ways in which the production of services and interactions with clients are organized as well as the way in which such activities may be deployed to generate new services (Bettencourt et al., 2002; Meyer and DeTore, 1999; Sundbo, 1997; Van der Aa and Elfring, 2002). Nevertheless, Tether (2005) puts forward a different perspective within which service sectors tend to innovate in ways different to transformation companies and suggesting service sector innovation tends to be smoother than in other company types.

One of the most important approaches to service innovation assumes that innovation in this sector ranges far beyond the firms making up the sector and impacts on all sectors of the economy with specific service types ending up conveying their innovation to other economic activities (Miles et al., 1995; Miles, 2003). Many of such studies have focused on the role played by Knowledge Intensive Business Services (KIBS) in innovation systems and how their engagement with companies in other sectors boosts performance levels, both of the latter companies and the regions (Miles, 2000; Leiponen, 2001).

Increasing knowledge in a region depends on its internal capacity to create innovation and likewise on its capacity to attain and to set into action the stock of knowledge produced in other areas. In this field, the leading factors in innovation are the internal production inputs, research and development (R&D) expenditure and human capital, as well as the diverse channels which enable the spread of the external knowledge towards the receiving region (Marrocu et al., 2011). According to Cohen and Levinthal (1990), human capital may be expected to play a decisive role in the external knowledge absorption process, which incorporates a core facet of innovative activities.

Howells (2000) concludes that the contracting of R&D services frequently falls within the framework of the technical specifications and design processes of the products launched by transformation companies. Furthermore, many KIBS studies have focused upon their relationships with clients (generally, companies in other sectors). This has evolved in parallel with KIBS beginning to be perceived as producers of innovation and drivers in the dissemination of knowledge through their very close relationships with their clients (den Hertog, 2000; Muller, 2001).

Recent contributions (Aslesen and Isaksen, 2007) have emphasized the extent to which KIBS prove crucial to the innovation and growth of other industries and regions. This position indirectly results from statistics demonstrating the sustained growth of the KIBS sector, which favors how KIBS, in cooperating with other companies may enhance the growth rates prevailing in their respective regions (Illeris, 1989). However, these findings have yet to be demonstrated by empirical evidence, in no small part due to the highly heterogeneous nature of the KIBS sector across its full extent (Wood, 2002). There is a corresponding need to carry out more empirical studies capable of proving whether such contributions actually exist in the field. Hence, this study seeks to empirically analyze if the human capital is the key factor in explaining business location differences on Portuguese knowledge intensive business services (KIBS) sector.

The paper is organized as follows. Section 2 describes the literature review. Section 3 presents the methods, based on the logistic regression model, and the sample. Section 4 presents and discusses the results with section 5 setting out the main conclusions and suggestions for further research.

2 Literature Review

In all sectors of economic activity, innovation proves fundamental to company survival and sustainability in an increasingly globalised market place. Innovation enables companies to respond to diverse levels of demand and constant change and evolution as well as fostering improvements across the domains of workplace health and safety, the environment, communication and quality of life in general. Hence, innovation is the motor of progress, competitiveness and economic development (Romer, 1994; Johansson et al., 2001).

Meanwhile, given the complexity inherent to innovation processes, companies often only prove able to innovate when cooperating with other companies thereby enabling all their partners to optimize their utilization of their own respective internal knowledge resources and deploying them in conjunction with the specific skills of their partners (Muller and Zenker, 2001). The knowledge bound up in innovation processes might be tacit or formally codified and either developed in-house or acquired from third parties, for example from innovation partnership networks (Muller and Zenker, 2001). Nonaka (1994) and Nonaka et al. (2000) set out how the knowledge transformation process best occurs within companies through the self-development of knowledge creation capacities. This approach stems from a neo-Schumpeterian viewpoint or from an evolutionary perspective on the economy that considers innovation as a gradual knowledge based process (Nelson and Winter, 1974, 1975, 1977; Freeman, 1982; Lundvall, 1992, Lundvall and Johnson, 1994).

Knowledge cycles able to actually produce innovation derive at least partially from interactions between different categories of actors. Additionally, innovation processes thrive in specific contexts, based on specific experiences, with core competences and purpose focused knowledge.

The role of KIBS as activities in support of the transformative industries and general purpose manufacturing companies has been raised by various authors (Cooke, 2001; Wood, 2005). According to Wood (2006), the growing importance of the KIBS sector to regional development has essentially been demonstrated by reference to: (i) the acquisition of fundamental knowledge by innovative small and medium sized enterprises (SMEs) and the public sector able to provide support to knowledge-based economies; (ii) increasing demand for external audits among SMEs; (iii) the importance attributed by each region to adaptability, to national and international production and to market norms and trends and all reflected in the level of demand for KIBS services and resources; (iv) the natural tendency for KIBS firms to adopt and shape new commercial and technical knowledge with such proving fundamental to meeting the new needs facing regions; (v) the way in which KIBS have been developing and fostering the appearance of a new fluidity in the exchange of knowledge; (vi) relationships involving the exchange of knowledge incorporating regional characteristics; and (vii) knowledge based technological innovation that depends on the prevailing local contexts although not necessarily in itself sufficient to boosting regional economic success. Such success depends more broadly on non-technological innovations and the capacity for adaptability, such as, for example, in terms of management and marketing methodologies, with KIBS providing ever more important inputs into this field.

According to Muller and Zenker, (2001) KIBS operate across two fundamental levels: (i) serving as an external source of knowledge and actively contributing towards client company innovation; and (ii) introducing internal innovations generated by highly qualified working environments contributing towards overall economic growth and development. According to Hipp (2000), KIBS produce innovations that convert into added value for other companies. Metcalfe and Miles (2000) portray KIBS as agents of innovation given that through their transferring knowledge, they become co-producers of innovation in conjunction with their clients (den Hertog and Bilderbeek, 1999; den Hertog, 2002). In addition, KIBS provide inputs of complementary knowledge able to facilitate innovation, organizing innovation processes at client companies and advising on the type and format of innovation that clients should undertake while simultaneously supervising and monitoring this process (Aslesen and Isaksen, 2007).

Some progress has been made regarding the general acceptance of services, including KIBS, as enhancing technological advancement and innovation (den Hertog, 2000; Haukness, 2000; Muller & Zenker, 2001; Gallouj, 2002). According to Miles (2001), KIBS are now recognized as playing a fundamental role as innovation system intermediaries. The relationship between KIBS firms and companies in other sectors has clearly proven a positive influence on those companies (Freel, 2006). This relationship has boosted recourse to R&D, raising the capacity of human resources and fostering cooperation to thereby collectively raise the innovation ratio.

KIBS are correspondingly deemed a strategically important sector for industrial and regional development (Aslesen and Isaksen, 2007). According to Fischer et al. (2001), KIBS are intensely concentrated in urban areas with the sector perceived as an essential component to the innovations systems existing in their surrounding environments. KIBS normally employ highly qualified human capital and combine general, scientific and technological information while accumulating experience and competences through ongoing research projects and the tacit knowledge acquired from clients. In sum, KIBS work towards resolving the problems posed by their clients (den Hertog, 2002). Sheamur and Doloreaux (2008) argue that KIBS contribute towards innovation and regional competiveness through the ways in which they interact with other local actors with the objective of producing the innovations that consequently foster regional development.

KIBS have also been analyzed as vectors for the interchange of information. Thus, their role is studied in terms of their innovative performance, as facilitators, transporters and sources of information between companies (den Hertog 2000; Muller and Zenker 2001; Wood 2005; Simmie and Strambach, 2006).

2.1 KIBS Knowledge Production and Dissemination

Strambach (2001) distinguishes between three stages defined as fundamental to the KIBS process of producing and spreading knowledge. Firstly, they acquire knowledge, whether of the tacit or the codified type, before moving onto the stage of recombining that knowledge (a phase involving the codification of knowledge) before finally arriving at the transfer of knowledge to client companies. Hauknes (1999) maintains that the creation and diffusion of innovation depend not only on new technological knowledge, generated both through the implementation of internal research processes and by third party laboratories and other research entities, but also and especially on the daily interaction, communication and exchange of information between companies and between these and other scientific institutions. KIBS increasingly act as bridges and convertors of technology to management professionals and tailored to the respective knowledge location and capacities in effect and enabling actors to resolve specialist problems through the input of complementary knowledge, in turn enabling the advent of innovation.

The innovation capacities of these companies are highly dependent on accessing external information resources (European Commission, 2000). Consequently, the capacity to combine internal and external resources should be interpreted as an improvement in their knowledge acquisition capacities (Cohen and Levinthal, 1990). In conjunction, this enables us to consider KIBS firms as potential co-innovators for small and medium sized companies. Teece (1986) provides insights into characterizing innovation stemming from the interactions between small and medium sized companies and KIBS firms. KIBS services result in highly interactive processes within the framework of which firm performance derives from its ability to adapt to the needs of their clients. Strambach (1998) highlights the depth and complexity of these relationships when pointing out how the procurement of knowledge intensive services is not the same as purchasing any other standardized product or service.

Hence, and in contrast to the traditional neoclassical and post-Keynesian approaches that take only the mobility of factors of production into account, regional competitiveness may be perceived as intrinsically bound up with endogenous resource effects leveraged through technological innovation when accepting that knowledge creation activities are key to regional development (Oakey et al, 1980; Jaffe et al, 1993; Paci and Usai, 2000; Kim and Knaap, 2001; Sohn and tal, 2003; Fischer and Varga, 2003).

Thus, through this literature review, we have ascertained that the strong performance of these firms, in conjunction with the contributions made towards innovation, is inherently related to how they incorporate qualified human resources into their working processes. This resource type then facilitates and drives the entire process of transferring knowledge and the service type provided by KIBS firms to companies in need (Miles et al., 1995).

This study seeks to empirically analyze both the impacts of location and the utilization of qualified human capital on KIBS firms.

3 Methods and Sample

The study sample stems from a database supplied by Coface Group containing the total population of KIBS in Portugal between 2004 and 2009. Based on the data, in 2004 Portugal hosted a total of 39,254 KIBS companies that declined to 34,644 in 2009.

We were also able to verify that 4,610 KIBS (13%) may be considered inactive in 2009, with only 87% actually operational. The sample was extracted from the data base according to the business codes CAE (REV.3) and NACE (REV 2), similar to the approaches made by other researchers (Frell, 2006, Miles et al., 1995; Doloreux and Muller, 2007, Shearmur, and Doloreux, 2008) so as to incorporate two KIBS groups into the sample: technological KIBS focused upon activities related to information and communication technologies, research and development, engineering and architecture and related activities, testing and analysis techniques (NACE codes: 62.01; 62.02; 62.03; 62.09; 63.11; 63.91; 63.99; 71.11; 71.12; 71.20; 72.1; 72.2) and professional KIBS firms operating in the legal, accountancy and bookkeeping sectors and auditing, fiscal consultancy, market studies activities as well as the entire public relations sector (NACE codes: 69.10; 69.20; 73.20; 70.22; 73.11; 73.12; 78.10; 78.30; 74.20; 74.90).

With the objective of analyzing the relationship existing between the number of employees with university level education, the KIBS type (professional vs. technological) and their location (rural vs. urban), we developed a questionnaire for a final sample of 500 KIBS. This sample was structured as follows: professional KIBS (p_KIBS) (65.6%, 328 companies) and technological KIBS (t_KIBS) (34.4%, 172 companies). Of the entire sample, 18.6% of companies were located in rural areas (r_KIBS) (93 companies) with 81.4% found in urban areas (u_KIBS) (407 companies). Of the 328 p_KIBS companies, 63 were located in rural regions with 265 in urban areas while the figures for technological KIBS came in at 30 and 142 respectively (Table 1). We define as rural, all locations containing fewer than 5,000 inhabitants (Kayser, 1990).

Table 1 Distribution of KIBS: typology and location

		KIBS Location		Total
KIBS Typology		Rural	Urban	
P_KIBS	N	63	265	328
	%	12.6%	53.0%	65.6%
T_KIBS	N	30	142	172
	%	6.0%	28.4%	34.4%
Total	N	93	407	500
	%	18.6%	81.4%	100.0%

4 Results

Regarding 2004, KIBS in the study return an average of around 80% (M = 0.80; DP = 0.28) of employees with an undergraduate degree or higher education qualification. In 2009, this proportion remained high (M=0.81; DP=0.26). Through the application of the non-parametric Mann-Whitney U test, we find the percentage of graduate workers in 2004 was higher on average in u_KIBS than their r_KIBS counterparts (given p=0.026<0.05) and hence rejecting the equal average null hypothesis). This finding does not hold for the 2009 figures given that the proportion of employees with higher education at rurally located KIBS rose significantly between 2004 and 2009 (up from 67% to 75%).

In fact, in 2009, the average proportion of employees with higher education did not differ significantly according to the KIBS location (p=0.152>0.05). As regards the KIBS typology, both the p_KIBS and the t_KIBS return high rates of professional employment with graduate levels of education (varying between 79% and 82%), with no statistically relevant differences between the two KIBS types (p2004=0.632 and p2009=0.702 >0.05). Analyzing the KIBS type separately to location (Table 2), we find that the urban p_KIBS return a higher level of graduate employment in 2004 than rural p_KIBS companies. In t_KIBS companies, this difference retains statistical significance in 2009.

Table 2 Comparison between graduate human resources (GHR) by KIBS

KIBS	Location (dummies)		GHR (2004)	GHR (2009)
P_KIBS	Urban	Average	0.82	0.82
		SD	0.25	0.24
	Rural	Average	0.65	0.78
		SD	0.42	0.31
Mann-Whitney U Test		p-value	0.039*	0.938
T_KIBS	Urban	Average	0.83	0.84
		SD	0.26	0.24
	Rural	Average	0.75	0.69
		SD	0.32	0.33
Mann-Whitney U Test		p-value	0.390	0.009*

* p< 0.05.

Finally, to evaluate the relationship between the KIBS location and the likelihood of employing graduate human resources, we applied a logistic regression model (Table 3).

Table 3 Logistic regression model

Independent Variables	B	EP	Sig.	Exp (B)
GHR04	-2.212	0.837	0.008**	.110
GHR09	2.386	1.001	0.017*	10.866
Undergraduate HR	-1.605	0.424	0.000***	.201
Gender (Female)	0.468	0.330	0.156	1.597
Age	-0.057	0.022	0.009**	.944
Intercept	1.684	1.078	0.118	5.390

* $p< 0.05$; ** $p< 0.01$; *** $p< 0.001$.

According to the Wald test (more specifically, the probability of significance) associated to the logit coefficients of the estimated model, the results do enable us to conclude that there is a statistically significant effect between employing higher education graduates ($p=0.008$ and $p=0.017<0.05$) and the age ($p=0.009<0.05$) and the academic background of owners ($p=0.000<0.05$) on the probability logit of companies locating in rural environments. Based upon the model's coefficients, we correspondingly find that the ratio of companies locating in rural communities rises in keeping with the level of employment of higher education graduates, with the owner having completed that level of study and when the business owner's age is lower.

Thus, we may conclude that rural professional and technological KIBS employ more graduate human resources than urban KIBS. As regards their urban professional and technological KIBS counterparts, we may state that statistically, the employment of higher education qualified professionals is not related to KIBS type. Furthermore, we find that the employment of graduate human resources, age, and the academic background of business owners do have a statistically significant impact on the logit probability of the KIBS company locating in a rural environment. This means that, while there is no direct cooperation between higher education institutions and KIBS companies, there is a transfer of knowledge generated by universities through the professionals employed by KIBS entities, as Delmar and Wennberg (2010) argued.

5 Conclusions

The main objective of our study was to seek if the human capital is the key factor in explaining business location differences on Portuguese knowledge intensive business services (KIBS) sector. To this end, we carried out a review of the literature on the relevance of innovation and the dissemination of knowledge by KIBS firms through recourse to employing graduate level human resources. The results of the logistic regression model verified how the employment of professionals with higher education in rural areas rises in proportion with the youthfulness of firm owners and the higher their own qualifications are at both p_KIBS and t_KIBS. This conclusion seems to evidence that one of the indicators fostering competitiveness is precisely the employment of individuals with higher levels of education (Roura, 2009).

Our results have important implications for entrepreneurs as well as for policy makers. There seems to be a gap between the support accessible and the entrepreneurs that should be getting this support. Hence, it becomes correspondingly important to strengthen the role played by this company type within the framework of overall regional economic performances. Although rural areas are lesser developed and consequently less attractive, we find that younger entrepreneurs prefer these regions. Therefore, the development of policies able to nurture and fund creativity at this company type proves of crucial importance. Furthermore, we should emphasize that while there are only 93 rurally located KIBS firms in Portugal, they employ individuals with higher levels of education. Correspondingly, within the recessionary context prevailing and the level of political priority attributed to youth unemployment, especially graduate level, enabling these companies to attract and retain these qualified resources may prove determinant.

Hence, and as a future line of research, we would propose undertaking new studies on other regional contexts internationally in order to ascertain whether this company type shares this behavioral pattern. Beyond many others, two questions in particular are still left unanswered: were we to adopt different methodologies, would we reach the same results? Do new KIBS start-ups display both the same need to employ specialist human resources and the same location preferences as already operational KIBS firms?

References

Aslesen, H., Isaksen, A.: Knowledge Intensive Business Services and Urban Industrial Development. The Service Industries Journal 27(3), 321–338 (2007)

Barras, R.: Towards a theory of innovation in services. Research Policy 15, 161–173 (1986)

Barras, R.: Interactive innovation in financial and business services: The vanguard of the service revolution. Research Policy 19, 215–237 (1990)

Bettencour, L.A., Ostrom, A.L., Brown, S.W., Roundtree, R.I.: Client Co-Production in Knowledge-Intensive Business Services. California Management Review 44(4), 100–128 (2002)

Cohen, W.M., Levinthal, D.A.: Absorptive capacity: a new perspective on learning an innovation. Administrative Science Quarterly 35, 128–152 (1990)

Cooke, P.: Strategies for Regional Innovation Systems. Policy paper, Vienna, United Nations Industrial Development Organization, UNIDO (2001)

Delmar, F., Wennberg, K.: Knowledge Intensive Entrepreneurship – the birth, growth and demise of entrepreneurial firms. Edward Elgar Publishing, Cheltenham (2010)

De Jong, J.P.J., Bruins, A., Dolfsma, W., Meijaard, J.: Innovation in service firms explored: what, how and why? EIM Business and Policy Research (2003),http://www.eim.net/pdf-ez/B200205.pdf

Den Hertog, P., Bilderbeek, R.: Conceptualising service innovation patterns. Research Programme on Strategic Information Provision on Innovation and Services (SIID), Ministry of Economic Affairs, Dialogic, Utrech (1999)

Den Hertog, P.: Knowledge intensive business services as co-producers of innovation. International Journal of Innovation Management 4(4), 491–528 (2000)

Den Hertog, P.: Co-producers of innovation: in the role of knowledge-intensive business services in innovation. In: Gadrey, J., Gallouj, F. (eds.) Productivity, Innovation and Knowledge in Services, pp. 223–255. Edward Elgar, Cheltenham (2002)

Doloreux, D., Muller, E.: The key dimensions of knowledge-intensive business services (KIBS) analysis. A decade of evolution. Working Paper Firms and Regions No. U1/2007, Fraunhofer-Institut für System-und Innovationsforschung-ISI, Karlsruhe (2007)

Edquist, C. (ed.): Systems of Innovation. Technologies, Institutions and Organizations. Pinter Publishers, London (1997)

Fischer, M., Diez, J., Snickars, F.: Metropolitan Systems of Innovation. Theory and evidence from three metropolitan regions in europe. Springer, Berlin (2001)

Fischer, M., Varga, A.: Spatial knowledge spillovers and university research: evidence from Austria. Ann. Reg. Sci. 37, 303–322 (2003)

Freeman, C.: The Economics of Industrial Innovation. Pinter Publishers, London (1982)

Freel, M.: Patterns of Technological Innovation in Knowledge-Intensive Business Services. Industry and Innovation 13(3), 335–358 (2006)

Gallouj, F., Weinstein, O.: Innovation in services. Research Policy 26, 537–556 (1997)

Gallouj, F.: Innovation in the Service Economy. Edward Elgar, Cheltenham (2002)

Hauknes, J.: Services in innovation — innovation in services. SI4S Final report to the European Commission, DG XII, TSER programme. STEP Group, Oslo (1999)

Haukness, J.: Dynamic innovation systems: what is the role of services? In: Boden, M., Miles, I. (eds.) Services and the Knowledge-based Economy, Continuum, London (2000)

Hipp, C.: Innovationsprozesse im Dientleistungssektor. Eine theoretisch und empirisch basiert Innovationstypologie. Physica, Heidelberg (2000)

Howells, J.: Research and Technology Outsourcing and Systems of Innovation. In: Metcalfe, J.S., Miles, I. (eds.) Innovation Systems in the Service Economy, pp. 271–295. Kluwer, Boston (2000)

Illeris, S.: Producer services: the key factor for future economic development? Entrepreneurship and Regional Development 1(3), 267–274 (1989)

Jaffe, A., Trujenberg, M., Henderson, R.: Geographic localization of knowledge spillovers as evidence by patent creations. Q. J. Econ. 108, 577–598 (1993)

Johansson, B., Karlsson, C., Stough, R.: Theories of Endogenous Regional Growth, Lessons for Regional Policies. Springer, Berlin (2001)

Johne, A., Storey, C.: New service development: a review of the literature and annotated bibliography. European Journal of Marketing 32(3/4), 184–252 (1998)

Kim, T., Knaap, G.: The spatial dispersion of economic activities and development trends in China: 1952-1985. An Reg. Sci. 35, 39–57 (2001)

Leiponen, A.: Knowledge Services in the Innovation System, Etla: Working Paper B185; SITRA 244, Helsinki (2001)

Lundvall, B.-Å.: National System of Innovation. Theory of Innovation and Interactive Learning. Pinter Publishers, London (1992)

Lundvall, B.-Å., Johnson, B.: The learning economy. Journal of Industry Studies 1(2), 23–42 (1994)

Malhotra, N.: Marketing Research: An Applied Orientation, 6th edn. Prentice-Hall (2010)

Marrocu, E., Paci, R., Usai, S.: The complementary effects of proximity dimensions on Knowledge spillovers, Working paper, 2011/21, Contributi di Ricerca Crenos (2011)

Meyer, M.H., DeTore, A.: Product development for services. Academy of Man-agement Executive 13(3), 64–76 (1999)

Miles, I., Kastrinos, N., Bilderbeek, R., den Hertog, P.: Knowledge-intensive Business Services: Their Role as Users, Carriers and Sources of Innovation, Report to the EC DG XIII. Sprint EIMS Programme, Luxembourg (1995)

Miles, I.: Services innovation: coming of age in the knowledge based economy. International Journal of Innovation Management 4(4), 371–389 (2000)

Miles, I.: Services in National Innovation Systems: from Traditional Services to Knowledge Intensive Business Services. In: Schienstock, G., Kuusi, O. (eds.) Transformation Towards a Learning Economy: the Challenge to the Finnish Innovation System. SITRA, Helsinki (2001)

Miles, I.: Innovation in Services, TEARI working paper no. 16, Policy Research in Engineering Science and Technology, University of Manchester (2003)

Muller, E.: Innovation Interactions between Knowledge intensive Business services and Small and Medium-Sized Enterprises: An Analysis in Terms of Evolution, Knowledge and Territories. Physica-Verlag, Heidelberg (2001)

Muller, E., Zenker, A.: Business services as actors of knowledge transformation: The role of KIBS in regional and national innovation systems. Research Policy 30, 1501–1516 (2001)

Nelson, R., Winter, S.: Neoclassical vs. evolutionary theories of economic growth. Critique and prospectus. Economic Journal 12, 886–905 (1974)

Nelson, R., Winter, S.: Growth theory from an evolutionary perspective: the differential productivity puzzle. The American Economic Review 65(2), 338–344 (1975)

Nelson, R., Winter, S.: In search of a useful theory of innovation. Research Policy 6, 36–76 (1977)

Nonaka, I.: A dynamic theory of organizational knowledge creation. Organization Science 5(1), 14–37 (1994)

Nonaka, I., Toyama, R., Nagata, A.: A firm as a knowledge-creating entity: a new perspective on the theory of the firm. Industrial and Corporate Change 9(1), 1–20 (2000)

Oakey, R., Thwaites, A., Nash, P.: The regional distribution of innovative manufacturing establishments in Britain. Reg. Stud. 44, 235–253 (1980)

Paci, R., Usai, S.: Technological enclaves and industrial districts. An analysus if regional distribution of innovative activity in Europe. Reg. Stud. 34, 97–114 (2000)

Pavitt, K.: Sectoral patterns of technical change: Towards a taxonomy and a theory. Research Policy 13, 343–373 (1984)

Romer, P.: The origins of endogenous growth. Journal of Economic Perspectives 8(1), 3–22 (1994)

Roura, J.: Towards new European peripheries?". In: Karlsson, C., Johansson, B., Stough, R. (eds.) Innovation, Agglomeration and Regional Competition, New Horizons in Regional Science, pp. 170–197. Edward Elgar Publishing, UK (2009)

Shearmur, R., Doloreux, D.: Urban Hierarchy or Local Buzz? In: High-Order Producer Service and (or) Knowledge-Intensive Business Service, The Professional Geographer, Canada, vol. 60, pp. 333–355 (2001)

Simmie, J., Strambach, S.: The contribution of KIBS to innovation in cities: And evolutionary and institutional perspective. Journal of Knowledge Management 10, 26–40 (2006)

Sohn, J., Kim, T., Hewings, G.: Information technology and urban spatial structure: a comparative analysis of Chicago and Seoul regions. Ann. Reg. Sci. 37, 447–462 (2003)

Strambach, S.: Innovation processes and the role of knowledge-intensive business services. In: Koschatzky, K., Kulicke, M., Zenker, A. (eds.) Innovation Networks — Concepts and Challenges in the European Perspective, pp. 53–68. Physica, Heidelberg (2001)

Strambach, S.: Knowledge-intensive business services (KIBS) as an element of learn-ing regions — the case of Baden–Württemberg. Paper presented at the ERSA Conference, Vienna (August 28-31, 1998)

Sundbo, J.: Management of innovation in services. The Service Industries Journal 17(3), 432–455 (1997)

Teece, D.J.: Profiting from technological innovation: implications for integration, collaboration, licensing and public policy. Research Policy 15, 285–305 (1986)

Tether, B.S.: The Sources and Aims of Innovation in services: variety between and within sectors. Economics of Innovation and New Technology 12(6), 481–505 (2003)

Tether, B.: Do Services Innovate (Differently)? Insights from the European Innobarometer Survey. Industry and Innovation 12(2), 153–184 (2005)

Van der Aa, W., Elfring, T.: Realizing innovation in services. Scandinavian Journal of Management 18(2), 155–171 (2002)

Wood, P.: Consultancy and Innovation: the business service revolution in Europe. Routledge, London (2002)

Wood, P.: A service-informed approach to regional innovation—Or adaptation? Service Industries Journal 25, 429–445 (2005)

Wood, P.: The regional significance of knowledge-intensive services in Europe. Innovation: The European Journal of Social Science Research 19, 151–166 (2006)

The Importance of Human Capital in Innovation: A System of Indicators

Rosa Mª Mariz-Pérez, Mª Mercedes Teijeiro-Álvarez, and Mª Teresa García-Álvarez

Department of Economic Analysis and Business Administration, University of A Coruna, Campus de Elviña, s/n. 15071. A Coruña, Spain
{rmariz,mteijeiro,mtgarcia}@udc.es

Abstract. This article seeks to clarify the impact of human capital on the innovation capacity of companies. For this purpose we employ the literature review and some personals interpretations. Results have relevant implications for managers of companies that are interested in promoting their innovation activity. By considering how human capital is related with the innovation process, this article attempts to provide a useful guide of human capital indicators within the intellectual capital framework. We consider that the basic contribution of our work is the development of a system of indicators for human capital management with the objective of allowing for a clear picture of links between strategic human resources and the innovation capacity of companies. Moreover, it can be easily adapted to a given type of firm and, therefore, serve to compare companies belonging to the same sector of activity.

Keywords: Human capital, innovation, intellectual capital, indicators guide.

1 Introduction

The nature of world economic growth is, in part, due to innovation speed. This is possible given the rapid technological evolution, shorter product life cycles and to the higher rate of development of new products (Plesis, 2007). In this sense, innovation can be defined as the process that allows companies to accumulate knowledge and technological capacities to improve productivity, cost reduction and prices while, at the same time, contributes to the creation of new products and to the quality increase of existing ones.

In the present moment, we must keep in mind that the capacity a company has to innovate depends, in great extent, on intangible assets and knowledge it possesses and, of course, on the manner it is able to employ these (Alegre and Lapiedra, 2005; Subramaniam and Youndt, 2005). This perspective of innovation depends on certain factors, such as the possession of adequate professional competencies, attitudes, aptitudes and intellectual agility, good relations within the workforce, adequate organizational technology, the capacity to bring in and retain the best professionals,

A.M. Gil-Lafuente et al. (Eds.): Soft Comput. in Manag. and Bus. Econ., STUDFUZZ 287, pp. 31–44.
springerlink.com

etc. These intangible assets are commonly called intellectual capital (IC) and most papers establish that it consists of three elements: human capital (HC), structural capital (SC) and relational capital (RC) (Edvinson and Malone, 1997; Bontis, 1998; Bontis et al., 2000).

In this paper we attempt to identify the elements of IC that can contribute to guaranteeing the innovation capacity of companies. With this aim, we pay special attention to HC (which includes experience, skills, employee professional development, teamwork, etc). We present a guide of indicators to manage this item taking into account aspects related both to the development and renovation of HC and to its efficiency and stability. From a business perspective, we consider that results have important implications for managers of companies committed with innovation, especially in knowledge intensive activities.

2 Relevance of Human Capital in Determining Company Innovation Capacity

Innovation is definitely one of the basic pillars of company competitiveness. We are all aware of the positive effects of technological innovation, given the possible productivity improvements, new product development, quality and differentiation increases, cost and price reductions, etc. Therefore, we can state that innovation is critical to increase company value (Tsent and Goo, 2005).

However, investments in research and development are necessary but not sufficient to develop innovation capacity (Martin et al., 2009). In this sense, it must be combined with investment in HC (this would allow for the transformation of innovation potential into productive realities), SC (this contributes to have the full command of processes related to assets owned) and RC (related to knowledge included in relations with stakeholders). Taking into account this definition, innovation can be considered as the most knowledge intensive organizational process, given that it depends both on individual employee know-how and internal and external company knowledge (Adamides and Karacapilidis, 2006).

Within the three components of intellectual capital, human capital, understood as both individual and group knowledge of company employees, is especially important in determining innovation capacity of firms. We therefore we consider a broader definition of HC to include not only individual knowledge, but also the part of knowledge that arises from relations between company personnel (Broking, 1996; Edvinson and Sullivan, 1996; Euroforum, 1998 and CIC, 2003).

Littlewood and Herman (2004) note the importance of HC when they establish that, nowadays, HC is one of the factors that determine organizational competitiveness, given competencies, knowledge, creativity, capacity to resolve problems, leadership and personal compromise are some of the assets required to meet the demands of turbulent environments and reach organizational goals. For Carson et al., HC includes tacit knowledge and communications skills, the entrepreneur spirit and other personal attributes such as disposition or aptitudes for lifelong learning.

Traditionally, management models have basically included tangible assets and have failed to capture the value of the intangible ones. However, over the last decades, models have highlighted the importance of incorporating intangible assets, such as HC, relational capital or structural capital. Furthermore, it has been estimated that IC accounts for most of the market value of a given organization although this is not recorded in financial statements (Nonaka and Takeuchi, 1995; Brennan and Connell, 2000; Fornell, 2000; Civi, 2000, Heng, 2001; Watson et al., 2005; Martínez-Torres, 2006). Difficulties related to the valuing of this type of resources are not a general impediment to finding increasing organizational proposals to manage and power them. Effective administration in this sense has great potential for value creation and, therefore, intangible assets cannot be ignored (Brennan and Connell, 2000; Bozbura et al., 2007). Several papers have focused on the importance of IC to create sustainable competitive advantages (Marr et al., 2002) and on the direct impact it has on business results (Ichniowski and Shaw, 1999; Carlucci et al., 2004; Luthans and Avolio, 2009).

The key to managing IC is to monitor its transformation from the beginning (Lynn, 1998), when it is simple information, until it provides organizational value. Knowledge (individual or organizational) can be considered part of IC only when it is used and shared to create value (Cannon-Bowers and Salas, 2001; Martínez Ramos, 2003; Hansen, et al., 2005).

One of the basic problems of HC management is the difficulty in measuring it, given it includes individual elements such as education, work experience, capacity, motivation, etc. Moreover, as with any other asset, obsolescence must be taken into account and it is rather surprising to find that very few empirical papers include it. De Grip and Van Loo (2002) distinguish, from an economic view, two types of depreciation:

- Technical depreciation: refers to decreases in the value of HC due to physical deterioration (skill atrophy, lack of new skills,...), given situations such as inactivity or unemployment, low motivation for work, etc.
- Economic depreciation: related to the loss in market value of employee qualification, basically as a consequence of specific skill obsolescence, given rapid technological change, changes in sector structure and lack of organizational skill adaptation.

In this paper, we focus on indicators to measure this type of capital and that are directly related to company innovation. Of course, we are presently working on the necessary empirical study in order to check the validity of the system of indicators proposed.

3 Components and Dimensions of Human Capital from an Intellectual Capital Perspective

In this section we outline the basic elements of IC models with respect to their contribution to HC, taking into account connections with company innovation.

Integral Credit Scorecard Model (Kaplan and Norton, 1992)

This model notes that intangible assets must be aligned with business strategy (those that are not will not create much value although investments made may be high) and that, at the same time, they must be integrated (they must be created using capacities built through other intangible and tangible assets and not through independent capacities that do not allow for synergetic effects to arise).

For these authors, HC represents the availability of skills, talent and know-how of employees to undergo fundamental internal processes that guarantee strategic success. Innovation is not considered as a capital on its own and it constitutes an internal process that increases the value of HC, informational capital and organizational capital. The influence of innovation on HC is centered on how strategic employees reach the specified objectives, given the human resources they have and the optimal personnel needs (Trillo and Rodríguez, 2007).

Skandia Navigator Model (Edvinsson, 1997)

The Skandia model constitutes a strategic and operational management system based on the fact that the performance of a company comes from its capacity to create sustainable value, given the strategic vision and mission of the firm. The concept Edvinsson and Malone includes within IC is that all assets are valuable for the organization, including HC and structural capital.

From a human approach, they include all individual capacities, knowledge, skills and experience of employees and managers as well as creativity and innovation capacity of the organization. This constitutes the core of the model, although it is perhaps the most difficult part to measure given it includes assets that are not owned by the company.

Innovation in this model is linked to the contribution of human resources to the organization. Professional experience and innovation are considered to be the foundations of the future while they support the burden of organizational structure

University of Western Ontario Model (Bontis, 1996)

Bontis considers that IC is the sum of HC, structural and relational capital.

With regard to HC, the author considers that innovation is the result of the communication and learning processes of individuals that take place in the organization. We can say that HC is a source of innovation, strategic renovation and value for the company and that it is made up of knowledge stock, both of tacit and explicit type, which members have. Bontis considers that HC explains the other two components of IC.

Intangible Assets Monitor (Sveiby, 1997)

In this model, intangible assets are valued with regard to business strategy. Sveiby considers that these resources include the internal and external structure of the company and the competencies of employees.

The problem related to measurement of intangible assets is related to the need to identify the specific flows that change or influence the market value of the company. The author establishes three types of indicators for each block of intangible assets. First, indicators of growth and renovation (considering the potential future of a company); second, efficiency indicators (inform about the productivity of intangibles) and, third, stability indicators (show the degree of permanence of these assets in the company). Innovation is considered to be a first-class factor in accordance with the previous indicators.

The author considers that not all employees must be taken into account within HC, only those that are true experts. Therefore, the remaining workers are left aside. He also establishes the existence of a dimension called employee competencies [1] that is defined as the capacity of organizational members to act consequently in different situations and create both material and immaterial assets.

Nova Model (Camisón et al., 2000)

The basic objective of this model is to measure and manage IC in any type of organization regardless of its size. The structure of IC is determined by four components (human, organizational, social, and innovation capital and learning). Innovation is considered as an individual capital on its own.

Human capital includes assets related to knowledge (tacit and explicit) that belong to people and it is divided into: technical knowledge, experience, leadership skills, teamwork skills, employee stability and managing skills with regard to prospective activities and advancing challenges.

Intellect Model (Euroforum, 1998)

Intellectual capital is integrated in three basic blocks: HC, structural capital and relational capital. Each must be measured and managed over time and towards the future.

Human capital is tacit and explicit knowledge that is useful for the organization and that people and teams belonging to a company possess, along with the capacity to generate it. They agree that although that which has this type of capital is each employee and not the organization, it is a part of company value and, therefore, must be considered organizational capital. For these authors HC is the basis for the generation of the other two types of IC.

Intangible assets belonging to HC are categorized in the present or future dimension. Within the present dimension, we would have personal satisfaction, employee typology, competencies of workers, leadership, teamwork, and stability or risk of losing stability. Within the future dimension, the model considers the improvement of employee competencies and the innovation capacity of people and teams.

For each element of HC, the model establishes several indicators to measure each element. This proposal can serve as a guide for each particular case of a given organization.

Organizational Learning Model of KPGM Consulting (Tejedor and Aguirre, 1998)

This model was developed as part of the "Logos Project: Research relating to the learning capacity of Spanish firms" by KPMG Peat Marwick Management Consulting. Three factors are considered relevant to explain organizational learning:

- The strong and conscience commitment of the organization to continuous learning at all levels of the company.
- The learning behaviors and mechanisms at all levels.
- The development of infrastructures that determine company operation and the behavior of people and groups to favor learning and permanent change.

According to this model, an organization can only learn if its employees learn. However, even if this happens, the knowledge gained may not be a useful asset for the organization. To achieve organizational learning, specific mechanisms should be developed to create, capture, store, transmit and interpret knowledge. This will make the most of and adequately use individual and team learning.

Stewart Model (1997)

Intellectual capital is seen as an intangible material that can be a source of wealth. It is the sum of HC, relational capital and customer capital.

The author establishes a series of principles that should guide IC administration. Those related to HC are the following:

- Firms do not own HC and customer capital. They share ownership of the first of these with employees and of the latter with suppliers and customers. To effectively manage these resources, the company must be aware of this sharing situation and think that inadequate managing will destroy value.
- To create usable HC, the organization must promote teamwork and other forms of social learning. Interdisciplinary teams learn, formalize and capitalize talent because they spread it and make it less dependent on the individual. This reduces the risk of losing talent if the individual leaves the firm.
- To administer and develop HC, consider that wealth is created through skills and talent. These are strategic resources for the firm due to their capacity to create value that, in the last stage, customers are willing to pay for. Talented people who do not create company value should be treated as costs to minimize, while those that do generate value must be seen as investment assets.
- Human capital and structural capital mutually reinforce themselves when the organization has its mission straight and the enterprising spirit surfaces. Symmetrically, these capitals also destroy each other when many of the organizational activities do not have value for customers or in situations in which the firm is concerned, to a greater extent, with conducts, instead of with strategies.

With respect to indicators, Stewart presents an open model that serves as a guide and allows each organization to elaborate a proposal closer to its own needs.

Intellectus Model (Bueno et al., 2003)

Intellectus is presented as an evolutionary model, based on social capital. The latter is the sum of present and potential resources that arise from individual or social relations.

Intellectual capital is subdivided into HC, structural (organizational and technological) and relational capital (business and social).

Each of these five components is integrated by basic elements that define the aspects that identify them. At the same time, each element is analyzed through a series of variables that represent the central object to measure and that must be managed with great efficiency, effectiveness and satisfaction for the parties involved. Each variable is associated with different indicators that help determine and create value.

Human capital makes reference to knowledge (explicit or tacit and individual or social) that people or groups possess and to the capacity of creating it and, therefore, it is useful for the organizational mission. This capital is integrated by what people and groups know and by their capacity to learn and share this knowledge in benefit of the organization.

Innovation is generated through technological capital and it acts as a link between internal values that arise from HC and organizational capital and external values related to business and social capital.

From these models, the following conclusions are reached:

They have many similarities. The concepts of HC are virtually identical and the real differences come from the emphasis given to some elements.

In whatever manner IC is categorized, HC is always present and this is due to its importance to business structure and function.

Related to innovation, signs of controversy become more visible. In this manner, some models consider this item as a part of the internal processes of the organization that increases firm value, while others see HC itself as a source of innovation.

4 A System of Indicators for Human Capital Management within an Innovation Context

The knowledge of an employee is based upon his or her skills and experiences and ability to absorb new knowledge. Therefore, while knowledge is a resource in its own right, the way in which knowledge is managed and used will affect the quality of services that can be leveraged from each resource owned by the firm (Darroch, 2005).

Although the potential value of HC is usually recognized by company managers, it is necessary to identify which of these investments actually have an effect on innovation capacity and how these connections take place.

This information, in the case of HC, is difficult to obtain (Golden and Raghuram, 2009) because we are working with individual virtues such as attitudes, motivation, learning capacities, etc. Moreover, keep in mind that these elements are not owned by the organization, so they can be lost if qualified employees leave the company. The organization must convert this knowledge into organizational knowledge. This leads us to include the value of all those elements that can favor value creating employee commitment apart from factors that help transform HC in structural capital, that is, assets belonging to the organization and not to individuals.

Nevertheless, the fundamental aim of a system of indicators for HC is not to value or measure knowledge directly, but to improve the capacity to create and exploit it. We will keep this in mind in our proposal and pursue clear connections between intangible resources and activities and consequent wealth (see Table 1). With regard to the frequency in the elaboration of indicators, it will depend on factors like the type of organization, the dynamics of the industry, or on strategy, although we recommend a minimum of once a year.

Table 1 A system of indicators for HC and innovation: a proposal

1. STRATEGIC EMPLOYEE COMPETENCIES	
	Percentage of strategic employees
	Level of education
	Experience
	Knowledge range and depth
	Training investment in the job
	Total investment in HC
	Employee rotation
	Effectiveness of personal selection sources
	Absenteeism ratio
	Cost of employee remuneration
	Variable retribution
	Employee skills and capacities
VALUES, APTITUDES AND ATTITUDES OF STRATEGIC EMPLOYEES	
	Employee satisfaction
	Degree of identification with corporate values
	Creativity and innovation
	Employees that have a professional development plan
	Competitive job creation

Our proposal classifies HC indicators into two large groups: those that belong to competencies of strategic employees and those that make reference to values and attitudes.

To discover strategic employee's competencies, it is necessary to take into account the following indicators:

1. *Percentage of strategic employees*: this measure informs of the level of HC concentration. Because HC only includes knowledge that has the potential to create organizational value, not all employees are taken into account [2].
2. *Educational level:* most models measure this through the number of professionals with higher education. However, high educational levels are relevant only in relation with specific aims and sector of each organization. Therefore, not all companies need professionals with higher education and each one should fix objectives taking strategy into account and measure educational level of HC in relation to those objectives.
3. *Experience:* this is a very important measure of HC, especially for organizations with high percentages of workplaces that require a wide range or specific knowledge regarding functions and technology used. It seems logical to think that more experienced employees will supply more new strategies or ideas for new products, etc., based on the experience acquired over time, compared to new workers.
   Experience must not be understood as learned knowledge in the present organization only, but also as experience acquired in other companies, workplaces, cultures, nationalities, etc. The emphasis is placed on the advantage of having a workforce with diverse experience because this should result in more flexible and productive organizations.
4. *Range and depth of knowledge:* one of the measures used to value this is HC value added, which reflects what part of the value added is due to each monetary unit invested in employees. It is calculated as the sum of all income minus all expenses not due to employees divided by the sum of wages and social fringe benefits.
5. *Investment in workplace training:* one way of influencing employee learning is to invest in formal training programs and processes. Within the different ways of measurement, we propose the investment in initial training (number of average hours of training needed to have productive workers) and the investment in workplace training (average number of hours of training per year).
6. *Total investment in HC:* this reflects the degree of organizational commitment to HC. It is the result of the sum of expenses of the human resource department and, therefore, includes cost of employee selection, hiring, professional development, motivation strategies, etc., plus salaries (including travel expenses and social services). The comparison of this indicator with that of other companies of the same sector can give us an idea of our relative HC allocation.
7. *Employee rotation:* as we explained in the second epigraph, one of the most threatening elements for HC deterioration is the voluntary abandonment of employees with high knowledge levels. Therefore, the index of personnel rotation must be included. We define it as the proportion of employees that leave the company each month.

8. *Effective sources of employee selection:* there are many ways to select employees and the use of one or another can significantly affect the adequacy of new employees to the workplace. We establish three indicators to measure this:

    - The comparison of a rotation index for the first six months or the first year for every selection source used; this period of time is usually sufficient to identify misunderstandings, conflicts and dissatisfaction.
    - The quality level of each employee will be measured by the hierarchical superior once the initial training period is concluded. This is done through a questionnaire regarding aspects related to personal and professional competencies of the new worker.
    - Efficiency of each selection alternative is measured as the time period needed to cover a given job position. The faster the process, the better the procedure, if it does not negatively affect quality and rotation.

9. *Absenteeism ratio:* the measure of absenteeism usually identifies workers with personal problems, such as dissatisfaction or lack of motivation. This has adverse effects for companies, especially if it affects key workers. This highlights the importance of its evaluation and tracking. We propose the percentage of unexpected absenteeism of employees as a measure for this.
10. *Remuneration cost per employee:* this variable represents the average cost of an individual employee. It is important to analyze which is the relative position of a professional's remuneration of each organization with respect to its sector. It is also possible to calculate this variable for each department or section.
11. *Variable remuneration:* one of the measures that have greatly improved performance is to connect salaries with yield. Our calculation proposal is the proportion of employees that have variable retribution, based on objectives.
12. *Skills and capacities of employees:* this refers to the knowledge related to the manner in which duties are performed, know-how, that is, employee skills acquired through work experience. To measure it, the immediate hierarchical superior or the human resources department must elaborate a questionnaire, based on a scale of 5 or 7 points, where employee skills and capacities are evaluated. The following items should be taken into account:

    - Employee skills related to the ability to cooperate, teamwork, interpersonal communication, problem-solving, information management, decision-making
    - Employee capacities to apply knowledge to practice, analyze and synthesize, organize and draw up plans, criticize and auto-criticize, work in interpersonal teams, learn, leadership, intellectually be able to adapt to dynamic changes in the organization.

To discover the values and attitudes of strategic employees, the following indicators must be taken into account:

1. *Employee satisfaction:* it is related to the employee's attitude at work and, of course, it is a rather subjective measure because it includes, along with other factors, the existence of equilibrium between contributions and compensations,

workplace recognition, the degree of connection and participation in strategic activities, the equilibrium between work and private life, work insecurity (due to the type of contractual arrangements), etc. Those workers with higher levels of satisfaction will, most probably, present higher performance measures and, therefore, are more productive. The relevance of this measure has grown over the past years because several researchers have demonstrated the link between this variable and the retention of workers within the organization and adequacy of customer services. To value employee satisfaction, we propose a personal survey with scales of 5 or 7 points.

2. *Identify with corporate values:* the degree of employee identification with organizational objectives, philosophy and practices, that is, employee commitment is closely related to workplace productivity. We consider that a personal survey, with a scale of 5 or 7 points, should be used to value personnel satisfaction.
3. *Creativity:* creativity is considered a precursor of innovation. For most organizations it entails a very important measure that determines product and service differentiation in the mid- and long-term. We suggest the following indexes:

    - Employee creativity through the analysis of suggestions and new ideas they present.
    - Innovation through the percentage of those new ideas that have succeeded.

4. *Number of employees that have a professional development plan:* the aim of professional development plans is to allow employees to have the possibility of being promoted within the organization. This directly influences motivation and commitment.
5. *Creation of competitive workstations:* as with personal development plans, the increase of strategic workplaces in the organization is important to retain HC and attract new competent employees. We establish the following measures:

    - Number of competitive workplaces created during the last year.
    - Percentage of employees promoted.

These indicators highlight the importance of HC in a given organization a along with its capacity for innovation.

5 Conclusions

Human capital management is a key organizational element for obtaining sustainable competitive advantages. Its effective administration sets up an enormous potential for value creation in the organization and, therefore, given the direct effect on innovation (Bozbura et al., 2007), special attention must be provided. Difficulties related to measurement of intangible resources are not an impediment for organizations to take them into account, manage and power their value increasingly. In this manner, employee competencies and their values and attitudes at work are prime elements to accomplish organizational mission in dynamic environments. Organizational capacities are, to a great extent, determined by the possession of these intangibles; to improve these, it is necessary to identify and manage the latter (Marr et al., 2004). Those related to knowledge resources can be considered a combination of processes, comprehension regarding the context and experience that is submitted to continuous accumulation. However, only those that add value and are rare, inimitable and do not have equivalent substitutes are able to generate sustainable competitive advantages (Barney, 1991, 2002).

In this paper, we have examined the most relevant contributions made by IC models with respect to HC and its relation to organizational innovation capacity. We have also gathered the most extended indicators used to measure it. Although the first research papers on IC date back to the middle of the previous century, the literature review reveals that there is still some confusion regarding the terms HC and its relation to innovation. In this sense, some models consider the latter as a consequence of investment decisions in HC, while others state innovation as another capital, individually considered, although most models accept the importance of HC as a powering factor of innovation in organizations.

From a strategic point of view, HC reveals employee knowledge as a key factor to develop innovation and commercialize it and, in this manner, it may determine the organization's competitive position, especially in technological intensive activities. Therefore, managers must have reliable, relevant and timely information regarding HC to make efficient management decisions and promote innovation.

In this context, we have developed a system of indicators for HC management with the objective of allowing for a clear picture of links between strategic human resources and wealth generation. The basic contribution of our work is its usefulness for managers; it may help them face up to existing challenges. We are, of course, aware of the ambitiousness of this idea although we also detect the need for organizations to have a common and easy to understand guide for HC and its implications on innovation.

We have taken into account several aspects related to HC. First, growth and renovation through indicators such as level of education, experience of strategic employees, the creation of competitive workplaces or the investment in HC. Second, we have also included factors regarding efficiency using indicators such as value added or the percentage of variable retribution. Third, we consider the importance of stability of HC and, therefore, include, for example, strategic employee rotation, level of satisfaction or identification with corporate values.

The originality of this system of indicators is that it is presented in a clear and brief manner and that it opens the possibility for decision makers to adapt it to a given organization and make comparisons between different companies. Of course, a future line of research should include an empirical study to check the importance of these indicators. In this sense, we are now working on the design of a survey and collecting data.

To conclude, we consider it is important for managers to take into account that not only HC has an effect on innovation, but also SC and RC. The innovation process of an organization depends, in great extent, on the incorporation of HC to productive realities and this, in turn, is supported by organizational structure and organizational external relations. Therefore, mastery of processes, organizational routines, customer accounts or ownership rights are a source of innovation success.

References

Alegre, J., Lapiedra, R.: Gestión del conocimiento y desempeño innovador: un estudio del papel mediador del repertorio de competencias distintivas. Cuadernos de Economía y Dirección de la Empresa 23, 117–138 (2005)

Aramides, E.D., Karacapilidis, N.: Information technology support for the knowledge and social process of innovation management. Strategic Management Journal 27, 621–639 (2006)

Barney, J.B.: Strategic factor markets: expectations, luck and business strategy. Management Science 32, 1231–1241 (1986)

Bontis, N.: There is a price on your head: managing intellectual capital strategically. Business Quarterly 60(4), 40–47 (1996)

Bontis, N.: Intellectual capital: An exploratory study that develops measures and models. Management Decision 36, 63–76 (1998)

Bozbura, F.T., Beskese, A., Kahraman, C.: Prioritization of human capital measurement indicators using fuzzy AHP. Expert Systems with Applications 32, 1100–1112 (2007)

Bueno, E., et al.: Hacia un modelo integrador de los procesos de negocio, conocimiento y aprendizaje en las organizaciones. In: Comunicación Presentada en el XIII Congreso Nacional ACEDE: Dirección de Empresas y Creación de Valor en Un nuevo Entorno Económico, Institucional y Cultural, Salamanca (septiembre 2003)

Camisón, C., Palacios, D., Devece, C.: Un nuevo modelo para la medición del Capital Intelectual en la empresa: El modelo Nova. X Congreso Nacional ACEDE, Oviedo (Septiembre 2000)

Carlucci, D., Marr, B., Schiuma, G.: The knowledge value chain –how intellectual capital impacts business performance. International Journal of Technology Management 27, 575–590 (2004)

Carmeli, A., Tishler, A.: The relationships between intangible organizational elements and organizational performance. Strategic Management Journal 25, 1257–1278 (2004)

Carson, E., Ranzijn, R., Winefield, A., Marsden, H.: Intellectual Capital. Mapping Employee and Work Group Attributes. Journal of Intellectual Capital 5, 443–463 (2004)

Darroch, J.: Knowledge management, innovation and firm performance. Journal of Knowledge Management 9, 101–115 (2005)

De Grip, A., Van Loo, J.: The economics of skills obsolescence: a review. In: de Grip, A., van Loo, J., Mayhew, K. (eds.) Research in Labor Economics, Understanding Skills Obsolescence, vol. 1, pp. 1–26. JAI Press, Amsterdam (2002)

Edvinsson, L.: Developing Intellectual Capital al Skandia. Longe Range Planning 30, 336–373 (1997)

Edvinsson, L., Malone, M.S.: Intellectual capital. Realizing your company's true value by finding its hidden brainpower, 1st edn. Harper Collins Publishers, Inc. (1997)

Euroforum: Proyecto Intelect. Medición del Capital Intelectual. Instituto Universitario Euroforum, Escorial, Madrid (1998)

Gallego, I., Rodríguez, L.: Situation of intangibles assets in Spanish firms: an empirical analysis. Journal of Intellectual Capital 6, 105–126 (2005)

Hatch, N., Dyer, J.: Human capital and learning as a source of sustainable competitive advantage. Strategic Management Journal 25, 1155–1178 (2004)

Hill, C.W.L., Rothaermel, F.T.: The performance of incumbent firms in the face of radical technological innovation. Academy of Management Review 28, 257–274 (2003)

Kaplan, R., Norton, D.: Balanced Scorecard-Measures that Drive Performance. Harvard Business Review, 71–79 (January-February, 1992)

Lev, B.: Intangibles: management, measurement and reporting. The Brookings Institution, Washington, DC (2001)

Littlewood, H.: Análisis factorial conformatorio y modelamiento de ecuación estructural de variables afectivas y cognitivas asociadas a la rotación de personal. Revista Interamericana de Psicologíca Ocupacional 23(1), 27–37 (2004)

Lynn, B.: Intellectual Capital. Key to value added success in the next millennium 72 (1), 10–15 (1998) CMA Magazine

Martín, G., Alama, M., Navas, J.E., López Sáez, P.: El papel del capital intelectual en la innovación tecnológica. Una aplicación a las empresas de servicios profesionales de España, Cuadernos de Economía y Dirección de la Empresa 40, 83–110 (2009)

Martínez Ramos, M.: De la contabilidad de los recursos humanos al capital intelectual. Una ampliación necesaria. Dirección y organización: Revista de Dirección, Organización y Administración de Empresas 29, 134–144 (2003)

Marr, B., Schiuma, G., Neely, A.: Intellectual capital –defining key performance indicators for organizational knowledge assets. Business Process Management Journal 10(5), 551–569 (2004)

Nonaka, I., Takeuchi, H.: The knowledge creating company: How Japanese companies create the dynamics of innovation. Oxford University Press, Oxford (1995)

Phillips, J.J.: Invertir en el Capital Humano. Estrategias para no gastar demasiado...o demasiado poco. Ed. Deusto, Barcelona (2006)

Plessis, M.: The role of knowledge management in innovation. Journal of Knowledge Management 11(4), 20–29 (2007)

Steward, T.A.: La nueva riqueza de las naciones: el Capital Intelectual, Granica, Buenos Aires (1998)

Subramaniam, M., Youndt, M.A.: The influence of intellectual capital on the types of innovative capabilities. Academy of Management Journal 48, 450–463 (2005)

Sveiby, K.E.: The intangible assets monitor. Journal of Human Resource Costing and Accounting 2(1), 73–97 (1997)

Tejedor, B., Aguirre, A.: Proyecto Logos: investigación relativa a la capacidad de aprender de las empresas españolas. Boletín de Estudios Económicos LIII (164), 231–249 (1998)

Trillo, A., Rodríguez, O.: La influencia de la innovación en el capital intelectual de una empresa. Propuesta de un modelo. Conocimiento, Innovación y Emprendedores; Camino al Futuro, 1419–1431 (2007)

Tseng, C., Goo, Y.J.: Intellectual capital and corporate value in an emerging economy: empirical study of Taiwanese manufacturers. R&D Management 35, 187–201 (2005)

Youndt, M.A., Subramaniam, M., Snell, S.A.: Intellectual capital profiles: an examination of investments and returns. Journal of Management Studies 41, 335–361 (2004)

Footnotes

[1] For the specific case of individual competencies, indicators proposed are: a) growth and renovation indicators (experience, educational level, training costs, personnel rotation, customers that promote competitiveness); b) efficiency indicators (proportion of professionals, value added by professionals) and c) stability indicators (average age, length of service, remuneration position, rotation of professionals).

[2] From here on, all indicators are based on strategic employees, that is, on personnel whose knowledge creates value or HC.

Use of Web 2.0 in the Audiovisual Series

Noelia Araújo Vila and José A. Fraiz Brea

Department of Business Administration and Marketing, University of Vigo,
Faculty of Business and Tourism, 32004, Ourense, Spain
{naraujo,jafraiz}@uvigo.es

Abstract. Web 2.0 is already a reality, a reality that every day increases the number of followers who use the tools that are part of it. Multiple sectors have glimpsed a possible way to promote and market their products or services through these channels (social networks, blogs, forums ...). An example is the audiovisual sector, a booming sector that gives alternatives to enjoy leisure time. Among these alternatives, audiovisual series have achieved a prominent place and have adapted themselves to the current environment being visible on the network and using the phenomenon 2.0.

Keywords: Web 2.0, Audiovisual series, The Internet, Social network, Forum, Blog.

1 Introduction: The New Digital and Multimedia World. Web 2.0

Current figures are far from 1980, a decade that Internet began its journey on a commercial basis after its inception as a tool of Defense Department of the United States (Kahn, Leiner, Pork, Clark, Kleinrock, Lynch, Postel, Roberts and Wolff, 1997:140-149). In the XXI century, more than 2,000 million people use Internet, being Asia the continent at the head, with 922,329,554 users (Nielsen Online, ITU and the Internet World Stats, 2011) (Table 1).

Table 1 World population and Internet users

Regions	*Population 2011*	*Internet users 2000*	*Internet users 2011*
Asia	3,879,740,877	114,304,000	922,329,554
Europe	816,426,346	105,096,09	476,213,935
North America	347,394,870	108,096,800	272,066,000
Latin America/Caribbean	597,283,165	18,068,919	215,939,400
Middle East	216,258,843	3,284,800	68,553,666
Oceania	35,426,995	7,620,480	21,293,830
Africa	1,037,524,058	4,514,400	118,609,620
TOTAL	6,930,055,154	360,985,492	2,095,006,005

Source: Nielsen online, ITU and Internet World Stats (2011).

A.M. Gil-Lafuente et al. (Eds.): Soft Comput. in Manag. and Bus. Econ., STUDFUZZ 287, pp. 45–59.
springerlink.com

Currently, the user can access a socio-cultural network where people are able to find the information they need. Internet use is increasing, becoming a medium used by most of the population, going in just a decade, from 360,985,492 users, to more than 2,000 million, which is a rise in its use of 480,36% (Nielsen Online, ITU and the Internet World Stats, 2011).

The main uses that until recently were given to the network of networks were seeking information and email access, activities that continue, although in recent decades, this medium has diversified its use, resulting in new trends, applications and functions, linked to problem solving, being entertained or to learn new things (Flanagin and Metzger, 2001: 162-163). To all of this is, it should be added to the world media and communication, either through forums, chats, blogs, instant messaging or calls, -Messenger, Skype-, and social networks, and all that it implies-photos, videos, podcasts...- in short, we can summarize as everything that surrounds the phenomenon of Web 2.0 or communication 2.0 (Zed digital, 2008).

Web 2.0 receives several definitions, being common characteristics the connectivity, participation, shared interests, belonging, identity and relationships (Castells, 2006) and its principles of collective intelligence, the role of customer relations between users and the collective knowledge generation (EconRed, 2009) (Figure 1).

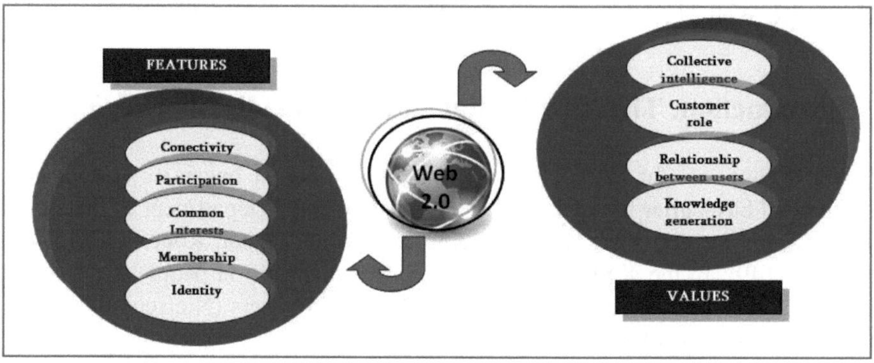

Source: personal compilation based on Castells (2009) and EconRed (2009).

Fig. 1 Web 2.0 features

2 Communication 2.0: Social networks, Forums, Chats and Weblogs

Web sites are an important source of information worldwide, with millions of queries that are made daily through search engines like Google or others (Spink, Wolfram, Jansen and Saracevic, 2000: 226). But this is not the only use the Internet gets, but we can quantify in a 45.8% (Orange Foundation, 2008) the number of Internet users who use the services related to Web 2.0, which can be segmented according to their activity within this phenomenon: from the most active, those who publish blogs and / or web sites, upload videos and / or photos, so-called "artists" who do not engage in any activity, so-called "inactives" (Figure 2).

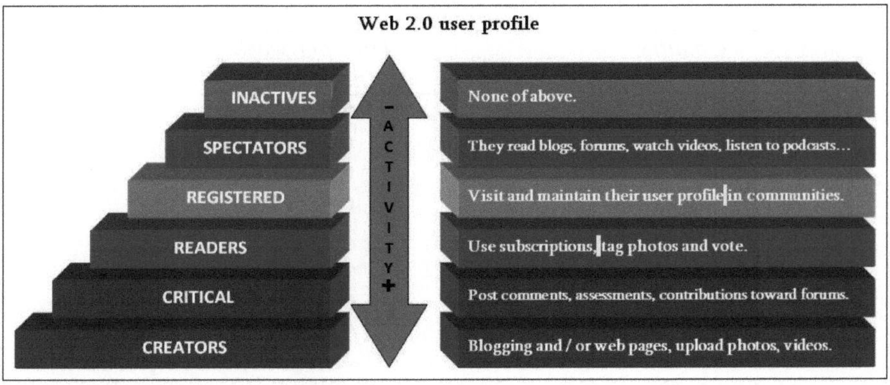

Source: personal compilation based on Li (2007:5).

Fig. 2 Web 2.0 User Profile

Within this world 2.0, stand out from other products or services, social networks, which have continued to win fans over the past years. For social network we understand "the ultimate online communications tool that allows users to create a public or semi-public profile, create and view their own networks, and other users, and interact with people in their networks" (Subrahmanyam, Reich, Waechter and Espinoza, 2008).

Despite its high degree of integration into society they are newly created, - decade of 2000- as the first social recognized network is Six Degrees, which was operational from 1997 to 2001. According Karinthy (1929). it only takes a small number of links to connect to any person for the rest of the population. This theory was supported and investigated by several authors, until 2001 Watts, after several experiments, concluded that "the average intermediate between any two people on a global scale is six" (Watts, 2006: 9).

Today, social networks are considered part of Web 2.0, as they have a number of characteristics of all products belonging to this phenomenon: connectivity, opinion, participation and relations, among others.

Most social networks we know and use today (Flicker, Myspace or Facebook), appeared in the early 2000s. 11 years later, interest in them has increased, both by sociologists and researchers, as the users themselves. It has been so successful and accepted in society, that 50% of Internet users are users of social networks, 38% do them but they do not use, and only 12%, do not know their existence (Zed Digital 2008).

The use of social networking has reached outstanding levels of consumption, with a global average of 5 hours and 27 minutes a week (The Nielsen Company, February 2010) - Table 2 -. Use in those under 25 years is highlighted, of which, with respect to the networks Facebook and Ttwitter, 19% stay connected 24 hours, 27% connect only occasionally and 32% as soon as they get up (Retrevo, 2010).

Table 2 Access to social networks by country / February 2011 Home and work

Country	*Time per week (hh:mm:ss)*
Average	5:27:33
Italy	6:27:53
Australia	6:25:21
United States	6:02:34
United Kingdom	5:50:56
Spain	4:50:49
Brazil	4:27:54
France	4:12:01
Germany	3:47:24
*Switzerland**	3:26:00
Japan	2:37:07
* only home	

Source: The Nielsen Company (2010).

Facebook has reached the top and is now the most advanced network, with over 800 million users (F8 Developer Conference, September 2011) and it is the one that has presence in more countries (Figure 3).

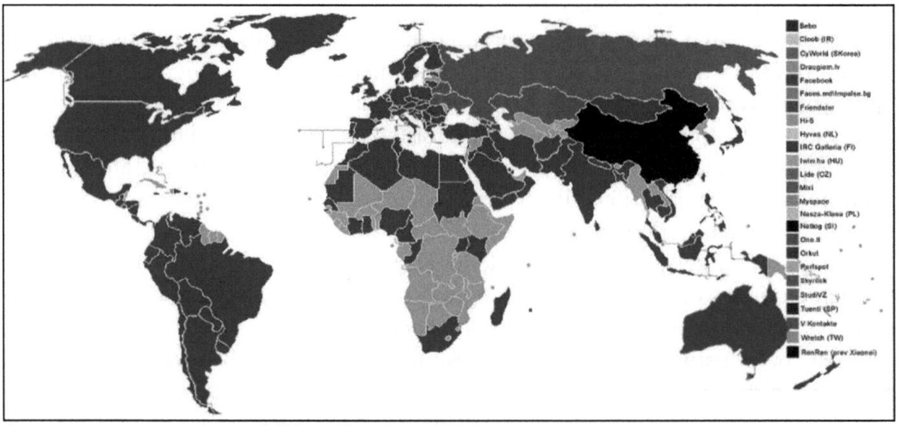

Source: www.oxyweb.co.uk/blog (October, 2011).

Fig. 3 Busiest social networks by country 2011

The figure above shows that social network Facebook is the most used in more than 80 countries (dark gray), followed by V Kontakte, present in several Asian countries. Only in China and Latvia, there are two networks more important than Facebook: Ren Ren and Draugien.lv. In just three years, Facebook has become the most used social network worldwide. The map below of 2008 shows the busiest networks in some countries, as Tuenti in Spain, Myspace in U.S. or Hi5 in various American, African, Asian countries as well as Portugal and Romania in Europe (Figure 4).

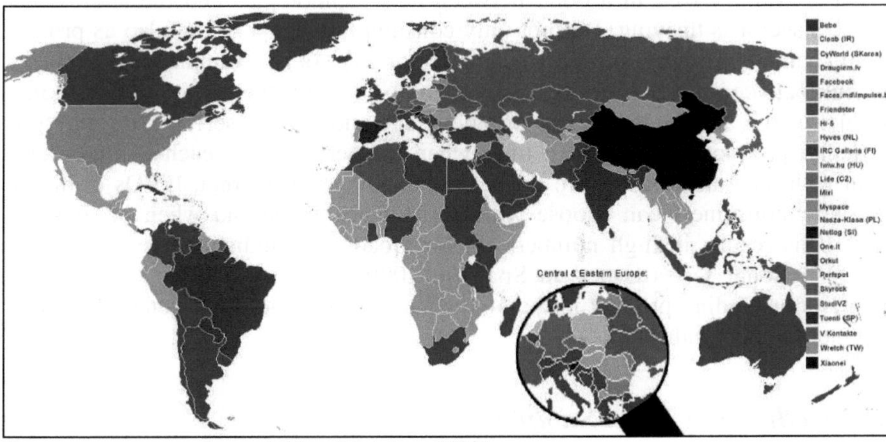

Source: www.oxyweb.co.uk/blog (October, 2008).

Fig. 4 Busiest social networks by country 2008

3 Application and Use of Tools 2.0 in the Audiovisual Series

It is therefore more than justified the proliferation of social networking and other Web 2.0 tools such as forums or blogs, and the importance attached by users of Internet and new technologies to them as a source of information and exchange of comments. There is an increasing in percentage of population that makes use of such tools. Therefore, various industries have not wanted to miss such an opportunity, including agencies and audiovisual companies, which have allocated resources so that their products are also present in these channels.

In recent decades, society has undergone notable changes in the workplace, forms of entertainment, communication or learning, all accompanied by a scanning process in which new technologies and evolution of computers have been protagonists (Franquet, 1999 : 1). Such changes have also affected in a direct way to audiovisual production, with examples such as the daily emergence of new formats, inventions or technologies that try to bring more simply, at any time, the products of this sector to the consumer. In the middle of 1990s, companies linked to the audiovisual industry, aware of these developments, began to invest in audiovisual media, such as Argentina, where they began to expand the bandwidth of television`s operators channels that required payment of this technology (Rossi, 1995).

As a result of this, the emergence of new media has happened continually, not only with the personal computer, but with the emergence of what might be called "new screens" defined as "PCs, mobile phones and PDAs, among others" (Murolo, 2009: 3). The Internet phenomenon, a network of networks, advances in technology come together in terms of media consumption and use of it, from computer to mobile phones of third and fourth generation, PDAs, iPods, MP4, digital TVs, consoles or terms which are beyond the reach of many market segments, such as tablet PC or webPads.

This is summarized in a bond between new technologies and the audiovisual sector, a sector is thriving. But not only consumption support, but also as promotional channels, coming into play in this sector 2.0. As an example, it is analyzed the link between the two fields, 2.0 and audio, through the use of tools 2.0 in a product of the audiovisual sector booming: the audiovisual series. But, why this particular product? We are in a time when the series have reached importance again with most accessibility to them: downloading via Internet, DVDs or digital TV. Now consumers can choose what series to consume and when to consume them. This results in high numbers of downloads and audiences, reaching only 41,099 fictional series issued in Spain in 2009 (EGEDA, 2010), which makes them an outstanding product in the audiovisual sector, and more generally within the entertainment industry.

3.1 Methodology and Sample

To check what is the use made by audiovisual series tools 2.0, we had chosen a qualitative research: the content analysis technique. To do this, we analyzed the presence of some series of international and national (Spanish) fiction in the social network Facebook, the most used worldwide (www.oxyweb.co.uk, 2011) and the use of tools 2.0 on the websites or blogs in the national series more valued by viewers in 2010. The chosen test specimen coincides with the most award-winning series, download or view in the years 2009 and 2010 (Tables 3 and 4).

Table 3 Sample international audiovisual series analyzed

Series	*Genre*	*Awards/ Ranking Downloads*
Heroes	Fiction	*Favorite Sci Fi/Fantasy Show* in *35th People's Choice Awards* and most downloaded series in the network (number 1 in the top ten)
Lost	Drama, suspense and fiction	Second most downloaded series in the network
24	Drama, suspense	Sixth most downloaded series in the network
Prison Break	Suspense and fiction	Fourth most downloaded series in the network
House	Drama	*Favorite TV Drama* in *36th and 35th People's Choice Awards*, and fifth most downloaded series in the network
Fringe	Suspense and fiction	Nominated for best television series in Saturn Awards 2009.

Table 3 *(continued)*

Series	Genre	Awards/ Ranking Downloads
Desperate Housewives	Comedy, intrigue	Seventh most downloaded series in the network
Grey's Anatomy	Drama	Eighth most downloaded series in the network
Gossip Girl	Drama, romantic comedy	Ninth most downloaded series in the network
Smallville	Drama, suspense and fiction	Number ten in BigChampagne ranking.
Dexter	Drama and suspense	Number four in TorrentFreak ranking (most downloaded series)
The Mentalist	Intrigue and drama	*Favourite New TV Drama* in *35th People's Choice Awards*
The Big Bang Theory	Comedy	*Favourite TV Comedy* in *36th People's Choice Awards y TCA Awards 2009*
True Blood	Drama and fiction	Best new series of the year in *TCA Awards 2009*
The mentalist	Intrigue and drama	*Favourite New TV Drama* in *35th People's Choice Awards*

Source: personal compilation based on BigChampagne, TCA Awards (2009), Saturn Awards (2009) and *Annual People's Choice Awards (*2010).

Table 4 Sample series analyzed national audiovisual

Audiovisual series	*Genre*	*Average audience (%)*
Águila Roja	Drama	*29,36*
Cuéntame cómo pasó	Drama	*25,22*
Hispania	Drama	*23,66*
Amar en tiempos revueltos	Drama	*22,33*
Gran reserva	Drama	*21,37*
La que se avecina	Comedy	*17,77*
Los protegidos	Fiction / suspense	*17,75*
El internado	Fiction / suspense	*15,71*
Aída	Comedy	*15,66*
Física o Química	Drama	*14,82*
Doctor Mateo	Drama/ Comedy	*13,15*

Source: personal compilation based on audience, laguiatv.com (2010).

3.2 Analysis and Results

All audiovisual series analyzed have their own group on Facebook and a considerable number of fans or followers (Table 5), and various applications such as tests to find out what is the character that you look over of a series (Facebook, 2010). In this space, as well as the information generated by the group managers (information from the plot of the series, players, pictures and even videos of a chapter), fans can make comments and replies to the comments of other users.

Table 5 Series presence in Facebook

Series	*Facebook Group*		
		International	
Heroes	Official HEROES Page	2.108.225	Information Links to news Protagonists 983 forums 44 photos 11 videos
Lost	Lost	6.873.620	Information Links to news Protagonists 7085 forums 122 photos 33 videos
24	24	2.774.815	Information Links to news Protagonists 60 photos More than 751 no official photos uploaded by fans 47 videos More than 25 no official videos uploaded by fans
Prison Break	Prison Break	3.221.410	Information Links to news Protagonists 28 photos More than 1.987 no official photos uploaded by fans More than 16 official videos More than 46 no official videos uploaded by fans

Table 5 *(continued)*

House	House	18.803.042	Information Links to news Protagonistas 1047 photos 51 videos More than 48 46 no official videos uploaded by fans Game: The HOUSE Daily Dose Game 810 forums
	Dr. House	16.795.495	2013 forums
Fringe	Fringe	2.156.615	Information Links to news Protagonists 588 photos 45 videos More than 24 no official videos uploaded by fans 261 forums
Desperate Housewives	Desperate Housewives	6.905.967	Information Links to news Protagonists 12 photos 16 videos Events: Desperate Housewives Season Premiere Sunday, September 26 on ABC 96 forums
Grey's Anatomy	Grey's Anatomy	12.024.114	Information Links to news Protagonists 43 videos 250 photos 599 forums
Gossip Girl	Gossip Girl	8.292.375	Information Links to news Protagonists 3 albums de fotos, 86 photos 105 videos 2775 forums "Play the game"

Table 5 *(continued)*

Smallville	Smallville	3.541.998	Information Links to news Protagonists 36 photo albums More than 1700 no official photos uploaded by fans 3032 foros 6 eventos
Dexter	Dexter	7.121.932	Information Links to news Protagonists 17 photo albums 1864 forums Event: Dexter at Comic-Con (San Diego) 80 videos More than 80 no official videos uploaded by fans
The Big Bang Theory	The Big Bang Theory	9.985.195	Information Links to news Protagonists Youtube links 13 photo albums More than 1466 no official photos uploaded by fans 3 events 554 forums
True Blood	True Blood	6.660.309	Information Links to news Protagonists 25 photo albums More than 19.134 no official photos uploaded by fans 3088 forums 26 videos
The Mentalist	The Mentalist	2.310.220	Information Links to news Protagonists 17 photos More than 295 no official photos uploaded by fans 327 forums 2 events: Premiere and Finale

Table 5 *(continued)*

		Spanish	
Águila Roja	Águila Roja	80.206	Information Links to news Protagonists 13 photos More than 153 no official photos uploaded by fans Game: my red eagle
Cuéntame cómo pasó	Cuéntame cómo pasó	3.911	Information Photos: 9
Hispania	Hispania	64.615	Information Photos: 10 albums, 109 photos 18 forums 3 events 8 videos
Amar en tiempos revueltos	Amar en tiempos revueltos	6.106	Information Links to news Protagonists Wikipedia link
Gran Reserva	Gran Reserva	8.159	Information Links to news Protagonists 2 photos More than 52 no official photos uploaded by fans 2 forums 1 video
La que se avecina	La que se avecina	53.208	Information Links to news Protagonists Wikipedia link
Los protegidos	Los protegidos	190.900	Information Links to news Protagonists 76 photos 46 forums
El internado	El internado	379.681	Information Links to news Protagonists 141 photos 102 forums 22 videos 11 events
Aída	Aída	103.859	Information Links to news Protagonists Wikipedia link

Table 5 *(continued)*

Física o Química	Física o Química	390.596	Information Links to news Protagonists 142 photos 7 events 42 forums
Dr. Mateo	Doctor Mateo	10.129	Information Links to news Protagonists 2 events: video-meetings 41 photos More than 47 no official photos uploaded by fans 3 forums

Source: personal compilation based on Facebook data (February, 2011).

In foreign series, known internationally, the number of followers or fans is more than two million in the majority of them, even some can reach the sixteen million fans as the group of *House* or twelve millions *Grey's Anatomy.* The profile of these groups follow the same model for all: information of the series (story and characters), a wall where the administrators of the group launched the series-related stories that can be commented by fans, official photos and videos and no official photos and videos uploaded by fans, forums and any section more like youtube links, events related to the series or even some application or game.

In the case of the Spanish shows, the number of followers is much lower, highlighting the cases of *Física o Química* and *El internado*, with 379,681 followers, and 390.5965, respectively, relevant figures for being much smaller series and less diffusion. Both series, along with *Los Protegidos* or *El Doctor Mateo,* follow the same format that foreign series, while the other series, like *Aida*, *Amar en tiempos revueltos* or *La que se avecina*,, follow a much simpler structure, information directly extracted from Wikipedia.In addition to social networks, there are other tools used by the audiovisual sector, and from 2000 a correlation analyzing study of terms through search engines, showed that a common inputs was *Buffy the Vampire Slayer* , a television series starring a vampire hunter, aimed at young people and teenagers (Spink et al, 2000: 232).

Today the series are still on the agenda in the rankings of search engines, an example is the Spanish case in the series *Física o Química*, third place of the most popular series in the ranking of most searched on Google in Spain (Google , 2009), followed by *Doctor Mateo* in sixth place and *La Señora* in tenth (Table 6).

It is therefore essential that the series have an official web site in which the viewer does not doubt the veracity of the information, or unofficial, where fans and supporters give their opinion and comments, through tools such as forums, blogs and chats.

As seen, the most successful series have a presence on social networks, in which these applications (forums or chat rooms) are enabled. Then we analyze the presence of the most successful Spanish series through their web sites, blogs or forums (Table 7).

Table 6 Ranking of most popular TVseries in the network in Spain in 2009

Position	*Series*
1	*Lost*
2	*Grey's anatomy*
3	*Física o Quimica*
4	*Doña Bárbara*
5	*Los Soprano*
6	*Doctor Mateo*
7	*Gossip Girl*
8	*Entre Fantasmas*
9	*The Wire*
10	*La Señora*

Source: personal compilation based on *Ranking Google* (2009).

Table 7 Spanish presence on the Internet series

	SPANISH		
Series	*Web site*	*Tools*	
Águila Roja	www.rtve.es/television/aguila-roja/ http://aguilaroja.mforos.com/	x	x
Hispania	http://www.antena3.com/series/hispania/	x	
Amar en tiempos revueltos	www.rtve.es/.../amarentiemposrevueltos.shtml blogs.rtve.es/amarologa/posts	x	x
Gran Reserva	www.rtve.es/television/gran-reserva/ blogs.rtve.es/granreserva/posts www.rutagranreserva.es	x	x
La que se avecina	www.telecinco.es/laqueseavecinalaqueseaveci natk.blogspot.com/	x	x
Los protegidos	www.antena3.com/series/los-protegidos/ www.losprotegidosweb.com/ www.losprotegidosfans.com/	x	x
El internado	www.antena3.com/series/el-internado/ www.elinternadofans.com/ www.elinternadoweb.com elinternadoblog.blogspot.com/ www.zonaelinternado.com/	x	x
Aída	www.telecinco.es/aida http://aida.mforos.com/	x	x
Física o Química	http://www.antena3.com/series/fisica-o-quimica/ http://www.fisicaoquimica.net/ http://fisicaoquimica.blogspot.com/	x	x
Dr. Mateo	http://www.antena3.com/series/doctor-mateo/ http://sanmartindelsella.es/	x	x

Source: own elaboration (2011).

4 Conclusions

Currently the tools of Web 2.0 (social networks, blogs, forums or chats, among others), have achieved outstanding consumption figures by the population, showing the number of followers of the popular social network Facebook, which reaches more than 800 million users (F8 Developer Conference, September 2011). Therefore, many sectors have glimpsed a possible way to promote and market their products or services through these channels.

One example is the audiovisual sector, integrated into the entertainment industry, industry with great weight on today's result of current trends and current consumer habits; a consumer interested in finding alternatives that make him enjoy his free time . Among these alternatives, an audiovisual product with its origins in the 1950s, emerges or re-emerges: the audiovisual series. The audiovisual series have reached the spotlight again, with the novelty of being more accessible to people: downloading via Internet, DVDs or digital TV. Now consumers can choose what to consume and when to consume it. They are doubly favoured from the emergence of new technologies with the emergence of new media consumption, and on the other hand with the creation of places or virtual meeting places (networks, forums or blogs) which can be annotated by his followers and promoted. Such is the importance of having a presence through these tools, that all the series with some recognition and hearing, make use of these tools.

Thus, at the international level, the highest-rated series and discharged *(House, Lost, Dexter, True Blood* and *Grey's Anatomy*, among others) have their own group on Facebook, groups of over two million followers (fans) in the analyzed sample , leading to significant figures as the 16 million followers of the group of *House* or the group 12 millions *Grey's Anatomy*. On the Spanish scene, the series also have a presence on Facebook, with much lower numbers of fans to the before mentioned, but relevant to the context in which these series are viewed: more than 390,000 followers *Física o Química* and over 370,000 followers *El Internado*. The structure of these groups is very similar in all cases, which is due to the limited uses within the application (Facebook):

- *A section with general information of the series*
- *Links to related news*
- *Albums for pictures and videos*
- *Forums*

Regarding the overall presence online audiovisual series, it shoud be mentioned the use of their own search engines (Google or others) as a source of information. A 2000 study of Google search engine itself shows how one of the most common entries is *Buffy the Vampire Slayer*, a television series starring a Vampire Slayer. These search engines redirect to web sites, forums, blogs or other online media, hence the importance of web presence. As an example, we analyzed the Spanish scene, in which all of the sample series have official web site (via the official web site), and other no official, blogs and forums.

In short, today, having web presence is inevitable for any industry that wants to be known and come to a large number of potential consumers, as Web 2.0, with the audiovisual sector, one that already benefit from the advantages that it entails.

References

Annual People's Choice Awards (2010), http://www.peopleschoice.com/pca/

BigChampagne (2010), http://bcdash.bigchampagne.com

Castells, M.: La sociedad red: una visión global. Editorial Alianza (2006)

ECONRED: Desarrollo de Comunidades Sociales. Canal Empresarial, Social Tec, ITH (2009)

EGEDA: Informe Panorama Audiovisual 2008-2009 (2010)

F8 DEVELOPER CONFERENCE 2011 (2011), https://f8.facebook.com/

Flanagin, A.J., Metzger, M.J.: Internet Use in the Contemporany Media Enviroment. Human Communication Research 27, 153–181 (2001)

Franquet, R.: Comunicar en la sociedad de la información. Revista de Estudios de Comunicación (7), 1–11 (1999)

Fundación Orange: Informe eEspaña: Informe anual sobre el desarrollo de la sociedad de la información en España (2008)

INTERNET WORLD STAT (2011), http://www.internetworldstats.com/

ITU (Organización de las Naciones Unidas para las tecnologías de la información y la comunica ción) (2011), http://www.itu.int/es/

Kahn, R., Leiner, B.M., Cerf, V.G., Clark, D.D., Postel, J., Roberts, L.G., Wolf, S.: The Evolution of the Internet as a Global Information System. En Intl. Inform & Libr. Rev. 29, 129–151 (1997)

Karinthy, F.: Chains (1929)

LA GUÍA TV: laguiatv.com (2010)

Li, C.: Social Technographics, p. 116. En Forrester Research, Inc. (2007)

Murolo, N.L.: Nuevas Pantallas frente al concepto de televisión. Un recorrido por usos y formatos. Razón y palabra (69) (2009)

NIELSEN ONLINE (2011), http://www.nielsen-online.com

OXYWEB (2011), http://www.oxyweb.co.uk/blog

RETREVO: Retrevo Gadget Census (2010)

Rossi, D.: La propiedad de los medios audiovisuales en la Argentina noliberal. En Causas y Azarares 3 (1995)

Spink, A., Wolfram, D., Jansen, B.J., Saracevic, T.: Searching the Web: The Public and their Queries. Journal of the American Society for Information Science and Technology 52(3), 226–234 (2001)

Subrahmanyam, K., Reich, S., Waechter, N., Espinoza, G.: Online and offline social Networks: use of social networking sites by emerging adults. Journal of Applied Development Psychology (6), 420–433 (2008)

TCA Awards (2009), http://teenchoiceawards.com/

Watts, D.J.: Seis grados de separación, Paidós Transiciones [traducción a español de, Six Degrees. W.W. Norton & Company, New Cork (2006)

ZED DIGITAL: El fenómeno de las redes sociales. Percepción, usos y publicidad (2008), http://www.zeddigital.es/Estudio%20de%20redes%20sociales_20_11_2008.pdf

Typologies of University Spin-Off Support Programmes: United Kingdom and Spain

José M. Beraza and Arturo Rodríguez

Department of Financial Economics, University of the Basque Country,
Plaza de Oñati 1, 20018 San Sebastián, Spain
{josemaria.beraza,arturo.rodriguez}@ehu.es

Abstract. This study identifies different models of programmes for supporting the creation of spin-offs in universities, and analyses their differentiated characteristics. The analysis was performed using data collected by way of a survey, targeted at the heads of spin-off support programmes in universities in the United Kingdom and Spain, to which we applied the cluster analysis technique, and then a Anova analysis to confirm their results. This enabled us to identify in these universities three types of spin-off support programmes, differing in terms of experience, resources, university´s commitment, proactivity, selectivity, number of spin-offs created and rate of survival. Among the various types of programmes identified, we found one that appear to be model for success. This type seems to opt for a policy intermediate of proactivity and selectivity. We have also found the existence of a certain "country effect" on the characteristics of the successful model. Finally, we have been able to confirm the importance given by the literature to university R&D activity and to the existence of a favourable environment to the success of spin-off programmes.

Keywords: University entrepreneurship, Knowledge transfer, Academic spin-offs, Technology transfer offices.

1 Introduction

Studies on the mission of the university have placed increasing importance on transferring research results to the market as a source of development and competitiveness (Etzkowitz et al., 2000; Geuna, 1999; Mowery and Sampat, 2001; Shane, 2004). However, this is a complex interactive process, involving a wide variety of agents (Etzkowitz et al., 2000). Conscious of the difficulties it poses, universities and governments have begun establishing policies for promoting this type of development (OECD, 1999).

The creation of knowledge-based firms has become particularly important in recent decades (Callan, 2001; European Commission, 2002; Shane, 2004), bringing with it a proliferation of university support programmes for setting-up spin-offs (Golob, 2003; Helm and Mauroner, 2007; Shane, 2004). Nonetheless, there are considerable differences in the way the various programmes are structured and

A.M. Gil-Lafuente et al. (Eds.): Soft Comput. in Manag. and Bus. Econ., STUDFUZZ 287, pp. 61–75.
springerlink.com

operate, in terms of their aims, strategies, functions and activities and the organisational structures and services they offer (Clarysse et al., 2002; Clarysse et al., 2005; Degroof, 2002; European Commission, 2002; Roberts and Malone, 1996; Wright et al., 2007). In addition, many of these programmes are newly-founded and consequently have neither a solid organisational structure nor clearly identified activities (Clarysse et al., 2005; Heirman and Clarysse, 2004). For this reason, different models of academic spin-off support programmes have been proposed (Clarysse et al., 2002; Clarysse et al., 2005; Degroof, 2002; Roberts and Malone, 1996).

Now, these works are intended to U.S. universities and some European countries like France, Belgium, United Kingdom, the Netherlands, Germany, Italy and Ireland. Therefore, it may be interesting to analyze the reality in other countries, and validate the models identified.

For this, the main aim of this work is to identify the different models of spin-off support programmes to be found in British and Spanish universities, and to analyse their differentiated characteristics. We believe that this study is of particular interest, since by studying universities from two European countries with a different degree of experience in developing spin-off support programmes we have a varied sample that may identify different types in supporting the creation of spin-offs and possible types characteristics of universities in the United Kingdom or Spain[1].

Using data collected by way of a survey, targeted at the people in charge of university programmes for supporting the creation of spin-offs, a statistic analysis was performed applying univariate and multivariate techniques. Firstly, we have applied the technique of factorial analysis to identify the most significant variables explaining the characteristics of these programmes. Secondly, using the cluster analysis technique, we have classified the universities. Using a one-factor analysis of variance (Anova) we went on to describe the differentiated characteristics of each of the clusters found.

The paper consists of an introduction and five other sections. In the second one, we review the literature on types of university spin-off support programmes, identifying the different models suggested by different authors. Thirdly, we explain the research methodology used. Fourthly, we perform a multivariate statistical analysis, using the cluster analysis technique, to identify the different existing models of spin-off support programmes in British and Spanish universities. Fifthly, we make a descriptive statistical analysis to characterise each of the models identified and highlight its main differentiating features. In the sixth section, we summarises the main conclusions obtained.

2 Typologies of University Spin-Off Support Programmes

We now proceed to review the main typologies of university spin-off support programmes developed in the literature. These mostly consist of qualitative research based on case studies.

---

[1] Choosing these two countries was due to a slightly higher response rate received in United Kingdom compared to the rest of Europe.

Roberts and Malone (1996) can be considered the pioneers in this area. For them, the two main dimensions of a policy of knowledge commercialization through the creation of spin-offs are *selectivity* and *support*, for each of which they distinguish two levels: high and low; they argue that there are only two viable models of support policy: *low selectivity/low support* and *high selectivity/high support.*

A *low selectivity/low support* policy means that many spin-offs are created to which little support is given. The choice of the projects eligible for support is left to external agencies (for example, venture-capital funds), which are seen as having greater experience and expertise in choosing possible "winners" and prevent possible conflicts of interest. This strategy makes sense in settings in which spinning-off is common practise and venture capital is in abundant supply.

By contrast, in a *high selectivity/high support* policy, a small number of spin-offs is created to which a large degree of support is given, giving them large possibilities of success. This strategy is more likely in environments where spinning-off is more unusual and venture capital is scarce. In this context, a university wanting to develop a support policy has no choice but to take the place of the financial market, playing the role of a financial investor, and encouraging a culture of entrepreneurship.

Degroof (2002) analyses how an environment unfavourable to entrepreneurship affects the type of spin-offs created in the academic institutions. Based on a study of spin-off creation in five organisations of this type, he distinguishes two models for spinning off in this type of environment. One model is that pursued by the specialist research institutions and the other is more typical of universities.

The process used by the specialist research institutes has a long incubation period, lasting several years. The new company is only incorporated when it can avail of protected technology, a business plan with strong market potential, a convincing business model to draw on and lastly an entrepreneurial team capable of leading the project with the help of venture-capital organisations, company directors from the industry and other advisers.

In contrast, the process pursued by universities does not normally include incubation or assistance in developing the business plan. The spin-offs are founded at a very early stage, when the project is still undefined. As a result, in most cases the business is developed after the spin-off's incorporation, when it is already up and running as a firm.

The author concludes that the process adopted by the specialist research institutes involves pursuing a high selectivity/high support policy, whereas the process identified in the universities involves a low selectivity/low support policy, thus contradicting Roberts & Malone's conclusions. However, the author notes this process is not static, but becoming increasingly more sophisticated as the institutions learn from their experience.

This approach sheds lights on one important practical aspect: the difficulty of establishing from the outset a high selectivity/high support policy in an environment that is unfavourable to entrepreneurship. Implementing such a policy requires considerable resources and skills, which may not exist in the universities; moreover, implementing them involves, among other factors, bringing about a considerable cultural and structural change.

Even in unfavourable environments, therefore, a university can start out in a position of low selectivity/low support and gradually move towards a position of greater selectivity and support, although it remains to be seen whether all universities are capable of following this path and at what speed.

For their part, Clarysse et al. (2002), Clarysse et al. (2005) and Wright et al. (2007) analyse the extent to which these institutions differ when it comes to organising and managing spin-off activities, the resources needed to undertake these activities appropriately, and whether the differences in organisation and management lead to different types of spin-off.

They initially distinguish three models of support for the creation of spin-offs: the *low* or *self selective* model, the *supportive* model and the *incubator* or *protective* model.

In the *low selective* model the aim is to generate as many spin-offs possible. To this end, the purpose is to stimulate entrepreneurial initiative, relegating to a second place the analysis of the economic or financial potential of the initiatives. This means that the main activities are opportunity search and awareness creation. The firms created may be based not only on technology, but also on skills developed within the university. Hence many companies are created, but only a few have the ambition to grow, and even fewer to actually succeed. In short, the companies created can be classed as *lifestyle* spin-offs.

In the *supportive* model, the aim is to create companies with economic potential and a will to grow, which will set up in the immediate territorial area of the institution, foster regional development and forge links with the institution that will encourage industry relations. In this model, management of intellectual property and business plan development, and public and private funding to allow the projects to develop in the initial stages are essential. Consequently, fewer spin-offs are created than in the previous model. Companies created in this model can be classed as *prospector* spin-offs.

The *incubator* model seeks to create solid companies, with the result that venture capital organisations are involved in them from the outset. Applying this model requires a research group that enjoys international recognition in a given technology. In addition, the technology transfer unit must be capable of incubating the project, facilitating the recruitment of outside managers, attracting international venture capital from the initial stages and forming the base of the company's intellectual property. Companies created along these lines can be classed as *venture capital backed* spin-offs.

In a later work, extending the study sample, the authors identified two further categories, differing from the three previous models: *resource deficient* organisations and *competence deficient* organisations.

In the *resource deficient* group they include spin-off programmes which have ambitious goals, but lack the necessary resources to achieve them. Because of these deficiencies, such programmes are positioned as being weak support and are therefore incapable of achieving the aims initially set out.

The *competence-deficient* group includes spin-off programmes that have sufficient resources, but lack the skills needed to perform the necessary activities, which can only be developed over time.

Summing up, we draw the following conclusions from the literature: firstly, there is no single model for a spin-off support policy; secondly, the two main dimensions of a policy of support for this type of company are *selectivity* and *support*; thirdly, it is very important that the universities, based on environment-specific conditions, know what their objectives are and clearly set out what resources and activities are required to achieve these objectives; and fourthly, establishing a high support and high selectivity policy requires considerable resources to which many universities rarely have access on an individual basis; and even if they can, they may require a certain period of time to develop the necessary skills to utilise these resources efficiently.

3 Research Methodology Used

3.1 Questions Asked

This empirical study used information taken in a survey on a series of variables related to resources and results, activities, organisational structure, the university's relations with the spin-offs, how long the programmes have been in place and the success achieved.

In selecting the questions used in the research we used a model of the linear process of valorisation via spin-offs. This model sets out the various stages a university needs to consider in supporting the creation of spin-offs, with particular emphasis on stages and activities at which the university's direct intervention can be decisively important. These can be grouped analytically into four basic stages: promotion of entrepreneurial culture, search for and detection of ideas, evaluation and valorisation of ideas, and creation of spin-offs. To these four stages we have added a further section corresponding to general information.

The boundary between these four stages is sometimes blurred, since there is a certain overlap between the different areas. Nonetheless, we feel that this division is useful as an analytical outline for deciding on the questions to be asked and indicating what aspects of the support programme they reflect.

3.1.1 General Information

The university itself is the most important element in the support programmes, as a source of marketable research results (Polt et al., 2001). For this reason, as well as the identifying data, three additional questions were included in the survey: the type of body performing the spin-off set-up support activities; the number of people in that body; and the activities related to this type of company carried out at the university.

3.1.2 Promotion of Entrepreneurial Culture

In unfavourable conditions for entrepreneurship, a prerequisite for creating spin-offs is the promotion of an entrepreneurial culture amongst university personnel (Henry et al., 2005; Klofsten, 2000; Trim, 2003). Four questions were included in the survey for this reason: the university's level of commitment to promoting entrepreneurial culture; the actions being taken to promote this culture; the number of people that will benefit from these activities and the degree of success obtained.

3.1.3 Search for and Detection of Ideas

Commercially exploitable ideas deriving out of university research do not normally arise spontaneously (McDonald et al., 2004; Shane, 2004; Siegel et al., 2003). Four questions were included in the survey for this reason: the university's level of proactivity in searching for and detecting ideas; the actions carried out to search for and detect ideas; the source of the entrepreneurial ideas and the degree of success obtained.

3.1.4 Evaluation and Valorisation of Ideas

The ideas initially detected need to be assessed to determine whether they meet a series of prerequisites for viable commercial exploitability (Wright et al., 2004). Likewise, the university, as an institution, and the people from whom the idea comes must support the project in order for it finally to become a spin-off (Vohora et al., 2004). Nine questions were included in the survey for this reason: the use of a specific methodology for evaluating and exploiting ideas; the use of external personnel in technological assessment; the use of external personnel in market assessment of the idea; the profile of these personnel; who assumes the leadership in promoting the spin-off; the role generally taken in the spin-off by the research group from which the idea originated; the number of exploitable ideas detected over a one-year period; the percentage of these ideas that are positively evaluated and, finally, the percentage of positively evaluated ideas that lead to the creation of a spin-off.

3.1.5 Creation of Spin-Offs

The ideas detected and supported lead to the creation of spin-offs, which can be of varying types and have varying degrees of success (Clarysse et al., 2005), and with which the university can have different links (Lockett et al., 2003). Nine questions were included in the survey for this reason: number of spin-offs created over the last five years; type of spin-offs created; average length of time from detection of the marketable idea to creation of the spin-off; most common source of financial resources; university's stake in the spin-off's capital; university's involvement in its management; survival rate of this type of company; percentage of firms that fail before three years and, finally the year in which the university began activities to support spin-off creation.

3.2 Population of European Universities, Selection of the Sample and Information-Gathering Technique

The study population comprises European universities in general, and Spanish and British universities in particular, that perform some type of activity involving spin-off creation.

The system used to identify the study population and select the sample was as follows: firstly, we searched in Google for a list of European universities, classified by countries; secondly, we visited the websites of each of these universities one by one; thirdly, we identified the body responsible for spin-off set-up support activities; and finally we identified the persons in charge of these bodies: name, position, telephone number and e-mail address.

Altogether a total of 74 universities were identified in the UK and 255 in the rest of Europe (not including Spanish universities). We sent these universities a letter of presentation by e-mail, inviting them to fill out the online questionnaire. Replies were received from 25 universities in the United Kingdom and 42 universities elsewhere in Europe, representing a response rate of 34% and 17%, respectively. In the case of Spanish universities, the letter of presentation was sent to practically all universities with a Technology Transfer Office (TTO), the great majority of which were public universities. 35 replies were received, representing a response rate of 58%. Altogether, a total of 389 letters were sent, and 102 replies were received, representing a response rate of 26%.

We then proceeded to discard universities with a certain number of incomplete replies or those that might be considered atypical, finally obtaining a database comprising a total of sixty-five universities, of which eighteen were in the UK, twenty-three in Spain and twenty-four elsewhere in Europe. The rate of complete replies is therefore 17% for the sample as a whole, 24% for British universities, 38% for Spanish universities and 9% for other European universities.

4 Multivariate Analysis

4.1 Factorial Analysis: Obtaining Representative Substitute Variables

We used the database drawn from the information obtained to make a factorial analysis to determine the most significant variables explaining the characteristics of the university spin-off support programmes.

Using this technique we proceeded first of all to identify factors. To do this, we used the *Principal Components Analysis* data reduction method. To determine the number of factors to be extracted, we used the *latent root criterion* technique. We then calculated the contributions of each variable to the different factors and selected the variables that contributed most to each one, in order to identify the variables that most appropriately described the university spin-off support programmes, and use them in the subsequent cluster analysis.

Forty-seven variables were originally used in this study, all the quantitative variables in the survey; and by applying the *latent root criterion* the final solution chosen was that formed by sixteen factors. In order to improve the solution, a Varimax rotation was used. This solution preserves 78.652% of total variability. We then selected the variables with the greatest load for each factor, as representative of each of the factors. In this way, we managed to group the original quantitative variables that are intended to represent different aspects corresponding to the different stages of support in the creation of spin-offs, into sixteen variables. Table 1 shows these variables classified according to the different phases of the process of exploitation by spin-off.

Table 1 Most significant variables classified according to the stages of the process of exploitation by spin-off

General information	• Number of people who are part of the body. • Relative importance of the promotion of entrepreneurial culture. • Relative importance of search for and detection of ideas.
Promotion of entrepreneurial culture	• Success in the promotion of entrepreneurial culture.
Search for and detection of ideas	• Importance of monitoring of the projects undertaken by research groups. • Importance of monitoring of business design/project competitions. • Relative importance of "Others" in the origin of ideas.
Evaluation and valorisation of ideas	• Relative frequency of postgraduate-PhD students as leaders in promoting spin-offs. • Relative frequency of external personnel hired as leaders in promoting spin-offs. • Appropiate for research groups to provide technological consultancy to the spin-offs. • Percentaje of ideas detected over a one-year period that are positively evaluated. • Percentaje of positively evaluated ideas that lead to the creation of a spin-off.
Creation of spin-offs	• Percentaje of knowledge-based spin-offs. • Percentaje of "Other" spin-offs. • University's involvement in the spin-off's management. • Percentaje of spin-offs who dies before 3 years.

4.2 Typology of Spin-Off Support Programmes in the Universities of United Kingdom and Spain: Cluster Analysis

Once the most significant variables had been detected, we have classified the British and Spanish universities using the cluster analysis technique, from the sixteen most significant variables identified in the previous sub-section. The distance used was the *Euclidean distance squared*, following standardisation of the variables by transforming them into *Z scores* with mean 0 and standard deviation 1. The clusters were formed using hierarchical clustering and *Ward's method*.

There is no standard procedure for determining the final number of clusters, and in this case the clustering coefficient criterion does not give clear results, since it does not experience relevant changes when the number of clusters is varied. We therefore opted for obtaining various different cluster arrangements, from two to four, and, by using the one-factor analysis of variance (Anova), check whether there were significant differences between the clusters obtained.

Table 2 Means of all variables with significant differences in clusters[a]

	Ward Method			
	1	2	3	Total
Number of people who are part of the body[b]	(-)1,05	(+)2,36	1,20	1,44
Relative importance of evaluation and valorisation of ideas	(-)21,75	24,00	(+)29,00	25,00
University´s commitment to the promotion of entrepreneurial culture	3,30	(+)4,36	(-)3,00	3,51
Importance of business design/project competitions	2,90	(+)4,00	(-)2,80	3,17
Importance of entrepreneurship courses	3,15	4,09	(-)2,90	3,34
Importance of promoting the spin-offs created	(-)3,05	(+)4,09	3,10	3,34
Number of people benefiting from these activities	(-)1,40	(+)2,91	2,00	1,95
Success in the promotion of entrepreneurial culture[b]	2,85	(+)3,82	(-)2,60	3,05
Proactivity in searching for and detecting ideas	(-)3,00	3,27	(+)3,80	3,27
Importance of monitoring of the projects undertaken by research groups[b]	(-)2,45	(+)3,55	3,50	3,00
Importance of monitoring of business design/project competitions[b]	(-)2,60	(+)4,09	3,70	3,27
Success in searching for and detecting ideas	(-)2,75	(+)3,73	3,10	3,10
Number of ideas detected over a year	(-)1,50	(+)2,82	1,70	1,90
Percentaje of ideas detected over a one-year period that are positively evaluated[b]	(-)1,75	2,18	(+)3,60	2,32
Percentaje of positively evaluated ideas that lead to the creation of a spin-off[b]	(-)2,00	2,18	(+)2,90	2,27
University's involvement in the spin-off's management[b]	(-)1,40	(-)2,00	1,70	1,63
Percentaje of spin-offs who dies before 3 years[b]	(-)7,10	(+)26,82	7,50	12,49
Year in which the university began activities to support spin-off	(+)2000,40	(-)1996,27	1999,90	1999,17
Average number of spin-offs created between 2000 and 2004	(-)1,72	(+)5,35	2,93	2,9895
Average of patents	(-)7,73	(+)23,05	10,8	12,56

[a] The cluster marked with a (+) or (–) has the highest or lowest mean for the corresponding variable.

[b] Representative variables showing significant differences between the three clusters.

The results obtained in the Anova analysis show that there are a number of variables with significant differences that are practically the same for the three- and four-cluster arrangements and that differences are smallest for the two-cluster arrangement. We have therefore opted for the three-cluster arrangement, since it allows greater differentiation between the programmes, without proving excessive. In this cluster arrangement, eight variables have significant differences. Table 2 shows these variables.

5 Characterisation of Spin-Off Support Models

Having identified the clusters and the variables with significantly different means, we should now explain in more detail the characterisation of the clusters identified, as representatives of different models or types of spin-off creation support, comparing and contrasting them, to highlight possible similarities and differences.

For the purposes of this characterisation, we are going to use all the quantitative variables included in the survey that show significant differences between clusters, as well as one new variable not included: number of patents. The reason for including this one variable is as follows: given the somewhat subjective nature of cluster analysis and in order to ensure the validity and practical relevance of the solution obtained, it is recommendable to incorporate variables that have not been used to form the clusters, but which are known to vary in value from one another, as is the case with the variable chosen; moreover, this variable is relevant in this respect, as we show below.

Patent applications by universities are an indicator of university R&D results and their commercial orientation. While not all academic spin-offs are based on patented knowledge, a relationship can also be expected between the number of patents and the business creation process in the universities. However, in some cases the number of patent applications by the universities varies greatly from one year to another. For this reason, the variable used in the analysis was the average number of patent applications during the period 2002-2005 in the case of British universities, and 2000-2005 in the case of Spanish universities.

Table 2 shows the means of all the quantitative variables with significant differences obtained in the survey for the clusters of spin-off support programmes.

Taking as our reference these variables we go on to describe the profiles of each of the three types of programme identified:

Cluster 1 is made up of twenty universities, seven in United Kingdom and thirteen in Spain, with a relatively small level of patent activity, which have spin-off programmes started up recently and have a small number of people devoted to spin-off support; they give relatively minor importance to the evaluation and valorisation of ideas and, conversely, a relatively greater importance to the promotion of entrepreneurial culture; their commitment to promoting an entrepreneurial culture and the importance given to business design/project competitions, entrepreneurship courses and promoting the spin-offs created is limited, with the result that their activities benefit a small number of people and they have little success in promoting an entrepreneurial culture; they are not very proactive in searching for and detecting ideas, they place little importance on monitoring the projects undertaken by research groups and the business design/project competitions, with the result that they have little success in searching for and detecting ideas and, therefore, the number of ideas detected is very limited; they give a positive rating to and spin-off a small percentaje of the ideas detected; they generate few ventures, but with a high survival rate, and do not intervene in their running.

Cluster 2 is in turn made up of eleven universities, seven in United Kingdom and four in Spain, with a relatively high level of patent activity, experience in spin-off support, and with a large number of people devoted to spin-off; they give relatively greater importance to the evaluation and valorisation of ideas, and to the spin-off support; their commitment to promoting an entrepreneurial culture and the importance given to business design/project competitions, entrepreneurship courses and promoting the spin-offs created is high, with the result that their activities benefit a large number of people and they have quite success in promoting an entrepreneurial culture; they are quite proactive in searching for and detecting ideas, lesser extent the cluster 3; moreover, they place quite importance on monitoring the projects undertaken by research groups and great importance on monitoring business design/project competitions, with the result that they have quite success in searching for and detecting ideas and, therefore, the number of ideas detected is relatively high; they give a positive rating to and spin-off a small percentage of the ideas detected, even exceed the percentage of cluster 1; they generate a large number of spin-offs –though their rate of mortality is also high–, and they tend to intervene in their running, but not actively.

Finally, cluster 3 is formed by ten universities, four in United Kigdom and six in Spain, with a relatively small level of patent activity, which have spin-off programmes started up recently and have a small number of people devoted to spin-off support; they give relatively greater importance to the evaluation and valorisation of ideas and, conversely, a relatively minor importance to the promotion of entrepreneurial culture; their commitment to promoting an entrepreneurial culture and the importance given to business design/project competitions, entrepreneurship courses and promoting the spin-offs created is limited, with the result that their activities benefit a small number of people and they have little success in promoting an entrepreneurial culture; they are quite proactive in searching for and detecting ideas, they place importance on monitoring the projects undertaken by research groups and the business design/project competitions, but they have little success in searching for and detecting ideas and, therefore, the number of ideas detected is very limited; they give a positive rating to and spin-off a high percentaje of the ideas detected; they generate an intermediate number of spin-offs, with a reduced mortality rate[2], and do not normally intervene in their running.

6 Conclusions

In this paper we have set out to identify the different models of support programme that exist in universities in Spain and in the United Kingdom. In doing so, we have applied multivariate statistical analyses –particularly the cluster analysis technique– to the results of a survey of the people in charge of these types of programme. Having identified the models of support programme, we performed a statistical analysis to characterise them clearly. This enabled us to reach a series of conclusions, summarised below. We identified three clusters of spin-off support programmes in Spanish and in the British universities, differing in terms of

[2] Should be taken with some caution the low rate of mortality among clusters 1 and 3 due to the recent implementation of policies to support the creation of spin-offs in many of the universities included in them..

experience, resources, university commitment, proactivity, selectivity, number of spin-offs created and rate of survival; all of these variables are identified by the literature as determining their characteristics and results. These results confirm that universities pursue different spin-off support policies, utilise different spin-off creation processes and generate different numbers of companies of different characteristics, as various authors have indicated (Clarysse et al. 2002; Clarysse et al. 2005; Wright et al. 2007).

Thus, Type 2 appear to be a successful model, implementing a policy that comes close to the *supportive* model proposed by Clarysse et al. (2002), Clarysse et al. (2005) and Wright et al. (2007), whose aim is to create companies with economic potential and a growth ambition.

Moreover, although types 1 and 3 appear to show that some spin-off support programmes lack resources or competences, follow different policies. Thus, Type 3 follows a low selectivity policy, while Type 1 follows a high selectivity policy, which taking into account that in both cases these programmes are relatively recent, it appears do not confirm the conclusion of Degroof (2002), whereby universities will initially be forced to start with a low selectivity/low support policy, to move gradually towards a position of greater selectivity and support.

Otherwise, the typology of the spin-off support programmes identified in this work enables us to make some recommendations for improving the least successful models.

Type 1 programmes must devote a greater quantity of resources and strengthen their competences. Universities that adopt such programmes appear to be clear on the support policy they wish to pursue, but lack the resources and competences needed to put them into practice. If they want to improve their results, therefore, they need to devote a greater quantity of resources to their programmes, and strengthen their competences through training and hiring of specialist personnel, as well as by establishing collaboration networks with external agents specialising in each of the activities in the process. One alternative for universities lacking sufficient scale could be to group together to create joint spin-off programmes.

Type 3 programmes must prioritise the establishment of the support policy they wish to pursue. Although universities that adopt such programmes might seem to be pursuing a policy with a high proactivity in searching for and detecting ideas and low selectivity of the ideas detected that are finally spun off, it actually appears to be the lack of necessary resources and competences that leads them to pursue this policy. They must first of all establish the support policy they wish to apply, and it is therefore essential to have prior commitment from the university management.

Finally, the results obtained in this paper enable us to offer three additional observations.

Firstly, Type 2 –i.e., the "successful model"– have the greatest number of patents. This result is in consonance with the literature, which sees R&D activity as one of the factors related to the business creation process in the universities; at the same time, it also shows that patent applications by universities are an indicator of the results of university R&D and its commercial orientation.

Secondly, Type 2 universities, with only three exceptions, two Spanish and one British, are found in regions with above-average innovation behaviour for their respective countries (Hollanders 2007). An innovating environment, therefore, appears to have a positive effect on the characteristics and results of university spin-off support programmes. This result backs the literature's insistence on the importance of a favourable environment to the success of spin-off programmes (European Commission 2002; Hague and Oakley 2000; Wright et al. 2007), but it also calls into question Roberts & Malone's assertion that a policy of high support/high selectivity is more likely in unfavourable environments[3].

Thirdly, according to the results of cluster analysis, cluster 2 can be decomposed into two subtypes which show the existence of a "country effect" on the characteristics of successful university programmes. Thus, subtype 1, made up of Spanish universities, with one exception, has a lot of experience in spin-off support, has plenty of resources for this task and enjoys great commitment from the university; it is quite proactive in searching for and detecting ideas and follows a intermediate selectivity policy; with the result that it generates a comparatively high number of spin-offs, technology- and knowledge-based, but with a high mortality rate, and do not intervene in their running. In turn, subtype 2, only made up of British universities, even though is similar to subtype 1 with regard to experience, resources, commitment and selectivity; however, it is little proactive in searching for and detecting ideas; with the result that it generates a smaller number of spin-offs, mostly technology-based, but with a reduced mortality rate, and it intervenes actively in their running.

The analyses made in this paper are faced by a series of limitations. Firstly, the methodology used is insufficient for reflecting the complexity of support for the creation of university spin-offs. This study is cross-sectionally in nature, and therefore has no dynamic perspective and, since the process of creating spin-offs is by its very nature longitudinal, more studies of this type are needed. Secondly, the technique used for collecting information -survey of heads of spin-off set-up support programmes in universities- does not allow a complete apprehension of the issues associated with support for the creation of spin-offs, due to a series of factors: the variety of agents involved in the different stages of the process, their different forms of participation, the need to limit the contents of the survey in order to achieve a sufficient number of answers and the difficulty of obtaining sensitive information, such as for example the financial resources used by the programme or its origin. For this reason, it might be advisable to complement this work with qualitative studies in order better to understand the nature of the spin-off activity. Finally, the sample is not sufficiently representative. While the Spanish sub-sample covers the majority of the support programmes in Spanish universities, the same cannot be said of the sub-sample from British universities.

---

[3] This last statement needs to be qualified, since when the authors speak of a favourable setting, they are referring to the USA, and more specifically to universities such as MIT and Stanford, which continue to be an international reference point for spin-off support programmes. However, these two cases are atypical, even in America.

References

Callan, B.: Generating Spin-offs. Evidence from across the OECD. Science Technology Industry Review 26, 13–56 (2001)

Clarysse, B., Lockett, A., Quince, T., Van de Velde, E.: Spinning off new ventures: a typology of facilitating services. Institute for the Promotion of Innovation by Science and Technology in Flanders, Brussels (2002)

Clarysse, B., Wright, M., Lockett, A., Van de Velde, E., Vohora, A.: Spinning out new ventures: a typology of incubation strategies from European research institutions. Journal of Business Venturing 20(2), 183–216 (2005)

Degroof, J.-J.: Spinning off new ventures from research institutions outside high tech entrepreneurial areas. PhD Thesis, Massachusetts Institute of Technology (2002)

Etzkowitz, H., Webster, A., Gebhardt, C., Cantisano, B.R.: The future of the university and the university of the future: Evolution of ivory tower to entrepreneurial paradigm. Research Policy 29(2), 313–330 (2000)

European Commission: University spin-outs in Europe. Overview and good practice, Office for Official Publications of the European Communities, Luxembourg (2002)

Fernández, J.C., Trenado, M., Ubierna, A., Huergo, E.: Las nuevas empresas de base tecnológica y la ayuda pública, evidencia para España. Economía Industrial 363, 161–177 (2007)

Golob, E.R.: Generating spin-offs from university-based research: an institutional and entrepreneurial analysis. PhD Thesis, The State University of New Jersey (2003)

Hague, D., Oakley, K.: Spin-offs and start-ups in UK universities, Committee of Vice-Chancellors and Principals of the Universities of the United Kingdom, London (2000)

Heirman, A., Clarysse, B.: How and Why do Research-Based Start-Ups Differ at Founding? A Resource-Based Configurational Perspective. Journal of Technology Transfer 29(3/4), 247–268 (2004)

Helm, R., Mauroner, O.: Success of research-based spin-offs. State-of-the-art and guidelines for further research. Review of Managerial Science 1(3), 237–270 (2007)

Henry, C., Hill, F., Leitch, C.: Entrepreneurship education and training: can entrepreneurship be taught? Part I. Education + Training 47(2), 98–111 (2005)

Hollanders, H.: 2006 European Regional Innovation Scoreboard (2006 RIS). European Commission, DG Enterprise (2007), `http://www.proinnoeurope.eu/ScoreBoards/Scoreboard2006/pdf` (accessed June 1, 2009)

Klofsten, M.: Training entrepreneurship at universities: a Swedish case. Journal of European Industrial Training 24(6), 337–344 (2000)

Lockett, A., Wright, M., Franklin, S.: Technology Transfer and Universities' Spin-Out Strategies. Small Business Economics 20(2), 185–200 (2003)

Mcdonald, L., Capart, G., Bohlander, B., Cordonnier, M., Jonsson, L., Kaiser, L., Lack, J., Mack, J., Matacotta, C., Schwing, T., Sueur, T., van Grevenstein, P., van den Bos, L., Vonortas, N.S.: Management of intellectual property in publicly-funded research organisations: Towards European Guidelines. Office for Official Publications of the European Communities, Luxembourg (2004)

Mowery, D.C., Sampat, B.N.: University Patents and Patent Policy debates in the USA, 1925-1980. Industrial and Corporate Change 10(3), 781–814 (2001)

OECD, University Research in Transition. OECD Publications, Paris (1999)

Polt, W., Rarner, C., Gassler, H., Schibany, A., Schartinger, D.: Benchmarking Industry Science Relations: the role of framework conditions. Science and Public Policy 28(4), 247–258 (2001)

Roberts, E.B., Malone, D.: Policies and structures for spinning off new companies from research and development organizations. R&D Management 26(1), 17–48 (1996)

Shane, S.: Academic Entrepreneurship. University Spinoffs and Wealth Creation. Edward Elgar Publishing Limited, Cheltenham (2004)

Siegel, D., Waldman, D., Link, A.: Assessing the impact of organizational practices on the productivity of university technology transfer offices: an exploratory study. Research Policy 32(1), 27–48 (2003)

Trim, P.R.J.: Strategic marketing of further and higher educational institutions: partnership arrangements and centres of entrepreneurship. The International Journal of Educational Management 17(2), 59–70 (2003)

Vohora, A., Wright, M., Lockett, A.: Critical junctures in the development of university high-tech spinout companies. Research Policy 33(1), 147–175 (2004)

Wright, M., Birley, S., Mosey, S.: Entrepreneurship and University Technology Transfer. Journal of Technology Transfer 29(3/4), 235–246 (2004)

Wright, M., Clarysse, B., Mustar, P., Lockett, A.: Academic Entrepreneurship in Europe. Edward Elgar Publishing Limited, Cheltenham (2007)

Information Technology: Potential in the Colombian Banking Sector

Álvaro F. Moncada N[1] and Cristina Isabel Dopacio[2]

[1] CESA – Colegio de Estudios Superiores de Administración
Calle 35 # 6-16, Bogotá, Colombia
amoncada@cesa.edu.co
[2] Universidad CEU San Pablo
Julián Romea, 23 - 28003 Madrid, España
dopacio.fcee@ceu.es

Abstract. The purpose of this work is to determine the importance of the interaction of Information Technology (IT) resources and capabilities, in complement of organizational resources and capabilities to explain the differences in generation of sustainable competitive advantages and superior performance among Colombian banking institutions. It uses a framework based on the resource-based-view (RBV) to set three sets of factors (Human, Managerials and Technological) that are analysed according to their potential complementary to IT, through the correlation analysis that is applied to representative sample of Colombian banks. Result shows that the capabilities to adapt to change, business knowledge, improvement, teamwork and alignment have relevant roles on the effective development of the competitiveness in this sector.

Keywords: Information Technology, Resource-based view, Sustainable competitive advantage, Organizational performance, Colombian banking system.

1 Introduction

The globalization of markets, the economic integration, the disintermediation and deregulation, the rapid growth of technological innovations and the convergence of information technology and telecommunications has drastically changed the way companies compete, increasing levels of dynamism and the competitive environment between them, redefining and renovating the way business are conducted (Loukis et al., 2009; Overby et al., 2006, Jeffers and Muhanna, 2008, Chen et al., 2010), demanding flexibility, adaptability, and responsiveness to the market and forcing them to respond more effectively to the conditions of the global business environment. Since its emergence in organizations, IT has demonstrated that impacts positively on the business performance, increasing its capacity, redefining the management and business models and modifying the conditions and structures of the markets in different ways (McFarlan, 1984; Porter and Millar, 1985, Cash and Konsynski, 1986, Applegate et al., 2004). The improvement in the efficiency

A.M. Gil-Lafuente et al. (Eds.): Soft Comput. in Manag. and Bus. Econ., STUDFUZZ 287, pp. 77–95.
springerlink.com

of processes, reducing costs, and development of new channels, differentiation of products and services, and value for customers is some of these results (Hitt and Brynjolfsson, 1996; Brynson and Ko, 2004; Melville et al., 2004; Liang and Tanniru, 2007, Kohli and Grover, 2008, Nevo and Wade, 2010).

The development of processes, human and IT resources have significantly influenced the concept of organizations and the economic sectors where they compete. IT has become a resource of strategic importance for the achievement of competitive advantages of the firms (Porter, 1982; Porter, 1985; Parsons, 1983; McFarlan, 1984; Ives and Learmonth, 1984; Porter and Millar, 1985; Levy et al., 1999, Jeffers and Muhanna, 2008). Initially as a source of operational efficiency, costs reducer and differentiation generator; evolving the concept of fundamental factor of corporate behavior and the results obtained. The potential effect of IT has been the subject of different approaches with the purpose of identifying the circumstances under which this contributes to business results, either by itself or through complementary factors.

This study is based on RBV, which presents a reference framework to analyze its contribution as agent responsible of the accumulation of resources and capabilities, allowing determining the set of factors of the firm, which enhance and complement the IT for superior performance (Barney and Clark, 2007). Productive impact, organizational transformation and contribution to business performance, makes from IT an essential element in the definition of business strategy, not only to achieve greater efficiency on the operation, but as a determining factor in the strategic direction of the business (Martinez, 2005, Kohli and Grover, 2008).

2 Theoretical Framework

2.1 Resource-Based View

Achieving and sustaining competitive advantage is a fundamental issue in strategic management (Teece et al., 1997, Grant, 2010, Nevo and Wade, 2010). Peteraf and Barney (2003) pointed out that a firm "has a competitive advantage if it is able to create more economic value than the marginal (breakeven) competitor in its product market" (p. 314). This concept suggests that a firm has a sustained competitive advantage when it creates more economic value than the marginal firm in the industry and "when others firms are unable to duplicate the benefits of this strategy" (Barney and Clark, 2007, p. 52).

From this perspective, firms can be understood as a bundle of unique resources (tangible and intangible) which competitive advantages can arise from resource[1] heterogeneity and immobility (Barney, 1991; Bingham and Eisenhardt, 2008), that can persist over time (Teece et al., 1997) and become a source of sustained

[1] Not all the assets controlled or hold by the firm has the potential to be a source of sustained competitive advantage. Most companies have many resources (both tangible and intangible) but few that are strategic in nature. Most strategic assets tend to be knowledge-based and are intangible.

competitive advantage when they are valuable, rare, difficult to imitate and non-substitutable. The benefits of these resources must be properly managed (Sirmon, Hitt, and Ireland, 2007). In this way, RBV explains why an organization can out-perform others (Penrose, 1962; Wernerfelt, 1984; Barney, 1991, Barney and Clark, 2007, Grant, 2010, Nevo and Wade, 2010).

Barney (1991) in his framework has made two fundamental assumptions: (1) firm resources and capabilities are heterogeneously distributed among firms; and (2) resources and capabilities are imperfectly immobile. The relationship between resource heterogeneity and immobility and sustained advantage competitive is summarized in figure 1.

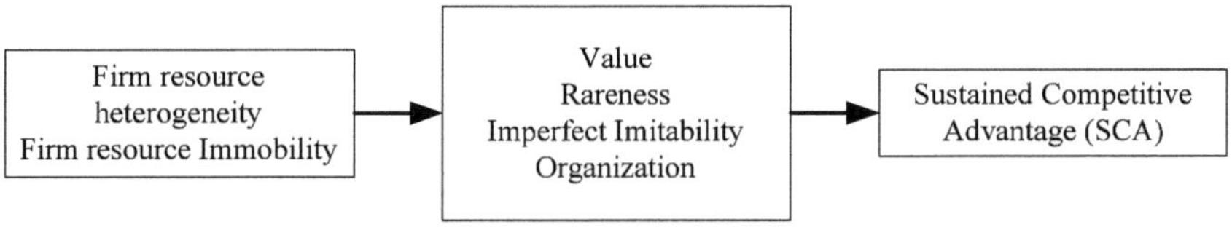

Source: Barney and Clark (2007, p. 69)

Fig. 1 Relationship between resource heterogeneity and immobility and SCA

The main principle for explaining the SCA in the RBV is the barrier to imitation (Mahoney and Pandian, 1992). According to Barney (1991) the firm's resources are imperfectly imitable for one or more of the following reasons: (1) resources obtained through unique historical conditions; (2) causal ambiguity and (3) social complexity. Bingham and Eisenhardt (2008) pointed out, SCA by inimitability can arise: (1) if a firm has property rights on resources that cannot be legally obtained by competitors; (2) if resource accumulation involves path dependencies and time compression diseconomies; or (3) if the linkages between resources and firm performance are causally ambiguous.

This concept has been developed by Barney (1991) in a Framework named VRIO[2], which represents the practical application of the RBV (Barney and Hesterly, 2008) which express the four characteristics that must be fulfilled by the firm resources in order to arise to SCA: (1) Valuable: valuable resources enable the firm to respond to environmental threats or opportunities, achieving at least the competitive parity, (2) Rare: valuable resources controlled by only a small number of competing firm that has potential to gain temporary competitive advantage, (3) Imperfectly imitable: if resources are valuable, rare, and inimitable, the firm has the potential to achieve long-term competitive advantage. Without this resource the firm will have a cost disadvantage, because they must obtain or develop, and

---

[2] Initially was known as VRIN: Valuable, Rare, imperfectly imitable and imperfectly substitutable (if competitors are able to find substitutable resources to support their competitive strategies).

(4) Organization, if the firm has policies, procedures and process oriented to support the exploitation of valuable, rare and costly to imitate resources.

The set of attributes of value, rarity, imitability and organization, provides a framework to understand the potential return associated with the exploitation of the resources and capabilities of the firm, as show in table 1.

Within the VRIO framework, if a resource is only valuable, it leads to competitive parity. Valuable and rare are required for a temporary competitive advantage. Value, rare and costly to imitate are essential for a sustained competitive advantage and an organizational focus is necessary to both develop a competitive advantage and sustain it. Barney and Clark (2007) pointed out that organization operates as an adjustment factor that if it fails some of its potential competitive advantage can be lost, obtaining only competitive parity or even competitive disadvantage.

Table 1 The VRIO framework

Valuable?	Rare?	Costly to imitate?	Exploited by Organization?	Competitive Implications	Economic Performance
No	--	--	No	Competitive disadvantage	Below Normal
Yes	No	--		Competitive parity	Normal
Yes	Yes	No		Temporary competitive advantage	Above Normal
Yes	Yes	Yes	Yes	Sustained competitive advantage	Above Normal

Source: Barney and Clark (2007, p. 70)

2.2 *Information Technology and RBV*

The researchers of the strategic role of IT and their potential for creating competitive advantage began in the 1980s (Cash and Konsynski, 1986; Ives and Learmonth, 1984; McFarlan, 1984; Parsons, 1983; Porter and Millar, 1985). From their works has emerged that IT can be used to create competitive advantage through efficiency improvements, differentiation and channel development (Piccoli and Ives, 2005). Later, in the mid-1990s, a wide variety of studies were developed, based on RBV (Mata et al., 1995; Ross et al., 1996 and Bharadwaj, 2000) to determine whether the firm´s efforts in the areas of IT can sustain a long-term competitive advantage. Many of these studies conclude that IT support the improvement of competitive position of firm (Dehning and Stratopoulos, 2003; Santhanam and Hartono, 2003; Ray et al., 2005). These works try to identify and define what constitutes IT capabilities (Jeffers and Muhanna, 2008; Tian et al., 2009).

The IT resources and capabilities are classified in different typologies Benitez,2009; Moncada,2011): Mata et al., (1995) defines five attributes of IT (Customer switching cost, access to capital, technology property, technical skills and managerial IT skills) to generate sustained competitive advantage; Ross et al., (1996) differentiate between IT assets (human, technological and relational assets) and IT process; Powell and Dent-Micallef (1997) classify IT resources into three categories (human resources, business resources and technological resources); Bharadwaj (2000) define six dimensions that categorizes in three areas: IT Infrastructure, Human IT resources and IT-enabled intangible; Wade and Hulland (2004) identify eight capabilities that grouped in three categories: inside-out (capabilities that are developed inside the firm), outside-in (capabilities that anticipate environment requirements) and spanning (capabilities that involve internal and external analysis and integrate the inside-out and outside-in capabilities).

Based on RBV, most of IT resources are valuable; however some of them are rare, less imitable and less substitutable than other IT resources, generating heterogeneity and imperfectly mobility in the firm, which causes differences in organizational performance (Benitez, 2009). However if IT can become a source of competitive advantage, it will be necessary to consider other resources that act as complementary of this (leveraging the value) to achieve and maintain the competitive advantage (Teo and Ranganathan, 2003, Jeffers and Muhanna, 2008).

Wade and Hulland (2004) pointed out that IT capabilities are repeated patterns used to exploit on a firm´s IT resources. Pavlou and El Sawy (2006) pointed out that IT capabilities are about acquiring, deploying, and leveraging IT resources to support and shape firm businesses. Benitez (2009) concludes that IT resources and capabilities should be complemented by other organizational resources and capabilities to enable improvements on the firm´s competitive position.

IT has transformed the business environment of the firm, suggesting that those organizations that have adequate IT resources and capabilities must obtain superior performances, either by a better manage of their costs, achieve higher profits, innovate their processes or products, or provide a better service to their customers (Maizlish and Handler, 2005).

2.3 The Colombian Banking System

In the last years, there have been a series of changes related to global trends in the supply of financial services. These trends include economic integration, technological change, increased competition, disintermediation, deregulation and financial crisis, which have forced financial institutions to constantly innovate their model to keep on "synchrony with the environment" (Palomo et al., 2010, p. 124) and to increase the business value.

"The search for operational efficiency, retention and growth of the customer base, acquisitions and mergers for expansion of markets, customer segmentation and custom product development" (Diaz, 2006, p. 1) emerge as major strategic choices of the sector, to achieve competitiveness in an increasingly complex and turbulent environment, where information technology is an indispensable platform for the development and implementation of the internal and external strategy in this sector of the economy.

Colombia has not been apart from this phenomenon in its financial market where it is assumed that this have led to an increase in competition during the last years (Murillo, 2009).

In particular, since the mid-1980s Colombia financial sector, has undertaken major changes in its structure, caused mainly by globalization, international competition and technological development that, together with economic cycles experienced by the country, has forced it to consider new strategies designing adjustments to the new business environment, in search of maintaining their competitiveness. In this environment, both for banks and nonbank financial institutions, Information Technology becomes a prerequisite for survival.

The Colombian banking system has been affected by a process of deregulation and consolidation. For this reason, Colombian financial institutions have reacted to the new market conditions. They were forced to reconsider their strategic options and to restructure. Between 1995 and 2009, the financial sector had a consolidation process, through mergers, takeovers and transformations; from 201 institutions to 65 financials intermediaries (18 of them are commercial Banks). Additionally, at the beginning of the 90's, the Colombian financial system was affected by an internationalization process with the incorporation of foreign banks, principally Spanish institutions, now there are only 7 foreign banks (Anif, 2006; Estrada, 2005; Garcia y Gomez, 2009).

In this century, the sector has a recovery period with positive growth rates and in many cases higher than the growth of the economy itself. In the recent global financial crisis has been affected in lower proportions than developed countries, in large part by the mechanisms of control and monitoring of risk, product of local crisis, were eventually implemented (Asobancaria, 2011).

Since its appearance in the Colombian banking sector, IT has demonstrated that it has impacted their business model to the extent, which has increased its operational capacity, its management, as well as the conditions and structures of the sector.

In Colombia, the advance of the banking sector has been product of the joint development of human resources and management as intangible factors act in combination with IT to produce improvements in the results of the banks, which can be observed in the competitive advantage and superior performance.

3 Research Methodology

The proposal model explores the relationships between Organizational resources and capabilities and IT resources and capabilities, based in three sets of factors that have been analyzed (and validated by factor analysis approach) according to their potential complementary and positive to IT.

This three complementary factors of resources and capabilities have been defined: human, management and technology, which according to previous studies (Ross et al., 1996; Powell and Dent-Micallef, 1997; Bruque et al., 2004 and Ravichandran and Lertwongsatien, 2005, among others) are considered as a whole, as elements that enhance the effect of IT on the results of the firm, which will be measured in this research work through generation of competitive advantage and Superior performance variables. The model developed in the research and which is intended to compare the proposed hypothesis, is shown in Figure 2.

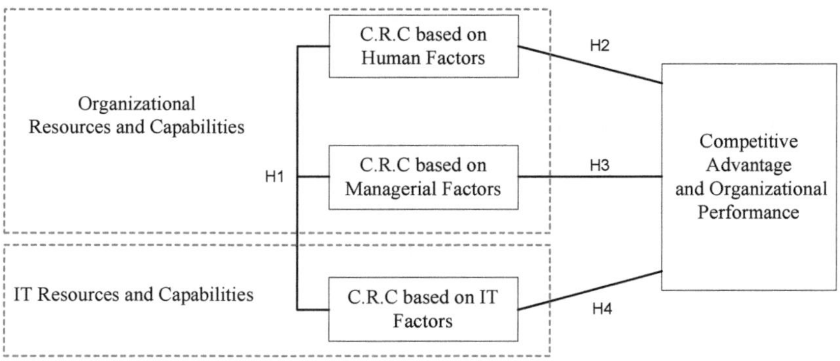

* C.R.C. – Complementary Resources and Capabilities

Source: Adapted from Bruque et al. (2004).

Fig. 2 Proposal Model

To reiterate, the main objective of this research is to determinate the significance of the interaction of IT resources and capabilities with organizational resources and capabilities, explain differences in performance among banking institutions, which in turn could contribute to explain firm-level-performance. To support this examination are used the researches and works indicated in table 2.

***Competitive advantage (VCO)*:** Porter pointed out that IT offers new innovative ways to compete through cost reduction and product differentiation. The use of IT to generate competitive advantages has been analyzed from different approaches such as strategic use of IT contribution to operational excellence and contribution to the strategic positioning (Ross et al., 1996, Tallon et al., 2000). Neo (1988) indicates that the analysis of market and customer needs can become creative elements capable of generating new strategic options based on IT to change the firm competitive position. Bhatt and Grover (2005) tested how knowledge of the business strategy by the IT teamwork and its relationship with other units of the organization, are positively associated with the competitive advantage of companies.

Table 2 Works and literature basis for questionnaire development

Definition	Literature Basis
IT as a strategic resource and as a complementary resource	Mata et al., 1995. Ross et al., 1996. Ross and Beath, 2002. Melville et al., 2004.
The technical and managerial skills	Mata et al., 1995.
The human assets, technology and related assets	Ross et al., 1996.
Factors boosters IT	Powell and Dent-Micallef, 1997.
Complementarity of IT resources	Ravichandran and Lertwongsa-tien, 2005
Model of the complementarity of resources	Jeffers, 2003.
Conditioning factors humans and managerials	Bruque et al., 2004.
Integrative model	Rivard et al., 2006.
Technical Capabilities in volatile environments	Tallon, 2007
Capabilities of IT Implementation	Tian et al., 2009.

Source: Moncada (2011, p. 229)

This construct is composed by the following variables: (1) Differentiation in products and services (VCO-DP); (2) Improved customer loyalty (VCO-LC); (3) Responsiveness to change (VCO-RC); (4) Products and Services of low cost (VCO-PS) and (5) Development of new segments (VCO-NS).

Superior Organizational performance (DSU): This construct examines results obtained by Colombian banks in implementing their IT strategies in their business plans, in order to understand the contribution that IT is building in various business areas and their interaction with the market and customers. It searches to establish the impact of IT on critical performance factors of the organization, determining the relationship between IT and superior performance and identifying the opportunities for some entities to generate value. The theoretical frameworks used in various studies to analyze the influence of IT on business performance (Powell and Dent-Micallef, 1997; Bharadwaj, 2000; Santhanam and Hartono, 2003, Bhatt and Grover, 2005; Oh and Pinsonneault, 2007) found that the positive effect of IT is greater when they are complementary to organizational resources (Brynjolfsson and Hitt, 2002). Tallon and Kraemer (2003), Avison et al. (2004) and Bergeron et al. (2004) have shown a positive contribution to organizational performance. Powell and Dent-Micallef (1997) find that when IT is exploited jointly with human resources and business resources there is a positive relationship. Bresnahan et al. (2002) and Galve and Gargallo (2005) confirm improvements in business performance.

This construct consists of six variables: (1) IT Contribution (DSU-CT); (2) Potential of IT (DSU-PT); (3) Improvement in Productivity (DSU-PR); (4) Improvement in competitive position (DSU-PC); (5) Increase in Sales (DSU-IV) and (6) Improved in profitability (DSU-MR).

Therefore are proposed the following hypothesis:

Hypothesis 1: IT resources and capabilities in the Colombian banking sector are complementary and their performance is leveraged in exploitation together with organizational resources and capabilities.

The Human Factor (FHU): refers to complementary resources and capabilities of human resources and their relationship to IT within the organization such as knowledge, skills, experience, training and learning capacity of the members of a bank, especially in their adaptive characteristics, communication and relationship (Araya et al., 2006) among others. Those elements related to the human factor inside the organization. Among these are the existence of an open and receptive climate in the organization, the fluidity of communication between management and technical staff and the leadership among supper management on the implementation of new technologies (Bruque et al, 2004).It also includes the core competencies and cross skills (functional skills) IT staff and the rest of the organization (Dorantes and Rao, 2006).

To develop the instrument, the factor FHU has referred mainly to the works of technical and managerial skills (Mata et al., 1995), human assets (Ross et al., 1996), human resources (Powell and Dent-Micallef, 1997), human conditioning factors (Bruque et al., 2004) and management skills in the use of IT (Bassellier and Benbasat, 2004).

This construct supports the following hypothesis:

Hypothesis 2: Complementary resources and capabilities based on the human factor are a source of sustainable competitive advantage and contribute significantly to higher organizational performance.

This construct (FHU) consists of six skills: (1) Adaptation and flexibility to change (FHU-OR); (2) Business Knowledge for IT (FHU-CN); (3) Technological competence of Senior Management (FHU-CT); (4) Incorporation of IT (FHU-AT); (5) IT Communication Organization (FHU-CA) and (6) Management commitment to IT (FHU-CD).

The management factor (FGO): This construct relates intangible assets and capabilities of the organization that are necessary for the development and successful implementation of IT. Also includes the application of business practices and management models that promote an organizational culture based on the use of IT as a strategic element of business. Encompass organizational culture, organizational structure, formal and informal relations between its members and their planning and control systems. The relationship and integration of IT with the rest of the business are core elements of this factor where business skills and interpersonal skills are positively related to business strategy for the use of IT in all areas of the organization (Bhatt and Grover, 2005).

To develop the instrument, the FGO factor have been based on works on teamwork skills (Mata et al., 1995), human assets and relationships (Ross et al ., 1996), complementary business resources (Powell and Dent-Micallef, 1997), human factors and management (Bruque et al., 2004).

This construct supports the following hypothesis:

Hypothesis 3: Complementary resources and capabilities based on the management factor are a source of sustainable competitive advantage and contribute significantly to higher organizational performance.

This construct (FGO) consists of eight variables: (1) Support and Sponsorship BB (FGO-RD); (2) Organizational Structure and Flexibility (FGO-FO); (3) Education and Training (FGO-FC); (4) Alignment and Planning (FGO-PI); Use of IT (FGO-UT); Interaction between teamwork (FGO-IE); (7) Process Improvement (FGO-DP) and (8) Shared Management Project (FGO-GP).

The IT Factor (FTE): This construct refers to the resources and capabilities of the IT infrastructure that are required for the development and implementation of this strategic element, in conjunction with human factors and management, allows the organization to adjust and reconfigure the environment (Beimborn et al., 2007). Models of planning and IT governance are the main component because through them the whole model set of IT management in the organization.

IT architecture with its components it is just one of its elements, given the fact that the generation of a competitive advantage and superior performance play an important role enhancing the effect that it is capable of providing as agility, flexibility, standardization, level of service and support to business processes.

The variables and factors used as support on the proposal model in this work are listed in table 3. Through them it is proposed confirming the generation of competitive advantage and superior performance by resources and capabilities complementary based on the human factor (FHU), resources and capabilities complementary based on the management (FGO) and resources and capabilities complementary based on Information Technology factors (FTE).

This construct supports the following hypothesis:

Hypothesis 4: The IT through complementary factors has positive effect on the generation of competitive advantages and superior performance in Colombian Banks.

Table 3 Constructs, Description and Variables

Constructs	Description	Variable
Competitive Advantage	Differentiation of products and services	VCO-DP
	Improvement of customer loyalty	VCO-LC
	Ability to respond to the change	VCO-RC
	Products and services lower cost	VCO-PS
	Development of new segments	VCO-NS
Superior Organizational Performance	IT Contribution	DSU-CT
	IT Potential	DSU-PT
	Improvement in Productivity	DSU-PR
	Improvement in competitive position	DSU-PC
	Increase in Sales	DSU-IV
	Improvement in profitability	DSU-MR
Based on the organization's human factor (FHU)	Adaptation and organizational flexibility to change	FHU-OR
	Business understanding for IT	FHU-CN
	Technological competition of senior management	FHU-CT
	Incorporation of IT	FHU-AT
	Open Communications between the Organization - IT	FHU-CA
	Executive commitment to IT	FHU-CD
Based on the organization's Managerial factor (FGO)	Senior Management Support and Sponsorship	FGO-RD
	Organizational Structure and Flexibility	FGO-FO
	Development and Training	FGO-FC
	Alignment and Planning	FGO-PI
	Utilization of IT	FGO-UT
	Interaction of teamwork	FGO-IE
	Improving processes	FGO-DP
	Shared Management of Projects	FGO-GP
Based on the organization's Information Technology factor (FTE)	IT Agility	FTE-AA
	IT Flexibility	FTE-FT
	IT Service Level	FTE-NS
	Standardization and Benchmarking	FTE-EP
	IT Architecture	FTE-AT
	Strategic value of IT	FTE-VE
	IT Investment Evaluation	FTE-IT

4 Data Analysis and Results

This study used surveys and personal interviews with senior level personnel (Managerial and technical directors) using a structured questionnaire which was validated by experts in Banking, with a pre-test, which helped to modify the language and rewrite items that were not clear. Initially the survey had 50 questions that were reduced to 32, following the expert suggestion. Each survey question requires a response based on 5-point Likert scale ranging from 1 to 5.

In the data collection process it sent banking institutions personalized invitations emails, which explained the study´s purpose, requested participation, assured the confidentialities of their responses and includes an excel questionnaire. To increase the response rate, it used a personal contact and networking efforts. By the end of the fieldwork, it has been obtained responses from 63.69% of banking institutions (the Colombian banking system is composed by 18 banks at 2010) for a total of 46 surveys of an initial objective of 72.

In order to demonstrate scale reliability, the Cronbach´s coefficient alpha was calculated as a reported in table 4.

Table 4 Reliability Analysis

Construct	Number of questions	Cronbach´s Alpha
Competitive Advantage	5	0.597
Superior Organizational Performance	6	0.527
Human Factor	6	0.804
Management Factor	8	0.831
Information Technology Factor	7	0.764

Although this index has not a minimum, some authors estimated at 0.35 the value for which ensures an acceptable level of consistency for each dimension (Van de Ven and Ferry, 1979, Powell and Dent- Micallef, 1997). Others recommend level above 0.70 (Hair et al., 1999). In this research there is a mean value of 0.705, so that expectations regarding the reliability of the scales were covered.

As indicated above, the three factors (FHU, FGO and FTE) were derived from previous studies and confirmed in the initial stage of this research through an analysis of factors approach which results were: (1) the determinant of correlation matrix is 1.53×10^{-13}, which is a very low value, indicating that the variables are correlated, (2) Barlett´s test of sphericity whose null hypothesis is Ho: the correlation matrix is the identity, which is rejected because the Chi-square is 1096.666

with 210 degrees of freedom and significance of 0, confirming the correlation between these variables, and (3) the Kaiser-Meyer-Oklin measure of sampling adequacy is 0.617 which is satisfactory for this analysis (Martin et al., 2008).

The analysis of correlation between human factor (FHU), management (FGO) factor and technological factor (FTE) is used for the test of hypothesis 1, found that the correlation is positive and meaningful among the three groups of factors: FHU-FTE (coefficient 0,829 with 99% confidence level) FHU-FGO (coefficient 0,451 with 95% confidence level) and FGO-FTE (coefficient 0,599 with 95% confidence level), with these coefficients and confidence levels it is accepted the proposed hypothesis. In this case were considered organizational resources and capabilities consisting on Human Factor and Managerial Factor.

The second hypothesis is contrasted by a comparison between the factor human (FHU) and the competitive advantage and superior performance (VCO_DSU). The correlation coefficient is 0,475 with a significance of 0.022, which shows that the level of significance is less than 0.05 (0.022 < 0.05) and confirms the existence of significant correlation.

The third hypothesis is contrasted by a comparison between the factor of management (FGO) and the competitive advantage and superior performance (VCO_DSU). The correlation coefficient is 0,543 with a significance of 0.007, which shows that the level of significance is less than 0.05 (0.007 < 0.05) and confirms the existence of significant correlation.

Finally the fourth hypothesis, correlational analysis bivariate between the technological factor (FTE) and the competitive advantage and superior performance (VCO_DSU), is used. Establishing the level of significance for measuring how relate these variables through the coefficient Spearman rho.

The correlation coefficient obtained is 0,455 with a significance of 0.029, which shows that the level of significance is less than 0.05 (0.029 < 0.05) and confirms the existence of correlation. These results are sufficient to test the hypothesis (positive and greater than 0.35) accepting the existence of the positive relationship between the technological factor and competitive advantage and superior performance for Colombian banking entities, with a 95% confidence level.

The table 5 contains a summary of findings by hypothesis.

Table 5 Summary of findings by hypothesis

Hypothesis	Factors	Spearman Coefficient	Significance	Supported Hypotheses
H1: IT resources and capabilities in the Colombian banking sector are complementary and their performance is leveraged in exploitation together with organizational resources and capabilities	FHU / FGO FHU / FTE FGO / FTE	0.451 0.829 0.599	$\rho < 0.05$ $\rho < 0.01$ $\rho < 0.05$	Confirm
H2: Complementary resources and capabilities based on the human factor are a source of sustainable competitive advantage and contribute significantly to higher organizational performance	FHU / VCO / DSU	0.475	$\rho < 0.05$	Confirm
H3: Complementary resources and capabilities based on the management factor are a source of sustainable competitive advantage and contribute significantly to higher organizational performance.	FGO / VCO /DSU	0.543	$\rho < 0.05$	Confirm
H4: The IT through complementary factors has positive effect on the generation of competitive advantages and superior performance in Colombian Banks.	FTE / VCO / DSU	0.455	$\rho < 0.05$	Confirm

5 Conclusions and Discussion

The results of theoretical and empirical development of this research to conclude that the generation of competitive advantages in the Colombian banking sector institutions are supported by the capabilities of responding to change, differentiation of products and services and developing new segments, while the superior performance, is supported on the capabilities of improved productivity, improved competitive position, improvement in profitability and potential of IT.

Similarly, states that in the Colombian banking, complementary resources and capabilities of the human factor are based on the capacity to adapt to change, business knowledge, integration of IT and strategic value of IT, resources and management capabilities are based on process improvement skills, teamwork interactions, alignment and IT planning and shared management of projects. All this, as complementary organizational capabilities of IT confirms the approaches developed in the theoretical and empirical, serving as a support for specific findings of this research.

Results of this work, there has been developed a theoretical framework and an empirical model that has shown the importance, potential and positive impact of Information Technology in the strategy of the Colombian banking and its impact on their competitiveness.

However, has been found that some complementary organizational capabilities and IT, despite its strategic impact, are not in sync with the rest of the development of organizational capabilities or have a low correlation with the factor they belong, thereby causing inconsistencies or differences with studies that have demonstrated their value or strategic importance: (1) IT-business knowledge or understanding of the nature of business processes and IT staff, which is not consistent with the work of Mata et al. (1995), Bhatt and Grover (2005) and Araya et al. (2006) who have shown that knowledge of the business by IT contributes to the effective use of IT and is positively related to competitive advantages, (2) Open Communications between the Organization and IT or the ability to establish fluid communication between the IT team and the organization for successful interaction, which is not consistent with the investigations of Neo (1988), Mata et al. (1995), Feeny and Wilcocks (1998), Basellier and Benbasat (2004), Bruque et al. (2004) and Dorantes and Rao (2006) who have pointed out the need of clear and accurate communication as a mechanism for building trust between members of IT and business areas as a basis for effective use of IT , (3) Education and Training or priority in the allocation of resources for training and skills development in the use of IT in the organization, which disagrees with the approach of Mata et al. (1995) and Bruque et al. (2004) who consider it as the basic element for the development of the culture of use and appropriation of IT, and (4) Organizational Structure and Flexibility understood as the adaptability of the organizational structure, processes and levels to suit the changing environment, which diverges with the studies of Beimborn et al. (2007) who determined the flexibility as critical enabler of the company's capabilities to respond to changes.

6 Limitations and Future Research

Although Colombian banking system represents 79.54%, of financial system, it should be noted that results of this research cannot be generalized to credit institutions (Financial Corporations and Cooperatives) as their business models are focused on institutional and corporate clients in the first and customer-members in the second. However, it is clear that the development of IT lags in these institutions, being their reasons the subject of a possible line of investigation.

The main lines of research suggested are: apply the same model to contrast and analyze the variables in similar investigations in other financial sub-sectors such as cooperatives, financial corporations, insurance companies and pension funds; apply the same model and contrast analysis for the variables in similar research elsewhere; Extending the Analysis Model by adding other variables that measure the impact of IT; Finally, it would be interesting to repeat periodically this study to analyze the changes in those institutions that have gradually introduced new IT processes.

References

Anif: Fusiones y Adquisiciones en el Sector Financiero Colombiano: Su Impacto sobre la Eficiencia (1990-2005). Reportes de ANIF, 1–50 (2006)

Applegate, L., Austin, R., McFarlan, F.: Estrategia y Gesti6n de la Información Corporativa, Retos de la economía en Red. McGraw Hill Internamericana de España, España (2004)

Araya, S., Orero, A., Chaparro, J.: Los Recursos y Capacidades y los Sistemas y Tecnología de Información: Una perspectiva organizativa integradora. In: Memorias X congreso de ingeniería de la organización, Valencia, España (2006)

Asobancaria: La banca colombiana: un ejemplo de innovación, confianza y expansión. Semana Económica 789, 1–12 (2011)

Avison, D., Jones, J., Powell, P., Wilson, D.: Using and validating the strategic Alignment model. Journal of Strategic Information Systems 13, 223–246 (2004)

Barney, J.: Firm Resources and Sustained Competitive Advantage. Journal of Management 17, 99–120 (1991)

Barney, J., Clark, D.: Resource-Based Theory – Creating and Sustaining Competitive Advantage. Oxford University Press, New York (2007)

Barney, J., Hesterly, W.: Strategic management and competitive advantage: Concepts and cases. Prentice Hall, Upper Saddle River (2008)

Bassellier, G., Benbasat, I.: Business competence of information technology professionals: Conceptual development and influence on IT-business partnerships. MIS Quarterly 28(4), 673–694 (2004)

Beimborn, D., Franke, J., Wagner, H., Weitzel, T.: The Impact of Operational Alignment on IT Flexibility - Empirical Evidence from a Survey in the German Banking Industry. In: Proceedings of Americas Conference on Information Systems (AMCIS), pp. 1–12 (2007)

Benitez, J.: Recursos de Tecnología de la Información y Desempeño Organizativo: El rol mediador de la capacidad de agilidad empresarial, Unpublished doctoral dissertation. Universidad de Granada (2009) `http://hera.ugr.es/tesisugr/18066343.pdf` (accessed September 12, 2011)

Bergeron, F., Raymond, L., Rivard, S.: Ideal patterns of strategic alignment and business performance. Information and Management 41(8), 1003–1020 (2004)

Bharadwaj, A.: A Resource-Based Perspective on Information Technology Capability and Firm Performance: An Empirical Investigation. MIS Quarterly 24(1), 169–196 (2000)

Bhatt, G., Grover, V.: Types of Information Technology Capabilities and Their Role in Competitive Advantage: An Empirical Study. Journal of Management Information Systems 22(2), 253–277 (2005)

Bingham, C., Eisenhardt, K.: Position, leverage and opportunity: a typology of strategic logics linking resources with competitive advantage. Managerial and Decision Economics 29, 241–256 (2008)

Bresnahan, T., Brynjolfsson, E., Hitt, L.: Information Technology, Workplace organization and the demand for skilled labor: Firm-level evidence. Quarterly Journal of Economics 117, 339–376 (2002)

Brynson, K., Ko, M.: Exploring the Relationship between Information Technology Investments and Firm Performance using Regression Splines Analysis. Information & Management, 1–13 (2004)

Bruque, S., Hernandez, M., Vargas, A.: Condicionantes humanos y de gestión en la implantación y desarrollo de las tecnologías de la información y de la comunicación: una aplicación al sector de distribución farmacéutica. Dirección y organización: Revista de dirección, organización y administración de empresas 30, 88–101 (2004)

Brynjolfsson, E., Hitt, L.: Computing Productivity: firm-Level Evidence. Review of Economics and Statistics 85(4), 793–808 (2003)

Cash, J., Konsynski, B.: Los sistemas de información establecen nuevas fronteras competitivas. Harvard Deusto Business Review 2, 45–58 (1986)

Chen, D., Mocker, M., Preston, D.: Information systems strategy: Reconceptualization, Measurement, and implications. MIS Quarterly 34(2), 233–259 (2010)

Dehning, B., Stratopoulos, T.: DuPont analysis of an IT-enabled competitive advantage. The International Journal of Accounting Information Systems 3(3), 165–176 (2002)

Díaz, G.: Los sistemas de información en las entidades bancarias: estrategias, escenarios y desafíos futuros 2007-2010) (2006), http://www.degerencia.com/articulo/los_sistemas_de_informacion_en_las_entidades_bancarias_estrategias_escenarios_y_desafios_futuros (accessed September 22, 2009)

Dorantes, C., Rao, S.: Competencias generadoras de ventaja competitiva en el uso de sistemas ERP. In: Proceedings of the Twelfth Americas Conference on Information Systems, Acapulco, Mexico (2006)

Estrada, D.: Análisis de las fusiones en el mercado bancario colombiano. Reporte de estabilidad financiera, Banco de la República, 69–79 (2005)

Feeny, D., Willcocks, L.: Core IS Capabilities for Exploiting Information Technology. Sloan Management Review 39(3), 9–21 (1998)

Galve, C., Gargallo, C.: Impacto de las Tecnologías de la Información en la Productividad de las Empresas Españolas. DT2004-05, Facultad de Ciencias Económicas y Empresariales, Universidad de Zaragoza, 1–26 (2005)

García, A., Gómez, J.: Determinantes de las fusiones y adquisiciones en el sistema financiero colombiano. 1990-2007 Revista de Economía del Rosario 121, 45–65 (2009)

Hair, J., Anderson, R., Tatham, R., Black, W.: Análisis Multivariante. Prentice Hall, Madrid (1999)

Hitt, L., Brynjolfsson, E.: Productivity, Business Profitability, and Consumer Surplus: Three Different Measures of Information Technology Value. MIS Quarterly 2(20), 121–142 (1996)

Ives, B., Learmonth, G.: The information system as a competitive weapon. Communications of the ACM 27(12), 1193–1200 (1984)

Jeffers, P.: Information Technology (IT) and Process Performance: an empirical investigation of the complementarities between it and non-it resources, The Ohio State University, Unpublished doctoral dissertation (2003), http://etd.ohiolink.edu/view.cgi?acc_num=osu1061404946, accessed (accessed January 06, 2008)

Jeffers, P., Muhanna, W.: Information Technology and Process Performance: An Empirical Investigation of the Interaction between IT and Non-IT Resources. Decision Series 39(4), 703–735 (2008)

Kohli, R., Grover, V.: Business Value of IT: An Essay on Expanding Research Directions to Keep Up with the Times. Journal of the AIS 9(1), 23–37 (2008)

Liang, T., Tanniru, M.: Customer-centric information systems. Journal of Management Information Systems 23(3), 9–15 (2007)

Levy, M., Powell, P., Galliers, R.: Assessing information systems strategy development frameworks in SMEs. Information & Management 36(5), 247–261 (1999)

Loukis, I., Sapounas, A., Milionis, E.: The effect of hard and soft information and communication technologies investment on manufacturing business performance in Greece. Telematics and Informatics 26(2), 193–210 (2009)

Mahoney, J., Pandian, R.: The Resource-based View within the Conversation of Strategic Management. Strategic Management Journal 13, 363–380 (1992)

Maizlish, B., Handler, R.: IT Portfolio management step-by-step: unlocking the business value of technology. John Wiley & Sons, New Jersey (2005)

Mata, F., Fuerst, W., Barney, J.: Information Technology and Sustained Competitive Advantage: A Resource-based Analysis. MIS Quarterly 19(14), 487–505 (1995)
Martin, Q., Cabero, M., De Paz, R.: Tratamiento estadístico de datos con SPSS. Thomson, Madrid España (2008)
Martínez, M.: Ideas para el cambio y el aprendizaje en la organización, una perspectiva sistémica, Ecoe Ediciones, Bogotá Colombia (2005)
Mcfarlan, W.: Information technology changes the way you compete. Harvard Business Review 62(3), 98–103 (1984)
Melville, N., Kraemer, K., Gurbaxani, V.: Review: Information technology and organizational performance: an integrative model of IT business value. MIS Quarterly 28(2), 283–322 (2004)
Moncada, A.: La Tecnología de la Información, potencial de utilidad en la estrategia del sector bancario Colombiano – Una perspectiva desde la teoría de Recursos y Capacidades, unpublished doctoral dissertation, Universidad San Pablo CEU (2011)
Murillo, G.: Sector Financiero Colombiano – Concentración, Internacionalización y Nuevas tecnologías en las organizaciones, Ecoe Ediciones, Bogotá Colombia (2009)
Neo, B.: Factors facilitating the use of information technology for competitive advantage: an exploratory study. Information and Management 15, 191–201 (1988)
Nevo, S., Wade, M.: The formation and value of IT-enabled resources: antecedents and consequences of synergistic relationships. MIS Quarterly 34(1), 163–183 (2010)
Oh, W., Pinsonneault, A.: On the Assessment of the Strategic Value of Information Technologies: Conceptual and Analytical Approaches. MIS Quarterly 31(2), 239–265 (2007)
Overby, E., Bharadwaj, A., Sambamurthy, V.: Enterprise agility and the enabling role of information technology. European Journal of Information Systems 15(2), 120–131 (2006)
Palomo, R., Sanchis, J., Soler, F.: Las Entidades Financieras de Economía Social ante la crisis financiera: un análisis de las cajas rurales españolas. Revista de Estudios Cooperativos, Universidad Complutense de Madrid 100, 101–133 (2010)
Parsons, G.: Information Technology: a new competitive weapon. Sloan Management Review 25(1), 3–14 (1983)
Pavlou, P., El Sawy, O.: From IT Leveraging Competence to Competitive Advantage in Turbulent Environments: The Case of New Product Development. Information Systems Research 17, 198–227 (2006)
Penrose, E.: Teoría del crecimiento de la empresa, Edición 1995, Aguilar, Madrid, España (1962)
Peteraf, M., Barney, J.: Unraveling the Resource-based Triangle. Managerial and Decision Economics 24, 309–323 (2003)
Piccoli, G., Ives, B.: Review: IT-dependent strategic initiatives and sustained competitive advantage: A review and synthesis of the literature. MIS Quarterly 29(4), 747–777 (2005)
Porter, M.: Estrategia competitiva: técnicas para el análisis de los sectores industriales y de la competencia, edición 1988, Cía. Editorial Continental, México (1982)
Porter, M.: Competitive Advantage: Creating and Sustaining Superior Performance. The Free Press, New York (1985)
Porter, M., Millar, V.: How information gives you competitive advantage. Harvard Business Review 63(4), 149–160 (1985)
Powell, T., Dent-Micallef, A.: Information Technology as Competitive Advantage: The Role of Human, Business, and Technology Resources. Strategic Management Journal 18(5), 375–405 (1997)

Ravichandran, T., Lertwongsatien, C.: Effect of Information Systems Resources and Capabilities on Firm Performance: A Resource-Based Perspective. Journal of Management Information Systems 21(4), 237–276 (2005)

Ray, G., Muhanna, W., Barney, J.: Information Technology and the performance of the customer service process: A resource-based analysis. MIS Quarterly 29(4), 625–652 (2005)

Rivard, S., Raymond, L., Verreault, D.: Resource-based View and Competitive Strategy: An integrated model of the contribution of information technology to firm performance. Journal of Strategic Information Systems 15, 29–50 (2006)

Ross, J., Beath, C.: New approaches to IT investment. Sloan Management Review 43(2), 51–59 (2002)

Ross, J., Beath, C., Goodhue, D.: Develop Long-term Competitiveness Through IT Assets. Sloan Management Review 38(1), 31–42 (1996)

Santhanam, R., Hartono, E.: Issues in Linking Information Tech¬nology Capability to Firm Performance. MIS Quarterly 27(1), 125–153 (2003)

Sirmon, D., Hitt, M., Ireland, D.: Managing Firm Resources in Dynamic Environments to Create Value: Looking Inside the black Box. The Academy of Management Review 32(1), 273–292 (2007)

Tallon, P.: Inside the Adaptive Enterprise: An Information Technology Capabilities Perspective on Business process agility, Center for Research on Information Technology and Organizations, pp. 1–35. University of California (2007)

Tallon, P., Kraemer, K.: Investigating the relationship between Strategic Alignment and IT Business Value: The discovery of a Paradox in Creating Business value with information technology: challenges and solutions, Namchul Shin, Idea Group Publishing, USA (2003)

Tallon, P., Kraemer, K., Gurbaxani, V.: Executives' Perceptions of the Business Value of Information Technology: A Process-oriented Approach. Journal of Management Information Systems 16(4), 145–173 (2000)

Teece, D., Pisano, G., Shuen, A.: Dynamic Capabilities and Strategic Management. Strategic Management Journal 18(17), 507–533 (1997)

Teo, T., Ranganathan, C.: Leveraging IT resources and capabilities at the housing and development board. Journal of Strategic Information Systems 12(3), 229–249 (2003)

Tian, J., Wang, K., Chen, Y., Johansson, B.: From IT deployment capabilities to competitive advantage: An exploratory study in China. Information Systems Frontiers 12(3), 239–255 (2009)

Ven de Ven, A., Ferry, D.: Measuring and assessing organizations. Wiley, Nueva York (1979)

Wade, M., Hulland, J.: Review: The Resource-Based View and Information Systems Research: Review, Extension, and Suggestions for Future Research. MIS Quarterly 23(1), 107–142 (2004)

Wernerfelt, B.: A Resource-Base view of the firm. Strategic Management Journal 5, 171–180 (1984)

A Double-Network Perspective on the Evolution of Subsidiary R&D Role: A Matter of Dual Embeddedness

Fariza Achcaoucaou and Paloma Miravitlles

Department of Business Administration, University of Barcelona,
Av. Diagonal 690, 08034 Barcelona, Spain
farizaa@ub.edu, paloma.miravitlles@ub.edu

Abstract. The International Business literature has recently sought to determine the drivers of R&D role changes in foreign subsidiaries over time. Yet, very few studies have examined network effects to explain this evolution. To fill this gap in the literature, the present work focuses on changes in subsidiary capabilities and on the dynamic mechanisms by which their R&D role might evolve, especially, as a consequence of their interaction with a variety of networks. Thus, the core aim of this paper is to summarize the findings from the previous literature as a first step for developing a general theoretical framework that integrates internal and external network embeddedness and to discuss its implications for the evolution in the R&D role of subsidiaries. The literature review highlights the need to investigate how subsidiary R&D roles evolve as a consequence of their being simultaneously engaged in both intra-organizational and local networks.

Keywords: R&D, technological innovation, role evolution, subsidiaries, multinational, embeddedness.

1 Introduction

The role played by subsidiaries, above all as regards their R&D activities, and their competitive position within their respective multinational corporations (MNCs) is perceived as being subject to change over time. Historically, headquarters was considered the only source of competitive advantage for an MNC and this was leveraged overseas by the transfer of knowledge to foreign subsidiaries (Vernon, 1966; Dunning; 1981). Recently, linked to the closer integration of subsidiaries into international networks, the latter have been able to generate new knowledge for the whole MNC. In fact, heterarchical (Hedlund, 1986) and transnational (Bartlett and Ghoshal, 1989) corporate models reflect the existence of an internal network within the MNC, where information flows freely in all directions. At the same time, the metanational corporate model (Doz, Santos, and Williamson, 2001) emphasizes the emergence of the company's external network. A subsidiary, thus, absorbs knowledge through its business linkages with local partners,

A.M. Gil-Lafuente et al. (Eds.): Soft Comput. in Manag. and Bus. Econ., STUDFUZZ 287, pp. 97–108.
springerlink.com

which represent an important source of technological competencies enabling it to contribute to the MNC's overall capabilities (Andersson, 2003). Thus, the ability to manage dispersed capabilities effectively within this 'double network' – comprising internal and external networks (Zanfei, 2000) – is seen as the key to an MNC's competitive advantage (Frost, Birkinshaw, and Ensign, 2002).

Taking the resource-based view (Prahalad and Hamel, 1990; Cantwell, 1991), many researchers have focused on this spread of knowledge as a basis for building subsidiary role typologies (e.g. Gupta and Govindarajan, 1991; Bartlett and Ghoshal, 1989). Shifting the focus from the MNC to the subsidiary, these studies analyze the latter as a unit with a unique, homogeneous role, while subsidiaries might operate within a narrow part of value chain (Roth and Morrison, 1992) and, thus, act as an incomplete unit that neither executes all the activities of the full value chain nor participates with the same intensity (Dörrenbächer and Gammelgaard, 2006). Thus, it is quite possible that a subsidiary will specialize in various or, indeed, just one value activity (e.g., production, marketing or R&D) (Bartlett and Ghoshal, 1986). In addition, capability development does not proceed at a uniform rate for every value activity (Kim, Rhee, and Oh, 2011), e.g. a subsidiary might play an active role in manufacturing but a receptive one in R&D.

Focusing on R&D activities, the International Business literature has recently identified the emergence of technologically advanced foreign subsidiaries (Blomkvist, Kappen, and Zander, 2010) performing a more creative R&D role (Cantwell and Mudambi, 2005). Today, we see foreign subsidiaries not only as knowledge receivers, or in the terminology of Cantwell and Mudambi (2005) performing a 'competence-exploiting' role, but also as knowledge creators in a fully integrated network (Di Minin and Zhang, 2010), fulfilling what Cantwell and Mudambi (2005) label as a 'competence-creating' role. This shift is important, as recent research highlights the more active role played by subsidiaries in the globalization of innovation, while examining their influence on MNC innovative ability (Phene and Almeida, 2008; Blomkvist et al., 2010).

A substantial body of the literature has analysed the drivers behind the configuration of subsidiary R&D roles (for example, Pearce, 1992; Bartlett and Ghoshal, 1990; Kuemmerle, 1997, 1999; Gerybadze and Reger, 1999; Gassmann and von Zedtwitz, 1999; von Zedtwitz and Gassmann, 2002; Sachwald, 2008). However, they present two major shortcomings: first, most of the studies take a static approach. Since they are primarily concerned with identifying the specialized roles adopted by overseas R&D laboratories, they neglect the prior evolution of capabilities within the subsidiary (notable exceptions are Cantwell and Mudambi, 2005; Kim et al., 2011). But as the specific R&D role of a subsidiary is a direct outcome of this evolution, the way in which these capabilities are created must first be analysed. Subsidiaries sustain and enhance their capabilities by learning (Zanfei, 2000; Andersson, Forsgren and Holm, 2002), which means that capability development is critically dependent on the subsidiary's ability to recognize, absorb, assimilate and combine value-adding knowledge over time (Andersson, 2003). This, in turn, ultimately determines its R&D role. Second, many of the studies analyse the drivers of a subsidiary's R&D role in isolation and so neglect any network effects. Specifically, they identify three main drivers of the configuration of strategic roles: task assignment by headquarters, the subsidiary's own choice and local environmental factors (Birkinshaw and Hood, 1998; Westney and Zaheer, 2001; Kim et al., 2011). However, less importance is attached to any underlying network effects, particularly those arising as a consequence of simultaneous engagement in the intra-organizational network and local embeddedness.

To fill this gap in the literature, this study examines the evolution in a subsidiary's capabilities and the dynamic mechanisms by which its R&D role can change over time. Thus, the core aim of this paper is to summarize the findings from the previous literature as a first step in developing a general framework that integrates internal and external network embeddedness and to discuss its implications for the evolution in the R&D role of subsidiaries.

The study makes what we consider to be three significant contributions: First, from a network-based perspective, we work from the premise that the relationship between headquarters and its subsidiaries is a driver of R&D evolution, which suggests that subsidiary initiative and headquarter mandates are at either end of the same continuum. The initiative taken by a subsidiary can be enhanced or inhibited by headquarters' stance (Ambos, Andersson, and Birkinshaw, 2010) and micro-political negotiation processes can modify headquarters' intended strategy (Birkinshaw, 1997; Dörrenbächer and Gammelgaard, 2006; 2011). Second, based on previous studies that consider the evolution in the R&D role as being driven by favourable and unfavourable environmental conditions (Frost, 2001; Benito, Grøgaard, and Narula, 2003), we introduce the degree of local embeddedness as a moderating factor of the effects of the environmental conditions. R&D units no longer absorb knowledge passively, but take initiatives to tap into external networks and, thus, increase their potential for the use and generation of knowledge (Zanfei, 2000). Third, based on our understanding of MNC dual networks, we analyse the impact of both internal and external network embeddedness on the evolution in subsidiary R&D roles. While some authors have examined the effect of headquarters-subsidiary relationships and knowledge transfer between units of the MNC (Pearce, 1992; Bartlett and Ghoshal, 1990; Kuemmerle, 1997, 1999; Gerybadze and Reger, 1999; Gassmann and Zedtwitz, 1999; von Zedtwitz and Gassmann, 2002), others have examined the impact of local embeddedness (Andersson, Forsgren, and Pedersen, 2001; Andersson et al., 2002, 2007). However, only a few recent studies have considered their simultaneous impact on subsidiary innovation, albeit not specifically on their evolving R&D roles (Di Minin and Zhang, 2010; Figueiredo, 2011; Yamin and Andersson, 2011). As a result, the framework proposed here allows us to evaluate the importance of various attributes involved in the underlying networks determining subsidiary R&D evolution. In this way the analysis gains in rigour and dynamism to better capture the complexity of reality.

The rest of the paper is structured as follows. Section two undertakes a review of the relevant literature examining internal and external MNC networks. The third section outlines the interrelation between the two networks and embeddedness. In the concluding section a conceptual framework of subsidiary R&D role evolution on the basis of dual embeddedness is developed. Finally, the paper is concluded with some implications for future research.

2 Theoretical Framework

2.1 Internal MNC Network

It is widely assumed that two of the key factors associated with subsidiary role development are subsidiary initiative-taking (Birkinshaw, 1997; Birkinshaw and Hood, 1998; Dörrenbächer and Gammelgaard, 2006), on the one hand, and parent company determinism in the allocation of mandates (Birkinshaw and Hood, 1998; Hood and Taggart, 1999), on the other.

The internal network perspective has recently been employed to conduct research on subsidiary initiative-taking. For example, Vernaik, Midgley, and Devinney (2005) find support for the importance of networking and autonomy in encouraging greater innovation and competitive advantage within MNCs. Gnyawali, Singal and Mu (2009) argue that internal network relationships boost subsidiary entrepreneurship. Andersson et al. (2007) conclude that the more valuable a subsidiary's initiatives are deemed to be for its peer subsidiaries within an MNC network, the better it will be able to influence these peers. Furthermore, the stronger the linkages that a subsidiary builds with its partners within that network, the greater its position of power will be (Young and Tavares, 2004). Yet, subsidiaries are not always able to increase their influence or central position by simply taking initiatives, unless they get headquarters' acknowledgement (Ambos et. al., 2010). Headquarters might either support the further development of subsidiary plans or might choose to strengthen their formal control mechanisms, which in turn, would threaten subsidiary autonomy. In addition, subsidiary initiatives have to contend with the 'corporate immune system' (Birkinshaw and Ridderstråle, 1999) and if the affiliate unit wishes to gain recognition and have its mandate upgraded, its actions must be in line with the corporate's dominant logic (Bettis and Prahalad, 1986; 1995). Consequently, subsidiary initiative must be taken within a corporate context that is shaped to a great extent by headquarters.

In the context of the internal network, the parent company's strategic vision, expressed through the assignment of mandates, is crucial for the development of the R&D role. Yet recent research has shed additional light on this assignment. Dörrenbächer and Gammelgaard's (2006) results reveal that headquarters' intended strategies might be modified by micro-political headquarters-subsidiary negotiations. In a subsequent study, Dörrenbächer and Gammelgaard (2010) present further findings supporting the effectiveness of a subsidiary's micro-political bargaining power, which depends on the resource and capability dependency of its sister affiliates, that is, on the internal network. Other studies, likewise, emphasize the role of parent company power on the configuration of the mandate network via the implementation of control mechanisms (Young and Tavares, 2004; Ambos and Schlegelmilch, 2007). In fact, headquarters has the authority to control its subsidiary network, and it can wield this power to further ensure subsidiary conformity (Yamin and Andersson, 2011). Consequently, drawing on Ambos et al. (2010), we consider headquarters-subsidiary relationships as a mixed-motive dyad governed by divergent and convergent interests.

2.2 External MNC Network

The International Business literature has tended to emphasise the importance of environmental factors in determining MNC subsidiary roles and evolution. For example, Birkinshaw and Hood (1998) refer to these factors as *local environmental determinism* and understand subsidiary evolution *as a function of the constraints and opportunities in the local market.* Kuemmerle (1999) and Pearce (1999) propose a typology for subsidiary-level R&D and consider the role of each subsidiary as being essentially determined by *the relative strength of a country's science base* and *the attributes of the location in which it is sited,* respectively. Cantwell and Mudambi (2005) allude to *location determinants* to explain that R&D development is conditioned by the *characteristics of the location in which the subsidiary is located* in terms of its quality and resource conditions.

However, most of the studies treat the external context quite generally, seeing environmental forces just as a driver to concentrate R&D where local conditions are most conducive to technology creation (Cantwell and Kosmopoulou, 2002). In other words, most studies confide their interest in location issues at a country level and neglect firm-location interactions as a potential platform for leveraging environmental effects. In its relationships with local actors a subsidiary is exposed to new knowledge outside the organization and this knowledge constitutes one of the key inputs for developing and accumulating the capabilities required for technological and organisational innovation (Anderson et al., 2002). For example, Andersson, Björkman, and Forsgren (2005) report that external embeddedness has a positive impact on the development of products and processes in the MNC. Almeida and Phene (2004) suggest that a subsidiary's knowledge linkages with the host country have a positive effect on innovation in the subsidiaries of the MNC. And Santangelo (2009) concludes that local linkages creation is greater when subsidiaries have 'competence-creating scope' within the corporate organizational structure.

In sum, the reason why some subsidiaries achieve better innovative performance than others operating in the same environmental context can be explained by the frequency, depth and quality of subsidiary linkages to local partnerships. Thus, arguably, improvements in a subsidiary's R&D role depend upon effective integration within the local host country's environment rather than simply on siting activities in a munificent location (Cantwell, 2009). In other words, the potential of environmental factors as a source of competitiveness lies in a subsidiary's awareness of how to benefit from the welfare effects of the country's science base through a certain degree of embeddedness.

The underlying idea is that maintaining strong ties of trust and cooperation with local actors potentially establishes a basis for learning, generating and transferring knowledge beyond the boundaries of the firm (Uzzi and Lancaster, 2003; Andersson, 2003). In turn this knowledge serves as the necessary basis for developing the technological competencies to undertake innovative activities (Figueiredo, 2011). When these technological competencies are subsequently transferred to other units, the overall level of competencies within the MNC rises (Andersson et al., 2001; Yamin and Andersson, 2011).

So while previous studies have considered the evolution in a subsidiary's R&D role as being driven by favourable and unfavourable environmental conditions (Frost, 2001; Benito et al., 2003), we argue that the degree of local embeddedness acts as a moderating factor of these conditions. Thus, the effects of favourable local conditions may be intensified by enhanced degrees of local embeddedness, while unfavourable local conditions may be restrained by lower degrees of local embeddedness.

3 A Double Network Perspective on Subsidiary R&D Evolution

As noted, subsidiary initiative and parent company determinism are more closely related than hitherto thought. Arguably, they are involved in a 'perpetual bargaining process' (Andersson et al., 2007). Subsidiary power in this relationship, as far as its R&D evolution is concerned, can be associated with the possession of knowledge-capabilities and a favourable host country environment (Dörrenbächer and Gammelgaard, 2006). Subsidiaries strengthen their competitive position within the corporate group by accumulating over time the competencies needed for innovation (Figueiredo, 2011). This is possible through their entrepreneurial undertakings that tap into new opportunities in the local environment, i.e. subsidiary initiative (Birkinshaw, 1997; Rugman and Verbeke, 2001) and the acquisition of value-adding resources, especially knowledge, on which the rest of the MNC can draw (Birkinshaw, Hood, and Young, 2005). When these resources are unique and valuable for other units in the corporate group, a subsidiary can occupy a central position within the MNC network (Bouquet and Birkinshaw, 2008) and upgrade its power situation vis-à-vis the parent company (Forsgren, Holm, and Johanson, 2005). For Dörrenbacher and Gammelgaard (2006; 2011), a subsidiary's influence on the allocation of headquarters' mandates often depends on ownership of valuable resources that can be used when bargaining with headquarters. Luo (2005) emphasises that it is the quality and rarity of these resources that determines the likelihood of the subsidiary gaining corporate support and parent mandate assignments. The result is an increasing capacity to influence headquarters' R&D strategic decision-making in favour of the subsidiary's own interests (Andersson et al., 2007; Ambos et al., 2010). This is positively associated with gaining mandates so as to increase the scope for R&D evolution.

Subsidiaries address their own future by balancing their own initiatives against requests from headquarters (Garcia-Pont, Canales and Noboa, 2009). Headquarters' power within internal network relationships depends on formal authority. The parent company managers have the recognized legitimacy to organize the activity of the MNC by delegating business areas and strategic responsibilities to its dispersed subsidiaries overseas (Dörrenbächer and Gammelgaard, 2010), i.e. the allocation of mandates. This formal authority can be exerted through the use of different planning and control mechanisms, including the distribution of decision-making rights and the allocation of resources (Ghoshal and Bartlett 1988), which constitute a major instrument in the hands of headquarters for changing subsidiary roles (Birkinshaw and Hood, 1998).

However, in the last decade, the shift towards 'supply-side' motivations to perform R&D operations overseas (Criscuolo, Narula, and Verspagen, 2005) has strengthened subsidiary autonomy to the detriment of headquarters control. MNCs have an increasing interest in the exploration of local knowledge and in accessing expertise complementary to the firm (Ivarsson and Jonsson, 2003; Santangelo, 2011). In such a situation, it is not easy for headquarters to manage and control knowledge development because of context specificity and information deficiencies (Ferner, 2000). Hence, subsidiary autonomy and initiative would appear necessary (Young and Tabares, 2004) to absorb knowledge effectively from the host country environment. Seen from this perspective, a subsidiary's external network can be considered a strategic source of capabilities and competitive advantage (Uzzi and Lancaster, 2003; Figueiredo, 2011) that can be exchanged with the parent company and sister subsidiaries (Ambos, Ambos, and Schlegelmilch, 2006). The logic of the arguments presented in these and other papers (see also Andersson et al., 2002, 2007; Andersson, 2003) implies that headquarters allocates different R&D mandates to specific subsidiaries so as to tap knowledge linked to the host environments of these subsidiaries.

Nevertheless, changes in a subsidiary's mandate depend not only on the endowment of the external environment but also on its potential to embed itself in the host country environment and to make local resources available to other MNC units (Andersson and Forsgren, 2000; Dörrenbächer and Gammelgaard, 2010). Thus, as Figure 1 illustrates, the subsidiary acts as a bridge for knowledge transfer between the host country environment and the international corporate network, including headquarters and peer subsidiaries (Giroud and Scott-Kennel, 2009). This means that subsidiaries are embedded, at one and the same time, in their own internal network, which includes headquarters and all the other MNC units, and in their external local network, which in the case of R&D activities involve other

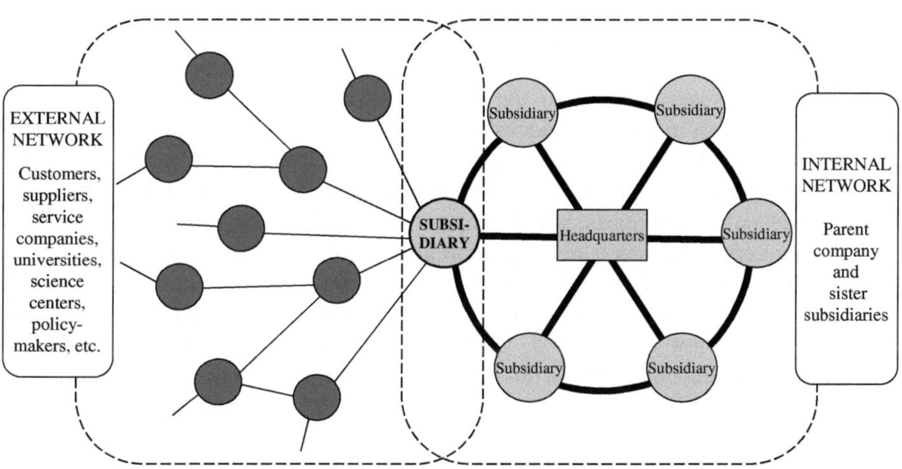

Source: Derived from the study.

Fig. 1 Subsidiary double-network embeddedness

actors besides customers, suppliers and service companies, such as universities,science centres or regulators and other policy-makers. In this respect, Andersson et al., (2005) have shown the degree of local embeddedness to be an important indicator of a subsidiary's ability to create new knowledge, while Andersson et al. (2002) have empirically demonstrated that high external embeddedenss can be correlated with an assignment of higher technological subsidiary mandates.

To summarize, it seems reasonable to expect that a subsidiary's R&D role is dependent on both its degree of external network embeddedness (so as to learn and assimilate knowledge from the host country environment) and its degree of intra-corporate embeddedness (allowing it to transfer its knowledge to the parent company and other subsidiaries). By focusing solely on the inter-organizational network, or only taking the intra-organizational network into account, is to see only half the picture.

4 Conclusions

In this study we have taken previous theoretical and empirical discussions of the drivers of the R&D role of subsidiaries as our starting point and placed them in a double-network context. On the one hand, studies of the R&D roles have traditionally placed greater emphasis on an MNC's internal conditions and relationships, specifically on headquarters-subsidiary relationships and the mechanisms by which a subsidiary transfers its capabilities and knowledge to other corporate units (Pearce, 1992; Bartlett and Ghoshal, 1990; Kuemmerle, 1997, 1999; Gerybadze and Reger, 1999; Gassmann and Zedtwitz, 1999; von Zedtwitz and Gassmann, 2002). On the other hand, studies examining the internationalisation of R&D activities emphasize relational embeddedness in external networks as a strategic source of knowledge and of technological capabilities outside the organization (Andersson et al., 2001; Andersson et al., 2002; 2007).

However, subsidiaries are clearly embedded in both external and internal networks at one and the same time (Forsgren et al., 2005; Yamin and Andersson, 2011) and, as is apparent from the above discussion, the double-network embeddedness paradigm makes it necessary to analyse effectively the evolution of a subsidiary's R&D role. Accordingly, a subsidiary's evolution towards a competence-creating mandate should combine stronger relationships with both its internal and external counterparts, which means that simultaneous internal and external network embeddedness is a necessary condition for achieving a more developed R&D role. However, few studies have yet to adopt such an approach in their analyses of the various issues associated with subsidiary innovation (but see, for example, Birkinshaw et al., 2005; Di Minin and Zhang, 2010; Figueiredo, 2011; Yamin and Andersson, 2011). Thus, there is a need to investigate more fully the effect of the interactions between internal and external embeddedness on subsidiary R&D roles.

The main limitation of this study concerns its overall scope. Beyond identifying a research opportunity in the literature, it is also necessary to identify specific patterns of R&D evolution from the dual embeddedness perspective and to undertake a more exhaustive analysis of the contextual factors and mechanisms that interact to determine these patterns. This would further our understanding of, one, the way in which subsidiaries acquire and transfer knowledge, thus generating major sources of competitive advantage, and, two, the way in which this knowledge impacts on processes of gaining or losing subsidiary mandates.

References

Almeida, P., Phene, A.: Subsidiaries and knowledge creation: The influence of the MNC and host country on innovation. Strategic Management Journal 8-9, 847–864 (2004)

Ambos, B., Schlegelmilch, B.B.: Innovation and control in the multinational firm: A comparison of political and contingency approaches. Strategic Management Journal 5, 473–486 (2007)

Ambos, T.C., Ambos, B., Schlegelmilch, B.B.: Learning from foreign subsidiaries: An empirical investigation of headquarters' benefits from reverse knowledge transfers. International Business Review 3, 294–312 (2006)

Ambos, T.C., Andersson, U., Birkinshaw, J.: What are the consequences of initiative-taking in multinational subsidiaries? Journal of International Business Studies 7, 1099–1118 (2010)

Andersson, U.: Managing the transfer of capabilities within multinational corporations: the dual role of the subsidiary. Scandinavian Journal of Management 4, 425–442 (2003)

Andersson, U., Björkman, I., Forsgren, M.: Managing subsidiary knowledge creation: The effect of control mechanisms on subsidiary local embeddedness. International Business Review 5, 521–538 (2005)

Andersson, U., Forsgren, M.: In Search of centre of excellence: network embeddedness and subsidiary roles in multinational corporations. Management International Review 4, 329–350 (2000)

Andersson, U., Forsgren, M., Pedersen, T.: Subsidiary performance in multinational corporations: The importance of technology embeddedness. International Business Review 1, 3–23 (2001)

Andersson, U., Forsgren, M., Holm, U.: The strategic impact of external networks: Subsidiary performance and competence development in the multinational corporation. Strategic Management Journal 11, 979–996 (2002)

Andersson, U., Forsgren, M., Holm, U.: Balancing subsidiary influence in the federative MNC: a business network view. Journal of International Business Studies 5, 802–818 (2007)

Bartlett, C.A., Ghoshal, S.: Tap your subsidiaries for global reach. Harvard Business Review 6, 87–94 (1986)

Bartlett, C.A., Ghoshal, S.: Managing Across Borders: The Transnational Solution. Harvard Business School Press, Boston (1989)

Bartlett, C.A., Ghoshal, S.: Managing innovation in the transnational corporation. In: Bartlett, C.A., Doz, Y., Hedlund, G. (eds.) Managing the Global Firm, pp. 215–255. Routledge, London (1990)

Benito, G.R.G., Grøgaard, B., Narula, R.: Environmental influences on MNE subsidiary roles: economic integration and the Nordic countries. Journal of International Business Studies 5, 443–456 (2003)

Bettis, R., Prahalad, C.: The dominant logic: Retrospective and extension. Strategic Management Journal 1, 5–14 (1995)

Bettis, R., Prahalad, C.: The dominant logic: A new linkage between diversity and performance. Strategic Management Journal 6, 485–501 (1986)
Birkinshaw, J.: Entrepreneurship in multinational corporations: The characteristics of subsidiary initiatives. Strategic Management Journal 3, 207–229 (1997)
Birkinshaw, J., Hood, N.: Multinational subsidiary evolution: Capability and charter change in foreign-owned subsidiary companies. Academy of Management Review 4, 773–795 (1998)
Birkinshaw, J., Hood, N., Young, S.: Subsidiary entrepreneurship, internal and external competitive forces, and subsidiary performance. International Business Review 2, 227–248 (2005)
Birkinshaw, J., Ridderstråle, J.: Fighting the corporate immune system: A process study of peripheral initiatives in large, complex organizations. International Business Review 2, 149–180 (1999)
Blomkvist, K., Kappen, P., Zander, I.: Quo vadis? The entry into new Technologies in advanced foreign subsidiaries of the multinational Enterprise. Journal of International Business Studies 9, 1525–1549 (2010)
Bouquet, C., Birkinshaw, J.: Managing power in the multinational corporation: how low-power actors gain influence. Journal of Management 3, 477–508 (2008)
Cantwell, J.: The international agglomeration of R&D. In: Casson, M. (ed.) Global Research Strategy and International Competitiveness, pp. 216–232. Basil Blackwell, Oxford (1991)
Cantwell, J.: Location and the multinational enterprise. Journal of International Business Studies 1, 35–41 (2009)
Cantwell, J., Kosmopoulou, E.: What determines the internationalisation of corporate technology? In: Havila, V., Forsgren, M., Hakanson, H. (eds.) Critical perspectives on internationalisation, pp. 305–334. Pergamon, Oxford (2002)
Cantwell, J., Mudambi, R.: MNE competence-creating subsidiary mandates. Strategic Management Journal 12, 1109–1128 (2005)
Criscuolo, P., Narula, R., Verspagen, B.: Role of home and host country innovation systems in R&D internationalisation: a patent citation analysis. Economics of Innovation and New Technology 5, 417–433 (2005)
Di Minin, A., Zhang, J.: An exploratory study on international R&D strategies of Chinese companies in Europe. Review of Policy Research 4, 433–455 (2010)
Dörrenbächer, C., Gammelgaard, J.: Subsidiary role development: The effect of micro-political headquarters-subsidiary negotiations on the product, market and value-added scope of foreign-owned subsidiaries. Journal of International Management 3, 266–283 (2006)
Dörrenbächer, C., Gammelgaard, J.: Multinational corporations, inter-organizational networks and subsidiary charter removals. Journal of World Business 3, 206–216 (2010)
Dörrenbächer, C., Gammelgaard, J.: Subsidiary power in multinational corporations: the subtle role of micro-political bargaining power. Critical Perspectives on International Business 1, 30–47 (2011)
Doz, Y., Santos, J., Williamson, P.: From Global to Metanational: How Companies Win in the Global Economy. Harvard Business School Press, Boston (2001)
Dunning, J.H.: International production and the multinational enterprise. Allen & Unwin, London (1981)
Ferner, A.: The underpinnings of "bureaucratic" control systems: HRM in European multinationals. Journal of Management Studies 4, 521–539 (2000)

Figueiredo, P.N.: The Role of Dual Embeddedness in the Innovative Performance of MNE Subsidiaries: Evidence from Brazil. Journal of International Business Studies 2, 417–440 (2011)

Forsgren, M., Holm, U., Johanson, J.: Managing the embedded multinational: A business network view. Edward Elgar, Chelterham (2005)

Frost, T.S.: The geographic sources of foreign subsidiaries' innovations. Strategic Management Journal 2, 101–123 (2001)

Frost, T.S., Birkinshaw, J.M., Ensign, P.C.: Centers of excellence in multinational corporations. Strategic Management Journal 11, 997–1018 (2002)

Garcia-Pont, C., Ignacio, J., Noboa, F.: Subsidiary Strategy: The Embeddedness Component. Journal of Management Studies 2, 182–214 (2009)

Gassmann, O., von Zedtwitz, M.: New concepts and trends in international R&D organization. Research Policy 2-3, 231–250 (1999)

Gerybadze, A., Reger, G.: Globalization of R&D: recent changes in the management of innovation in transnational corporations. Research Policy 2-3, 251–274 (1999)

Ghoshal, S., Bartlett, C.A.: Creation, adoption, and diffusion of innovations by subsidiaries of multinational corporations. Journal of International Business Studies 3, 365–388 (1988)

Giroud, A., Scott-Kennel, J.: MNE linkages in international business: A framework for analysis. International Business Review 6, 555–566 (2009)

Gnyawali, D., Singal, M., Mu, S.: Knowledge ties among subsidiaries in MNCs: a multi-level conceptual model. Journal of International Management 4, 387–400 (2009)

Gupta, A.K., Govindarajan, V.: Knowledge flows and the structure of control within multinational corporations. The Academy of Management Review 4, 768–792 (1991)

Hedlund, G.: The hypermodern MNC: a heterarchy? Human Resource Management 1, 9–35 (1986)

Hood, N., Taggart, J.H.: Subsidiary development in German and Japanese manufacturing subsidiaries in the British Isles. Regional Studies 6, 513–528 (1999)

Ivarsson, I., Jonsson, T.: Local technological competence and asset-seeking FDI: an empirical study of manufacturing and wholesale affiliates in Sweden. International Business Review 3, 369–386 (2003)

Kim, K., Rhee, S., Oh, J.: The strategic role evolution of foreign automotive parts subsidiaries in China: a case study from the perspective of capabilities evolution. International Journal of Operations & Production Management 1, 31–55 (2011)

Kuemmerle, W.: The drivers of foreign direct investment into research and development: an empirical investigation. Journal of International Business Studies 1, 1–24 (1999)

Kuemmerle, W.: Building effective R&D capabilities abroad. Harvard Business Review 2, 61–70 (1997)

Luo, Y.D.: Toward coopetition within a multinational enterprise: a perspective from foreign subsidiaries. Journal of World Business 1, 71–90 (2005)

Pearce, R.D.: World product mandates and MNE specialisation. Scandinavian International Business Review 2, 38–58 (1992)

Pearce, R.D.: Decentralised R&D and strategic competitiveness: globalised approaches to generation and use of technology in MNEs. Research Policy 2-3, 157–178 (1999)

Phene, A., Almeida, P.: Innovation in multinational subsidiaries: The role of knowledge assimilation and subsidiary capabilities. Journal of International Business Studies 5, 901–919 (2008)

Prahalad, C.K., Hamel, G.: The core competence of the corporation. Harvard Business Review 3, 79–91 (1990)

Roth, K., Morrison, A.J.: Implementing global strategy: Characteristics of global subsidiary mandates. Journal of International Business Studies 4, 715–735 (1992)

Rugman, A.M., Verbeke, A.: Subsidiary-Specific Advantages in Multinational Enterprises. Strategic Management Journal 3, 237–250 (2001)

Sachwald, F.: Location choices within global innovation networks: the case of Europe. Journal of Technology Transfer 4, 364–378 (2008)

Santangelo, G.D.: MNCs and linkages creation: Evidence from a peripheral area. Journal of World Business 2, 192–205 (2009)

Santangelo, G.D.: The tension of information sharing: Effects on subsidiary embeddedness. International Business Review (2011) (in press)

Uzzi, B., Lancaster, R.: Relational embeddedness and learning: the case of bank loan managers and their clients. Management Science 4, 383–399 (2003)

Venaik, S., Midgley, D.F., Devinney, T.M.: Dual paths to performance: the impact of global pressures on MNC subsidiary conduct and performance. Journal of International Business Studies 6, 655–675 (2005)

Vernon, R.: International investments and International trade in the product cycle. Quarterly Journal of Economics 2, 190–207 (1966)

von Zedtwitz, M., Gassmann, O.: Market versus technology drive in R&D internationalization: four different patterns of managing research and development. Research Policy 4, 569–588 (2002)

Westney, D.E., Zaheer, S.: The multinational enterprise as an organization. In: Rugman, A.M., Brewer, T.L. (eds.) The Oxford Handbook of International Business, pp. 349–379. Oxford University Press, Oxford (2001)

Yamin, M., Andersson, U.: Subsidiary importance in the MNC: What role does internal embeddedness play? International Business Review 2, 151–162 (2011)

Young, S., Tavares, A.: Centralization and autonomy: back to the future. International Business Review 2, 215–237 (2004)

Zanfei, A.: Transnational firms and the changing organisation of innovative activities. Cambridge Journal of Economics 5, 515–542 (2000)

Finance

A Fuzzy Random Variable Approach to Life Insurance Pricing

Jorge de Andrés-Sánchez[1] and Laura González-Vila Puchades[2]

[1] Department of Business Administration, Faculty of Economics and Business Studies, Rovira i Virgili University, Av. de la Universitat 1, 43204 Reus, Spain
jorge.deandres@urv.net

[2] Department of Economic, Financial and Actuarial Mathematics, Faculty of Economics and Business, University of Barcelona, Av. Diagonal 696, 08034 Barcelona, Spain
lgonzalezv@ub.edu

Abstract. This paper develops life insurance pricing with stochastic representation of mortality and fuzzy quantification of interest rates following the methodology by Andrés and González-Vila (2012). We show that modelling the present value of life insurance contracts with fuzzy random variables allows a well-founded quantification of their fair price and the risk resulting from the uncertainty of mortality and discounting rates. So, we firstly describe fuzzy random variables and define some associated measures: the mathematical expectation, the variance, distribution function and quantiles. Subsequently the present value of life insurance policies is modelled with fuzzy random variables. We finally show how an actuary can quantify the price and the risk of a life insurance portfolio when the contracts present value is given by fuzzy random variables.

Keywords: Fuzzy finance, Fuzzy random variables, Life insurance mathematics.

1 Introduction

Life insurance pricing has to model the uncertainty of demographic events and financial variables. From its beginning, actuarial science has paid much attention to quantifying demographic phenomena and its stochastic uncertainty. In fact, its probabilistic behaviour is commonly accepted and practitioners obtain the corresponding probabilities from life tables. Nonetheless, recent research in actuarial science has focused on formalizing the uncertainty related to the economic parameters by means of random variables and stochastic processes. The most important of those parameters is, undoubtedly, the discount rates used to price contracts.

In the actuarial field, Fuzzy Sets Theory (FST) has been used to model problems that require a great deal of actuarial subjective judgement and problems for which the information available is scarce or vague. Some surveys of FST applications in actuarial science can be found in Derrig and Ostaszewski (2004), Ostaszewski (1993), Shapiro (2004) and Yakoubov and Haberman (1998). The

A.M. Gil-Lafuente et al. (Eds.): Soft Comput. in Manag. and Bus. Econ., STUDFUZZ 287, pp. 111–125.
springerlink.com

abovementioned reasons explain why one of the applications of FST in actuarial science is actuarial pricing of insurance contracts with fuzzy interest rates. In this respect, the papers published by Andrés and Terceño (2003), Betzuen et al. (1997), Lemaire (1990) and Ostaszewski (1993) are particularly noteworthy in a life-insurance context as are those of Andrés and Terceño (2003), Cummins and Derrig (1997) and Derrig and Ostaszewki (1997) within the context of property-liability insurance.

Practically all papers on fuzzy actuarial pricing reduce the randomness of the behaviour of claiming processes to predefined frequencies – i.e. the randomness of the present value of premiums and benefits is reduced to its mathematical expectation – and, therefore, these processes become deterministic. On one hand, this approach allows insurance contracts to be priced by automatically applying the financial mathematics with fuzzy parameters developed in Buckley (1987), Gil-Aluja (1998), Kaufmann (1986) and Li Calzi (1990). On the other, the information that provides the complete statistical description of claiming is lost, making it hard to rigorously introduce the uncertainty of claiming when fitting magnitudes like reserves for deviations of mortality or premium surplus. In this paper we develop an approach to pricing life insurance contracts that combines the stochastic approach to life insurance mathematics (see Gerber (1995) under deterministic interest rates) and the quantification of interest rates with fuzzy numbers, following the developments by Andrés and González-Vila (2012). Our approach will therefore allow us to maintain stochastic and fuzzy sources of uncertainty throughout all of the valuation processes. Related to our fuzzy methodology, Shapiro (2009) describes fuzzy random variables with actuarial modelling in view and Huang et al. (2009) develop a non-life individual risk model where the number of claims follows a Poisson process whereas their value is estimated with a triangular fuzzy number (FN).

We have structured this paper as follows. In section 2 we describe the concepts and instruments of FST on which our approach is based. In section 3 we calculate price life insurance policies with a fuzzy random approach whereas in section 4 we evaluate life insurance portfolios.

2 Fuzzy Random Variables

In many real situations the uncertainty is the result any one of numerous different causes: randomness, hazard, inaccuracy, incomplete information, etc. As Kaufmann and Gil-Aluja (1990) points out stochastic variability is described by the use of probability theory and other types of uncertainties such as incomplete information or imprecision can be captured with the use of fuzzy subsets. The concept of Fuzzy Random Variable (FRV) combines both random and fuzzy uncertainty (Krätschmer, 2001; Kruse and Meyer, 1987; Kwakernaak, 1978; Kwakernaak, 1979; Puri and Ralescu 1986; Zhong and Zhou, 1987), but there is not a unique definition for it. This paper uses the concept of FRV introduced by Puri and Ralescu (1986) because it is very suitable for modelling the present value of life insurance contracts. When pricing life insurances the randomness is due to the demographical phenomenon in such a way that the moment in which the benefit is paid can be described with a conventional random variable (RV). Likewise, the outcomes of the present value of life insurance will not be real but a FN because we suppose that discount rates used to calculate present values are estimated by means of generalised intervals.

Let $\{\Omega, \mathrm{A}\}$ be a measurable space, $\{\Re, \mathrm{B}\}$ the Borel measurable space and $F(\Re)$ denote the set of FNs. The fuzzy set valued mapping $\widetilde{\boldsymbol{X}}$:

$$\begin{aligned} \widetilde{\boldsymbol{X}}: \quad \Omega \quad &\rightarrow \quad F(\Re) \\ \forall \omega \in \Omega \quad &\rightarrow \quad \widetilde{X}(\omega) = \left\{ \left(z, \mu_{\widetilde{X}(\omega)}(z) \right) \right\} \in F(\Re) \end{aligned}$$

is called a *fuzzy random variable* if:

$$\forall B \in \mathrm{B},\ \forall \alpha \in [0,1],\ \ \left\{ \omega \in \Omega \mid X(\omega)_\alpha \cap B \neq \varnothing \right\} \in \mathrm{A}$$

Where $\tilde{X}(\omega)$ is a FN that must be viewed as a generalized interval with membership function $\mu_{\tilde{X}(\omega)}(z)$ and α-level representation $X(\omega)_\alpha$:

$$X(\omega)_\alpha = \left\{ z \in \Re \mid \mu_{\tilde{X}(\omega)}(z) \geq \alpha \right\} = \left[\underline{X(\omega)_\alpha}, \overline{X(\omega)_\alpha} \right]$$

Guangyuan and Yue (1992) demonstrate that any FRV $\widetilde{\boldsymbol{X}}$ defines, $\forall \alpha \in [0,1]$, an infima RV $\underline{\boldsymbol{X}_\alpha}$ and a suprema RV $\overline{\boldsymbol{X}_\alpha}$ whose realizations are, respectively, the lower and upper extremes of α -cuts of $\tilde{X}(\omega), \forall \omega \in \Omega$, $\underline{X(\omega)_\alpha}, \overline{X(\omega)_\alpha}$.

Let $\{\Omega, \mathrm{A}, P\}$ be a probability space. Given that in our paper we will price discrete life insurances, the next definitions are referred to discrete FRVs that come from the set of elemental outcomes $\Omega = \{\omega_i\}_{i=1,\ldots,n}$ with $P(\omega_i) = p_i, \forall i = 1,\ldots,n$.

Let $\tilde{\boldsymbol{X}}$ be a discrete FRV on $\{\Omega, \mathrm{A}, P\}$, being $F_{\underline{\boldsymbol{X}_\alpha}}$ and $F_{\overline{\boldsymbol{X}_\alpha}}$, $\forall \alpha \in [0,1]$, the distribution functions of the RVs $\underline{\boldsymbol{X}_\alpha}$ and $\overline{\boldsymbol{X}_\alpha}$ obtained from $\tilde{\boldsymbol{X}}$. Then, ∀α, we define *the couple of the distribution functions* of the RVs infima and suprema for that membership level $F_{\tilde{X}}(x)_\alpha = \left\{ \underline{F_{\tilde{X}}(x)_\alpha}, \overline{F_{\tilde{X}}(x)_\alpha} \right\}$:

$$\underline{F_{\tilde{X}}(x)_\alpha} = P\left(\overline{\boldsymbol{X}_\alpha} \leq x\right) = F_{\overline{\boldsymbol{X}_\alpha}}(x) \text{ and } \overline{F_{\tilde{X}}(x)_\alpha} = P\left(\underline{\boldsymbol{X}_\alpha} \leq x\right) = F_{\underline{\boldsymbol{X}_\alpha}}(x) \quad (1)$$

Likewise, for a discrete FRV $\tilde{\boldsymbol{X}}$ with $F_{\underline{\boldsymbol{X}_\alpha}}$ and $F_{\overline{\boldsymbol{X}_\alpha}}$, $\forall \alpha \in [0,1]$, being the distribution functions of the probability of the RVs $\underline{\boldsymbol{X}_\alpha}$ and $\overline{\boldsymbol{X}_\alpha}$ obtained from $\tilde{\boldsymbol{X}}$, we define *the couple of εth quantiles* of the RVs infima and suprema for that membership level $Q^{\varepsilon}_{\tilde{X}_\alpha} = \left\{ \underline{Q^{\varepsilon}_{\tilde{X}_\alpha}}, \overline{Q^{\varepsilon}_{\tilde{X}_\alpha}} \right\}$:

$$\underline{Q^{\varepsilon}_{\tilde{X}_\alpha}} = \min\left\{x \mid F_{\underline{X}_\alpha}(x) \geq \varepsilon\right\} \text{ and } \overline{Q^{\varepsilon}_{\tilde{X}_\alpha}} = \min\left\{x \mid F_{\overline{X}_\alpha}(x) \geq \varepsilon\right\} \tag{2}$$

In the case that $\forall \omega_i \in \Omega, i = 1,\ldots,n$, the FNs $\tilde{X}(\omega_i)$ satisfy, $\forall \alpha \in [0,1]$, $\underline{X(\omega_i)}_\alpha \leq \underline{X(\omega_{i+1})}_\alpha$ and $\overline{X(\omega_i)}_\alpha \leq \overline{X(\omega_{i+1})}_\alpha$, $i = 1,\ldots,n-1$ for $Q^{\varepsilon}_{\tilde{X}_\alpha}$ we find:

$$\underline{Q^{\varepsilon}_{\tilde{X}_\alpha}} = \min_i \left\{ \underline{X(\omega_i)}_\alpha \mid \sum_{j \leq i} p_j \geq \varepsilon \right\} \text{ and } \overline{Q^{\varepsilon}_{\tilde{X}_\alpha}} = \min_i \left\{ \overline{X(\omega_i)}_\alpha \mid \sum_{j \leq i} p_j \geq \varepsilon \right\} \tag{3}$$

Given the probability space $\{\Omega, \mathrm{A}, P\}$ with $\Omega = \{\omega_i\}_{i=1,\ldots,n}$ and $P(\omega_i) = p_i, \forall i = 1,\ldots,n$, the mathematical expectation of a discrete ordinary RV X is a function of its crisp realizations $\{x_1, x_2,\ldots,x_n\}$: $E(X)(x_1, x_2,\ldots,x_n) = \sum_{i=1}^{n} x_i p_i$. So, given a FRV $\tilde{X}$ its *mathematical expectation*, $\tilde{E}(\tilde{X})$, is the FN induced by the FNs $\tilde{X}(\omega_1), \tilde{X}(\omega_2),\ldots,\tilde{X}(\omega_n)$ through $E(X)$. Concretely, following Puri and Ralescu (1986) we can compute the extremes of the α-cuts of $\tilde{E}(\tilde{X})$, $E(\tilde{X})_\alpha = \left[\underline{E(\tilde{X})}_\alpha, \overline{E(\tilde{X})}_\alpha \right]$, $\forall \alpha \in [0,1]$ as:

$$E(\tilde{X})_\alpha = \left[\sum_{i=1}^{n} \underline{X(\omega_i)}_\alpha \cdot p_i, \sum_{i=1}^{n} \overline{X(\omega_i)}_\alpha \cdot p_i \right] = \left[E(\underline{X}_\alpha), E(\overline{X}_\alpha) \right] \tag{4}$$

Regarding the variance of FRVs some authors propose fuzzy definitions, as in the case of mathematical expectation, whereas other authors such as Feng *et al.* (2001) and Körner (1997) propose using scalar (crisp) values for the variance since it is a dispersion measure. This dichotomy in the definition makes that a choice of one definition must be done (for a more detailed discussion of this topic see Couso et al. (2007)). Due to the choice we have made of the FRV concept we will expose the concept of fuzzy variance contained in Feng *et al.* (2001) that is built up from the variance of the infima and suprema RVs $\underline{X}_\alpha$ and $\overline{X}_\alpha$ obtained from $\tilde{X}$.

So, for a discrete FRV $\tilde{X}$ with infima and suprema discrete RVs $\underline{X}_\alpha$ and $\overline{X}_\alpha$, $\forall \alpha \in [0,1]$, the *variance of* $\tilde{X}$, $V(\tilde{X})$, is the real number:

$$V(\tilde{X}) = \frac{1}{2} \int_0^1 \left(V(\underline{X}_\alpha) + V(\overline{X}_\alpha) \right) \cdot d\alpha \tag{5}$$

Of course, from this definition of the variance of a FRV we can derive a crisp standard deviation as $D(\tilde{\boldsymbol{X}}) = \sqrt{V(\tilde{\boldsymbol{X}})}$.

Notice that we use the superscript " ~ " to symbolise fuzzy magnitudes and we write random variables with bold letters. So, the symbols corresponding to fuzzy random variables will be in bold and contain the superscript " ~ ".

3 Pricing Life Insurance with Fuzzy Random Variables

Following Andrés and González-Vila (2012) we propose adapting the stochastic approach to life insurance contracts to the use of fuzzy discount rates. In this case, the RV present value of premiums and present value of benefits turn into FRVs that will allow us to maintain all the uncertainty associated with discount rates but also with mortality. Considering that the discount rates are given via FNs, the value of discount function for 1 monetary unit (m.u.) payable at t is a FN, $\tilde{d}_t$, with α-cut representation $d_{t_\alpha} = \left[\underline{d_t}_\alpha, \overline{d_t}_\alpha\right]$. Notice that Andrés and González-Vila (2012) expose several ways to estimate actuarial discount rates with FNs.

Now let us consider a temporary life insurance. The insured party aged x will receive 1 m.u. at the end of the year of his death if this happens within the next n years. Otherwise he does not receive any quantity. Notice that our definition also includes whole life insurances for $n = \varpi - x + 1$ where ϖ is the maximum attainable age in the mortality tables. The space of events is $\Omega=\{\omega_0, \omega_1, \ldots, \omega_{n-1}, \omega_n\}$ where ω_0= "the insured survives n years (and so perceives no amount of the insurance)" and ω_j= "the insured dies within the jth year (and so perceives the m.u. at the end of this year)", j=1,2,…,n.

From the discount function $\tilde{d}_t$,we can generate the FRV *present value* of a life insurance temporary n years associated to a person aged x years $_n\tilde{\boldsymbol{A}}_{\boldsymbol{x}}$. The outcomes of the present value of life insurance are random because they depend on the insurer's death age. But these outcomes are also FNs since they are calculated with discount rates that are generalized intervals. This FRV adopts as values the following FNs, with respective probabilities P:

outcomes	P
$\tilde{d}_{r+1}$	$_{r\mid}q_x, r = 0,1,\ldots,n-1$
0	$_n p_x$

The FRV $_n\tilde{\boldsymbol{A}}_{\boldsymbol{x}}$ defines, $\forall \alpha \in [0,1]$, the infima and suprema RVs $\underline{_n\boldsymbol{A}_{\boldsymbol{x}}}_\alpha$ and $\overline{_n\boldsymbol{A}_{\boldsymbol{x}}}_\alpha$ as:

$$\underline{{}_{n}A_{x}}_{\alpha} \qquad \begin{array}{c|c} \text{outcomes} & P \\ \hline \underline{d_{r+1}}_{\alpha} & {}_{r|}q_x \\ 0 & {}_{n}p_x \end{array}, r=0,1,\ldots,n-1 \qquad\qquad \overline{{}_{n}A_{x}}_{\alpha} \qquad \begin{array}{c|c} \text{outcomes} & P \\ \hline \overline{d_{r+1}}_{\alpha} & {}_{r|}q_x \\ 0 & {}_{n}p_x \end{array}, r=0,1,\ldots,n-1$$

We want to remark that the outcomes of these two RVs are not in increasing order.

Based on the concepts defined in section 2, contained in (1) to (5), we can determine the next magnitudes.

α-cuts of the mathematical expectation of the FRV ${}_{n}\tilde{A}_{x}$, $\forall\alpha\in[0,1]$, $E\left({}_{n}\tilde{A}_{x}\right)_{\alpha}=\left[\underline{E\left({}_{n}\tilde{A}_{x}\right)}_{\alpha},\overline{E\left({}_{n}\tilde{A}_{x}\right)}_{\alpha}\right]$ with:

$$\underline{E\left({}_{n}\tilde{A}_{x}\right)}_{\alpha}=E\left(\underline{{}_{n}A_{x}}_{\alpha}\right)=\sum_{r=0}^{n-1}\underline{d_{r+1}}_{\alpha}\cdot{}_{r|}q_x \tag{6a}$$

$$\overline{E\left({}_{n}\tilde{A}_{x}\right)}_{\alpha}=E\left(\overline{{}_{n}A_{x}}_{\alpha}\right)=\sum_{r=0}^{n-1}\overline{d_{r+1}}_{\alpha}\cdot{}_{r|}q_x \tag{6b}$$

The variances of the RVs $\underline{{}_{n}A_{x}}_{\alpha}$ and $\overline{{}_{n}A_{x}}_{\alpha}$ are:

$$V\left(\underline{{}_{n}A_{x}}_{\alpha}\right)=\sum_{r=0}^{n-1}\left(\underline{d_{r+1}}_{\alpha}\right)^2\cdot{}_{r|}q_x-\left(\sum_{r=0}^{n-1}\underline{d_{r+1}}_{\alpha}\cdot{}_{r|}q_x\right)^2 \tag{7a}$$

$$V\left(\overline{{}_{n}A_{x}}_{\alpha}\right)=\sum_{r=0}^{n-1}\left(\overline{d_{r+1}}_{\alpha}\right)^2\cdot{}_{r|}q_x-\left(\sum_{r=0}^{n-1}\overline{d_{r+1}}_{\alpha}\cdot{}_{r|}q_x\right)^2 \tag{7b}$$

So the variance and standard deviation of the FRV ${}_{n}\tilde{A}_{x}$ are :

$$V\left({}_{n}\tilde{A}_{x}\right)=\frac{1}{2}\int_0^1\left[V\left(\underline{{}_{n}A_{x}}_{\alpha}\right)+V\left(\overline{{}_{n}A_{x}}_{\alpha}\right)\right]\cdot d\alpha \text{ and } D\left({}_{n}\tilde{A}_{x}\right)=\sqrt{V\left({}_{n}\tilde{A}_{x}\right)} \tag{8}$$

To determine the couple of distribution functions defined in (1), $\forall\alpha\in[0,1]$, $F_{{}_{n}\tilde{A}_{x}}(y)_{\alpha}=\left\{\underline{F_{{}_{n}\tilde{A}_{x}}(y)}_{\alpha},\overline{F_{{}_{n}\tilde{A}_{x}}(y)}_{\alpha}\right\}$ we must consider that $\underline{F_{{}_{n}\tilde{A}_{x}}(y)}_{\alpha}$ $\left(\overline{F_{{}_{n}\tilde{A}_{x}}(y)}_{\alpha}\right)$ can be obtained from the distribution function of the RV $\overline{{}_{n}A_{x}}_{\alpha}$ $\left(\underline{{}_{n}A_{x}}_{\alpha}\right)$. So for $r=0,1,...,n-2$:

$$\underline{F_{{}_n\tilde{A}_x}(y)}_{\alpha} = \begin{cases} 0 & \text{if} \quad y < 0 \\ {}_n p_x & \text{if} \quad 0 \le y < \overline{d_n}_{\alpha} \\ {}_n p_x + \sum_{s=0}^{r} {}_{n-(s+1)|}q_x & \text{if} \quad \overline{d_{n-r}}_{\alpha} \le y < \overline{d_{n-(r+1)}}_{\alpha} \\ 1 & \text{if} \quad y \ge \overline{d_1}_{\alpha} \end{cases} \tag{9a}$$

$$\overline{F_{{}_n\tilde{A}_x}(y)}_{\alpha} = \begin{cases} 0 & \text{if} \quad y < 0 \\ {}_n p_x & \text{if} \quad 0 \le y < \underline{d_n}_{\alpha} \\ {}_n p_x + \sum_{s=0}^{r} {}_{n-(s+1)|}q_x & \text{if} \quad \underline{d_{n-r}}_{\alpha} \le y < \underline{d_{n-(r+1)}}_{\alpha} \\ 1 & \text{if} \quad y \ge \underline{d_1}_{\alpha} \end{cases} \tag{9b}$$

From (9a) and (9b) the couples of εth quantiles of the RVs infima ($\underline{{}_nA_x}_{\alpha}$) and suprema ($\overline{{}_nA_x}_{\alpha}$), $\forall \alpha \in [0,1]$, $Q^{\varepsilon}_{{}_n\tilde{A}_x\,\alpha} = \left\{ \underline{Q^{\varepsilon}_{{}_n\tilde{A}_x}}_{\alpha}, \overline{Q^{\varepsilon}_{{}_n\tilde{A}_x}}_{\alpha} \right\}$ are:

- If $0 < \varepsilon \le {}_n p_x$: $Q^{\varepsilon}_{{}_n\tilde{A}_x\,\alpha} = 0$
- If ${}_n p_x < \varepsilon \le {}_n p_x + {}_{n-1|}q_x$: $Q^{\varepsilon}_{{}_n\tilde{A}_x\,\alpha} = \left\{ \underline{d_n}_{\alpha}, \overline{d_n}_{\alpha} \right\}$
- If ${}_n p_x + \sum_{s=0}^{r} {}_{n-(s+1)|}q_x < \varepsilon \le {}_n p_x + \sum_{s=0}^{r+1} {}_{n-(s+1)|}q_x$:

$$Q^{\varepsilon}_{{}_n\tilde{A}_x\,\alpha} = \left\{ \underline{d_{n-(r+1)}}_{\alpha}, \overline{d_{n-(r+1)}}_{\alpha} \right\}, \; r = 0,1,\ldots,n-2$$

Numerical application

We will analyze a life insurance for a person aged 60 years with n = 5. To price the life insurance we use the mortality tables GRM-80. We consider a fuzzy discount rate given by the triangular FN $\tilde{i}$ =(0.03, 0.04, 0.055) that will be applied throughout all the duration of the contract. Its α-cuts are:

$$\forall \alpha \in [0,1], \quad i_{\alpha} = \left[\underline{i}(\alpha), \overline{i}(\alpha) \right] = [0.03 + 0.01 \cdot \alpha, 0.055 - 0.015 \cdot \alpha]$$

So, the α-cuts of the discount function $\tilde{d}_t = (1+\tilde{i})^{-t}$ are $\forall \alpha \in [0,1]$:

$$d_{t\alpha} = \left[\underline{d_t}_{\alpha}, \overline{d_t}_{\alpha} \right] = \left[(1.055 - 0.015 \cdot \alpha)^{-t}, (1.03 + 0.01 \cdot \alpha)^{-t} \right]$$

The FRV *present value of the life insurance* ${}_5\tilde{A}_{60}$, will adopt as values the FNs with their respective probabilities reflected in Table 1.

Figure 1 shows these FNs as well as the mathematical expectation of the FRV, whose α- cuts are calculated as indicated in (6). Using (8) the standard deviation of this FRV is $D\left({}_5\tilde{A}_{60}\right)=0.0021$.

Table 1 FRV present value of the life insurance ${}_5\tilde{A}_{60}$

outcomes	α – cuts of the outcomes	P
$(1+\tilde{i})^{-1}$	$\left[(1.055-0.015\cdot\alpha)^{-1},(1.03+0.01\cdot\alpha)^{-1}\right]$	0.0108
$(1+\tilde{i})^{-2}$	$\left[(1.055-0.015\cdot\alpha)^{-2},(1.03+0.01\cdot\alpha)^{-2}\right]$	0.0115
$(1+\tilde{i})^{-3}$	$\left[(1.055-0.015\cdot\alpha)^{-3},(1.03+0.01\cdot\alpha)^{-3}\right]$	0.0123
$(1+\tilde{i})^{-4}$	$\left[(1.055-0.015\cdot\alpha)^{-4},(1.03+0.01\cdot\alpha)^{-4}\right]$	0.0132
$(1+\tilde{i})^{-5}$	$\left[(1.055-0.015\cdot\alpha)^{-5},(1.03+0.01\cdot\alpha)^{-5}\right]$	0.0142
0	$0=[0,0]$	0.9380

With expressions (9) the couples of distribution functions of probability and εth quantiles of the RVs $\underline{{}_5A_{60}}_{\alpha}$ and $\overline{{}_5A_{60}}_{\alpha}$ associated to ${}_5\tilde{A}_{60}$ can also be obtained. Moreover, it is possible to calculate the probability $P(B)$ for different Borel sets of the real line which will depend on the value considered for $\alpha\in[0,1]$.

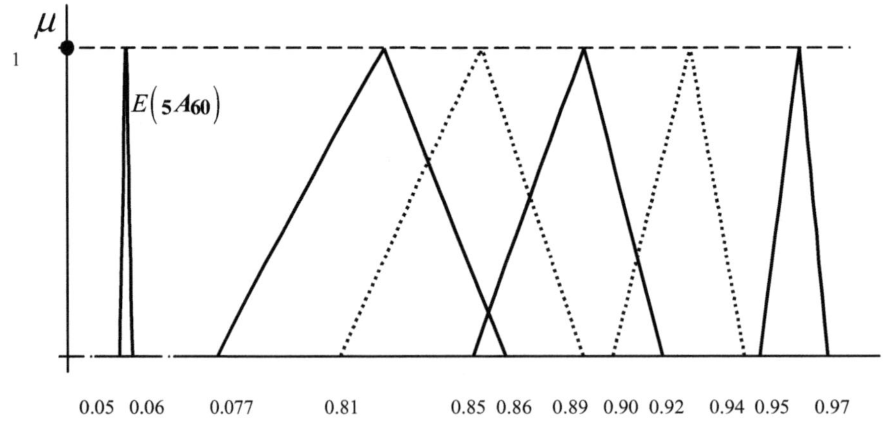

Fig. 1 Present value of the life insurance and its mathematical expectation

4 Pricing Portfolios of Life Insurances with Simulated Fuzzy Random Variables

This subsection introduces the fuzzy stochastic quantification of the present value of a group of life insurances. The FRV *present value of the portfolio of life insurances,* $\widetilde{PV}_P$, is obtained by adding the individual present values (${}_{n_j}\tilde{A}_{x_j}$), j=1,2,...,J, with J the total number of insured parties, i.e., $\widetilde{PV}_P = \sum_{j=1}^{J} {}_{n_j}\tilde{A}_{x_j}$.

In our analysis we will suppose, as is commonplace, independence in the mortality among insured parties. Of course, the dependence of the interest rate behaviour is complete, i.e. the path of interest rates throughout pricing horizon is common for all contracts. Any case, in order to obtain operational results, we must use the infima and suprema RVs $\underline{PV_P}_\alpha$ and $\overline{PV_P}_\alpha$, obtained from $\widetilde{PV}_P$ $\forall\alpha\in[0,1]$, defined as $\underline{PV_P}_\alpha = \sum_{j=1}^{J} \underline{{}_{n_j}A_{x_j}}_\alpha$ and $\overline{PV_P}_\alpha = \sum_{j=1}^{J} \overline{{}_{n_j}A_{x_j}}_\alpha$ where $\underline{{}_{n_j}A_{x_j}}_\alpha$ and $\overline{{}_{n_j}A_{x_j}}_\alpha$ are the RVs obtained from the FRV ${}_{n_j}\tilde{A}_{x_j}$ $\forall\alpha\in[0,1]$, j=1,2,...,J.

The α-cuts of the mathematical expectation of $\widetilde{PV}_P$ and its variance are easily obtained using the results of section 3. Specifically, for $E\left(\widetilde{PV}_P\right)_\alpha$ we obtain, $\forall\alpha\in[0,1]$:

$$E\left(\widetilde{PV}_P\right)_\alpha = \left[E\left(\underline{PV_P}_\alpha\right), E\left(\overline{PV_P}_\alpha\right)\right] = \left[\sum_{j=1}^{J} E\left(\underline{{}_{n_j}A_{x_j}}_\alpha\right), \sum_{j=1}^{J} E\left(\overline{{}_{n_j}A_{x_j}}_\alpha\right)\right] \tag{10}$$

where $E\left(\underline{{}_{n_j}A_{x_j}}_\alpha\right)$ and $E\left(\overline{{}_{n_j}A_{x_j}}_\alpha\right)$ are calculated as depicted in (6a) and (6b).

Regarding the variance, taking into account (7a) and (7b):

$$\begin{aligned} V\left(\widetilde{PV}_P\right) &= \frac{1}{2}\int_0^1 \left[V\left(\underline{PV_P}_\alpha\right) + V\left(\overline{PV_P}_\alpha\right)\right]\cdot d\alpha = \\ &= \frac{1}{2}\int_0^1 \left[\sum_{j=1}^{J} V\left(\underline{{}_{n_j}A_{x_j}}_\alpha\right) + \sum_{j=1}^{J} V\left(\overline{{}_{n_j}A_{x_j}}_\alpha\right)\right]\cdot d\alpha \end{aligned} \tag{11}$$

Now we can determine the fair price of life insurance (net premiums or net premium reserves). On the other hand, fixing stability surpluses for mortality deviations is difficult because the risk can only be quantified with the variance of present value of portfolio. To make an accurate estimate of cost of risk magnitudes it is also necessary to obtain the quantiles of $\widetilde{PV}_P$ using the distribution functions of probability of the infima and suprema RVs $\underline{PV_P}_\alpha$ and $\overline{PV_P}_\alpha$. However, it is not possible to find an exact analytical expression of these distribution functions.

Our approximation is based on the random simulation for pricing life insurances in Pitacco (1986) and Alegre and Claramunt (1995).

This means the results of the simulations will be FNs, due to the fuzziness in discount rates, instead crisp values.

To simulate the FRV $\widetilde{PV}_P$ we consider the RVs "moment when the insured amount will be paid", ${}_{n_j}T_{x_j}$ $j=1,2,\ldots J$. For the jth member of the collective the realizations of ${}_{n_j}T_{x_j}$ are $\{1,2,\ldots,n_j,\infty\}$ and their probabilities: $\left\{{}_{0/}q_{x_j}, {}_{1/}q_{x_j}, \ldots, {}_{n_j-1/}q_{x_j}, {}_{n_j}p_{x_j}\right\}$. Subsequently we implement the following steps:

Step 1. We will simulate S times the RVs ${}_{n_j}T_{x_j}$, $j=1,2,\ldots,J$. We suppose that those RVs are stochastically independent. So, the sth simulation of ${}_{n_j}T_{x_j}$, $j=1,2,\ldots,J$, generates a vector for the moment of payment $\vec{T}_s = \left(t_1^s,\ldots,t_j^s,\ldots,t_J^s\right)$, $s=1,2,\ldots,S$. Of course, t_j^s is the moment when the insured amount will be paid for the jth contract in the sth simulation.

Step 2. For the sth simulation we can now calculate the present value of the life insurance for the jth insured, that is the FN $\tilde{d}_{t_j^s}$, whose α-cuts, $\forall\alpha\in[0,1]$, are

$$d_{t_j^s\alpha} = \left[\underline{d_{t_j^s}}_\alpha, \overline{d_{t_j^s}}_\alpha\right].$$

Step 3. For the sth simulation we calculate the present value of the portfolio, $\widetilde{PV}_P^s$, by adding the present value of the J policies. It is the FN $\widetilde{PV}_P^s = \sum_{j=1}^{J}\tilde{d}_{t_j^s}$, where the α-cuts, $PV_{P\alpha}^s$, are:

$$PV_{P_\alpha}^s = \left[\underline{PV_P^s}_\alpha, \overline{PV_P^s}_\alpha\right] = \left[\sum_{j=1}^{J}\underline{d_{t_j^s}}_\alpha, \sum_{j=1}^{J}\overline{d_{t_j^s}}_\alpha\right].$$

Notice that in this step the original FRV $\widetilde{PV}_P$ has been approximated by a simulated FRV $\widetilde{PV}_P^*$ whose realizations are the FNs $\left\{\widetilde{PV}_P^1, \widetilde{PV}_P^2, \ldots, \widetilde{PV}_P^s, \ldots, \widetilde{PV}_P^S\right\}$ with the same probability of occurrence $\frac{1}{S}$. This simulated FRV defines, $\forall\alpha\in[0,1]$, the infima and suprema RVs $\underline{PV_P^*}_\alpha$ and $\overline{PV_P^*}_\alpha$.

Step 4. We describe $\widetilde{\boldsymbol{PV}}^{*}_{\boldsymbol{P}}$ from its infima and suprema RVs $\underline{\boldsymbol{PV}^{*}_{\boldsymbol{P}}}_{\alpha}$ and $\overline{\boldsymbol{PV}^{*}_{\boldsymbol{P}}}_{\alpha}$. To do this, the values of these RVs, $\forall\alpha\in[0,1]$, are ordered increasingly in such a way that the outcomes of $\underline{PV^{*}_{P}}_{\alpha}$ are $\underline{PV^{(1)}_{P}}_{\alpha}\leq\underline{PV^{(2)}_{P}}_{\alpha}\leq\ldots$ $\ldots\leq\underline{PV^{(s)}_{P}}_{\alpha}\leq\ldots\leq\underline{PV^{(S)}_{P}}_{\alpha}$ and analogously for $\overline{PV^{*}_{P}}_{\alpha}$: $\overline{PV^{(1)}_{P}}_{\alpha}\leq\overline{PV^{(2)}_{P}}_{\alpha}\leq\ldots$ $\ldots\leq\overline{PV^{(s)}_{P}}_{\alpha}\leq\ldots\leq\overline{PV^{(S)}_{P}}_{\alpha}$. With the parentheses we symbolize that the realizations of the RVs are ordered increasingly and not from their position in the simulation. Of course, in this case $\underline{PV^{(s)}_{P}}_{\alpha}$ and $\overline{PV^{(s)}_{P}}_{\alpha}$ may be the extremes of the α-cuts of two different realizations of $\widetilde{\boldsymbol{PV}}^{*}_{\boldsymbol{P}}$ that were obtained in step 3. Now we can obtain , $\forall\alpha\in[0,1]$, the couple $F_{\widetilde{\boldsymbol{PV}}^{*}_{\boldsymbol{P}}}(y)_{\alpha}=\left\{\underline{F_{\widetilde{\boldsymbol{PV}}^{*}_{\boldsymbol{P}}}(y)}_{\alpha},\overline{F_{\widetilde{\boldsymbol{PV}}^{*}_{\boldsymbol{P}}}(y)}_{\alpha}\right\}$ of the FRV $\widetilde{\boldsymbol{PV}}^{*}_{\boldsymbol{P}}$:

$$\underline{F_{\widetilde{\boldsymbol{PV}}^{*}_{\boldsymbol{P}}}(y)}_{\alpha}=F_{\overline{\boldsymbol{PV}^{*}_{\boldsymbol{P}}}_{\alpha}}(y)=\begin{cases}0 & y<\overline{PV^{(1)}_{P}}_{\alpha}\\ \dfrac{s}{S} & \overline{PV^{(s)}_{P}}_{\alpha}\leq y<\overline{PV^{(s+1)}_{P}}_{\alpha},\ s=1,2,\ldots,S-1\\ 1 & y\geq\overline{PV^{(S)}_{P}}_{\alpha}\end{cases}$$

$$\overline{F_{\widetilde{\boldsymbol{PV}}^{*}_{\boldsymbol{P}}}(y)}_{\alpha}=F_{\underline{\boldsymbol{PV}^{*}_{\boldsymbol{P}}}_{\alpha}}(y)=\begin{cases}0 & y<\underline{PV^{(1)}_{P}}_{\alpha}\\ \dfrac{s}{S} & \underline{PV^{(s)}_{P}}_{\alpha}\leq y<\underline{PV^{(s+1)}_{P}}_{\alpha},\ s=1,2,\ldots,S-1\\ 1 & y\geq\underline{PV^{(S)}_{P}}_{\alpha}\end{cases}$$

From these expressions we obtain the couple of εth quatiles $Q^{\varepsilon}_{\widetilde{\boldsymbol{PV}}^{*}_{P\alpha}}$:

$$Q^{\varepsilon}_{\widetilde{\boldsymbol{PV}}^{*}_{P\alpha}}=\left\{\underline{Q^{\varepsilon}_{\widetilde{\boldsymbol{PV}}^{*}_{P}}}_{\alpha},\overline{Q^{\varepsilon}_{\widetilde{\boldsymbol{PV}}^{*}_{P}}}_{\alpha}\right\}=\left\{\underline{PV^{(s)}_{P}}_{\alpha},\overline{PV^{(s)}_{P}}_{\alpha}\right\}\text{ for }\frac{s-1}{S}<\varepsilon\leq\frac{s}{S},s=1,2,\ldots,S$$

Numerical application

We will analyse the liability of a portfolio comprised of 12 whole life insurance contracts with an insured amount of 1.000 m.u. The insured parties for j=1,2,…,5 are 65 years old whereas for j=6,7,…,12 the insured parties are 75 years old. We price the contracts with the technical basis used in section 3.

The maximum age in our mortality tables is 114, therefore for people aged x_j=65 years, n_j=50 and when x_j=75 years, n_j=40. The possible results of ${}_{50}T_{65}$ are {1,2,...,50} and their probabilities: $\left\{ {}_{0|}q_{65}, {}_{1|}q_{65}, \ldots, {}_{49|}q_{65}, {}_{50}p_{65} \right\}$. Likewise ${}_{40}T_{75}$ can take {1,2,...,40} with the probabilities: $\left\{ {}_{0|}q_{75}, {}_{1|}q_{75}, \ldots, {}_{39|}q_{75}, {}_{40}p_{75} \right\}$

Figure 2 shows the shape of the fuzzy numbers expectation of the present value for the two life insurances, which are denoted by ${}_{40}\tilde{A}_{75}$ and ${}_{50}\tilde{A}_{65}$. Likewise, $D\left({}_{50}\tilde{A}_{65} \right)$=177.49 and $D\left({}_{40}\tilde{A}_{75} \right)$=161.09.

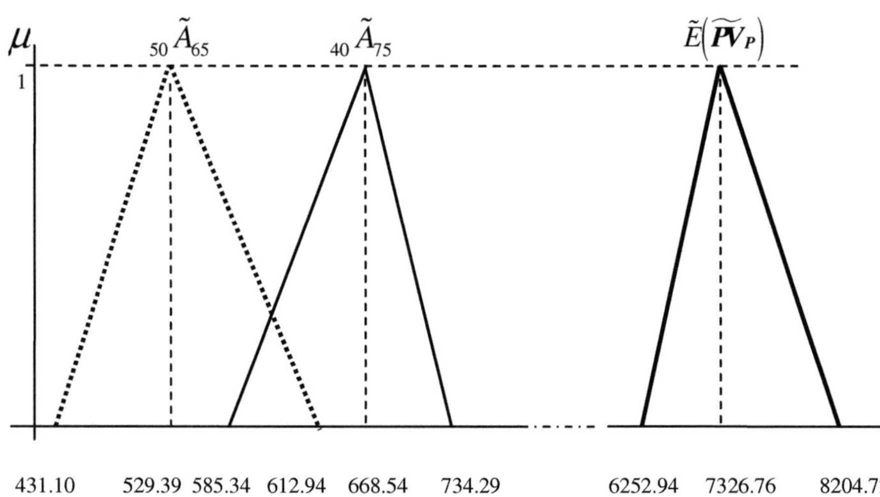

Fig. 2 Expectation of present value for the two types of contracts and for the portfolio

From (10) we deduce the mathematical expectation of $\widetilde{PV}_P$, $\tilde{E}\left(\widetilde{PV}_P\right)$, which can be interpreted as the fuzzy fair price of the portfolio of life insurances and which is also represented in Figure 2. Moreover, using (11), $D\left(\widetilde{PV}_P\right)$=2015.11.

We approximate the FRV $\widetilde{PV}_P$ from S=5000 simulations of the present value of the portfolio. Table 2 shows the approximate to its 90th, 95th and 99th quantile couples.

Table 3 shows the values of the infima and suprema distribution functions evaluated in the mode of the 90th, 95th and 99th quantiles in Table 2; i.e. y=8062.72, 8275.06 and 8679.66. The infima and suprema distribution functions are evaluated for α=0, 0.25, 0.5, 0.75, 1. We can check that the accumulated probability in the most feasible value of the quantile is its own probability level (e.g. 90% for y=8062.72) but the distribution functions may take other values. When y=8062.72, the accumulated probability can oscillate between 39.2% and 99.6%.

Table 2 Couples of several present value quantiles of the portfolio

	$Q^{90\%}_{\widetilde{PV}^*_{P\alpha}}$		$Q^{95\%}_{\widetilde{PV}^*_{P\alpha}}$		$Q^{99\%}_{\widetilde{PV}^*_{P\alpha}}$	
α	$\underline{Q^{90\%}_{\widetilde{PV}^*_{P}}}_{\alpha}$	$\overline{Q^{90\%}_{\widetilde{PV}^*_{P}}}_{\alpha}$	$\underline{Q^{95\%}_{\widetilde{PV}^*_{P}}}_{\alpha}$	$\overline{Q^{95\%}_{\widetilde{PV}^*_{P}}}_{\alpha}$	$\underline{Q^{99\%}_{\widetilde{PV}^*_{P}}}_{\alpha}$	$\overline{Q^{99\%}_{\widetilde{PV}^*_{P}}}_{\alpha}$
1	8062.72	8062.72	8275.06	8275.06	8679.66	8679.66
0.75	7802.02	8242.54	8023.87	8449.55	8450.39	8840.83
0.5	7553.92	8429.34	7785.22	8630.68	8234.61	9005.88
0.25	7321.14	8625.89	7557.38	8816.70	8039.00	9173.62
0	7094.34	8829.53	7344.54	9009.41	7837.01	9351.56

Table 3 Values of the couple $F_{\widetilde{PV}^*_P}(y)_\alpha$ for y=8062.72, 8275.06 and 8679.66.

	y=8062.72		y=8275.06		y=8679.66	
α	$\underline{F_{\widetilde{PV}^*_P}(y)}_\alpha$	$\overline{F_{\widetilde{PV}^*_P}(y)}_\alpha$	$\underline{F_{\widetilde{PV}^*_P}(y)}_\alpha$	$\overline{F_{\widetilde{PV}^*_P}(y)}_\alpha$	$\underline{F_{\widetilde{PV}^*_P}(y)}_\alpha$	$\overline{F_{\widetilde{PV}^*_P}(y)}_\alpha$
1	0.9	0.9	0.95	0.95	0.99	0.99
0.75	0.832	0.957	0.911	0.98	0.98	0.996
0.5	0.733	0.981	0.839	0.99	0.959	0.999
0.25	0.574	0.99	0.734	0.996	0.915	1
0	0.392	0.996	0.559	0.999	0.837	1

If total reserves are priced (or the premiums if the contracts are new) at 8062.72 m.u., the probability that this value will be sufficient is 90% with a membership level equal to 1. This situation arises when the profitability of assets is the most feasible, i.e. 4%. However, if the ex-post interest rate of assets is not 4% (notice that we have predicted that it may oscillate between 3% and 5.5%) the solvency level attained with 8062.72 m.u. changes. When the profit of assets is greater than 4%, 8062.72 m.u. can sustain 99.6% of mortality deviations (this solvency level is associated with an investment profit equal to 5.5%). However, if the assets profits are under 4% (in our example between 3% and 4%), 8062.72 m.u. units allows a solvency probability of less than 90%. In the worst interest rate scenario, the level of solvency may be 39.2%, which is reached when the insurer's profit is 3%.

5 Conclusions

Following the developments by Andrés and González-Vila (2012) for life annuities, we model the present value of life insurance contracts with FRVs because they allow quantifying their expected price and risk resulting from the uncertainty sources considered.

As several authors mentioned above have done, in this paper we use FNs to quantify uncertain insurance discount rates. We show how the use of FRV not only allows the fair price of the policy to be quantified, but also measures for the risk of mortality, both of which are fundamental for fitting surplus over pure premiums or reserves for deviation of mortality. It is important to consider that, to the contrary, "traditional" fuzzy life insurance quantification reduces random cash flows to their expected values, thereby making the risk of mortality difficult to quantify.

Notice that the most representative value of a FRV, its mathematical expectation, is a FN. However, a crisp quantification of this relevant magnitude is required, for example, to state fair prices or account reserves in financial statements. In section 4, the value of the 90th quantile for the present value of premiums has a 1-cut equal to 8062.72 whereas its 0-cut is {7094.34, 8829.53}. If this quantile quantifies net reserves of the portfolio it can be understood that "the reserves must be approximately 8062.72 m.u." but they may fluctuate between 7094.34 and 8829.53 m.u. To obtain the definitive value of the magnitude, it must be transformed into a crisp value. To do this we need to apply a *defuzzifying* method (see Zhao and Govind (1991) for a wide discussion of fuzzy mathematics, and Cummins and Derrig (1997), for applications in fuzzy-actuarial analysis). Another way of doing this, which is very consistent with practice in the real world, is to consider the fuzzy quantification as a first approximation that allows a margin for the "actuarial subjective judgement" or upper and lower bounds for acceptable market prices (the 0-cut). Finally, the actuary must use his/her experience and attempt to establish the crisp value of the fuzzy estimate.

References

Alegre, A., Claramunt, M.M.: Allocation of solvency cost in group annuities: Actuarial principles and cooperative game theory. Insurance: Mathematics and Economic 17, 19–34 (1995)

de Andrés, J., González-Vila, L.: Using fuzzy random variables in life annuities pricing. Fuzzy Sets and Systems 188, 27–44 (2012)

de Andrés, J., Terceño, A.: Applications of Fuzzy Regression in Actuarial Analysis. Journal of Risk and Insurance 70, 665–699 (2003)

Betzuen, A., Jiménez, M., Rivas, J.A.: Actuarial mathematics with fuzzy parameters. An application to collective pension plans. Fuzzy Economic Review 2, 47–66 (1997)

Buckley, J.J.: The fuzzy mathematics of finance. Fuzzy Sets and Systems 21, 57–73 (1987)

Couso, I., Dubois, D., Montes, S., Sánchez, L.: On various definitions of the variance of a fuzzy random variable. In: 5th International Symposium on Imprecise Probabilities and Their Applications, Prague, Czech Republic (2007)

Cummins, J.D., Derrig, R.A.: Fuzzy financial pricing of property-liability insurance. North American Actuarial Journal 1, 21–44 (1997)

Derrig, R.A., Ostaszewski, K.: Managing the tax liability of a property liability insurance com-pany. Journal of Risk and Insurance 64, 695–711 (1997)

Derrig, R.A., Ostaszewski, K.: Fuzzy Sets. In: Encyclopaedia of Actuarial Science, vol. 2, pp. 745–750. John Wiley & Sons, Chichester (2004)

Feng, Y., Hu, L., Shu, H.: The variance and covariance of fuzzy random variables and their applications. Fuzzy Sets and Systems 120, 487–497 (2001)

Gerber, H.U.: Life Insurance Mathematics. Springer, Berlin (1995)

Gil-Aluja, J.: Investment on uncertainty. Kluwer Academic Publishers, Dordretch (1998)

Guangyuan, W., Yue, Z.: The theory of fuzzy stochastic processes. Fuzzy Sets and Systems 51, 161–178 (1992)

Huang, T., Zhao, R., Tang, W.: Risk model with fuzzy random individual claim amount. European Journal of Operational Research 192, 879–890 (2009)

Kaufmann, A.: Fuzzy subsets applications in O.R. and management. In: Jones, A., Kaufmann, A., Zimmermann, H.J. (eds.) Fuzzy Set Theory and Applications, pp. 257–300. Reidel, Dordrecht (1986)

Kaufmann, A., Gil-Aluja, J.: Las matemáticas del azar y de la incertidumbre. Ceura, Madrid (1990)

Körner, R.: On the variance of fuzzy random variables. Fuzzy Sets and Systems 92, 83–93 (1997)

Krätschmer, V.: A unified approach to fuzzy random variables. Fuzzy Sets and Systems 123, 1–9 (2001)

Kruse, R., Meyer, K.D.: Statistics with vague data. Reidel, Dordrecht (1987)

Kwakernaak, H.: Fuzzy random variables I: Definitions and Theorems. Information Sciences 15, 1–29 (1978)

Kwakernaak, H.: Fuzzy random variables II: Algorithms and Examples for the Discrete Case. Information Sciences 17, 253–278 (1979)

Lemaire, J.: Fuzzy insurance. Astin Bulletin 20, 33–55 (1990)

Li Calzi, M.: Towards a general setting for the fuzzy mathematics of finance. Fuzzy Sets and Systems 35, 265–280 (1990)

Ostaszewski, K.: An investigation into possible applications of fuzzy sets methods in actuarial science. Society of Actuaries, Schaumburg (1993)

Pitacco, E.: Simulation in insurance. In: Goovaerts, M., De Vylder, F., Haezendonck, J. (eds.) Insurance and Risk Theory, pp. 43–44. Reidel, Dordrecht (1986)

Puri, M.L., Ralescu, D.A.: Fuzzy random variables. Journal of Mathematical Analysis and Applications 114, 409–422 (1986)

Shapiro, A.F.: Fuzzy logic in insurance. Insurance: Mathematics and Economics 35, 399–424 (2004)

Shapiro, A.F.: Fuzzy random variables. Insurance: Mathematics and Economics 44, 307–314 (2009)

Yakoubov, Y.H., Haberman, S.: Review of actuarial applications of fuzzy set theory. Actuarial Research Paper n. 105. Department of Actuarial Science and Statistics of the City University, London (1998)

Zhao, R., Govind, R.: Defuzzification of fuzzy intervals. Fuzzy Sets and Systems 43, 45–55 (1991)

Zhong, C., Zhou, G.: The equivalence of two definitions of fuzzy random variables. In: Proceedings of the 2nd IFSA Congress, Tokyo, pp. 59–62 (1987)

Customer Loyalty Strategies and Tools in the Spanish Banking Sector

Carlos del Castillo Peces, Carmelo Mercado Idoeta, and Camilo Prado Román

Department of Business Administration, Rey Juan Carlos University,
Paseo de los Artilleros, s/n, 28032 Madrid, Spain
carlos.delcastillo@urjc.es, carmelo.mercado@urjc.es,
camilo.prado.roman@urjc.es

Abstract. The present study sets out to analyze the pervasiveness and importance attached to customer loyalty strategies in the Spanish banking sector. In addition, the authors evaluate the effects of scientific marketing and customer relationship management (CRM) tools on customer loyalty. Results of a survey completed by fifty-three Spanish financial institutions show that the implementation of customer loyalty strategies is widespread: all participating institutions have put in place client segmentation practices and differentiated marketing strategies for each segment. The use of specific support tools is not widespread, although results point toward their positive effect on customer loyalty.

Keywords: Customer loyalty, Scientific Marketing, CRM, Financial sector.

1 Introduction

In the current climate of increased competitiveness, firms are (re)structuring their organizations to generate more value for their clients in what Porter (1985) defines as a differentiation strategy. Excess supply in most sectors of developed economies makes it more difficult and expensive to attract customers. Therefore, companies realize that customer loyalty is crucial in order to assure the desired levels of growth, profitability, and long-term survival (Heskett et al., 1997).

The Spanish financial industry shares many of the aforementioned characteristics of mature markets, with elevated levels of competition among participants. Fainé and Tornabell (2001) indicate that survival in the financial sector is determined by the capacity to anticipate clients' needs and provide quality service, thereby augmenting customer loyalty. In this paradigm, the traditional 'P's of the marketing mix: product, price, place, and promotion give way to the three 'R's of relationship marketing: relate, retain, and recovery.

Numerous strategies have been developed in order to systematically encourage customer loyalty. This paper studies how common and effective these strategies are within the context of Spanish financial institutions. On a more detailed level, the authors evaluate whether the implementation of certain scientific marketing tools (such as datawarehouse, data mining, or neural networks) or customer

A.M. Gil-Lafuente et al. (Eds.): Soft Comput. in Manag. and Bus. Econ., STUDFUZZ 287, pp. 127–135.
springerlink.com

relationship management (CRM) tools ameliorate the effectiveness of customer loyalty strategies. The methodology of the study consists of a questionnaire delivered by mail to 53 Spanish banks and savings banks. Results and conclusions are presented in sections 5 and 6, respectively.

2 Customer Loyalty

According to Huete et al. (2003), developing customer loyalty implies reducing drop-out rates (or inversely, increasing retention rates). Reinares and Ponzoa (2002) develop this idea by insisting on the fact that loyalty is a measure of the intensity of the client's relationship with the brand or company. It also represents a barrier for brand-switching to competitor's offerings. Therefore, as Bigné and Andreu (2004) point out, customer loyalty is a concept that does not only encompass repeat purchases but also satisfaction and a sense of affective engagement with the brand.

Studies on the relationship between customer loyalty and company profits were collected by Fainé and Tornabell (2001). According to them, a 2000 study on US banks shows that a 2% increase in the retention rate has the same effect as a cost reduction of 10%. Furthermore, a 20-year old client contributes 85% more income than a 10-year old client. Daemon Quest (2005) calculates that retaining 25% of high-value clients increase operating profits by 15%.

An explanation of the relationship between loyalty and profitability is provided by Reichheld (1990), who posits that there exist six "high tension cables" that convert increased loyalty into higher profits. The connectors that explain these loyalty-driven sources of profitability are: a) repeat sales of the same product or service; b) cross-selling or up-selling practices; c) referrals processes and/or member-get-a-member (MGM) programs, which generate up to 50% of new clients in certain sectors; d) diminished price sensitivity in loyal clients; e) reduced use of company resources by loyal clients, since they are familiar with the company's processes and products; f) the fact that loyal clients require a smaller marketing expense, in comparison with that required to acquire new clients. Ashley and Varki (2009) identify another profitability driver for loyalty: loyal clients are prone to complain directly to the company instead of spreading negative word-of-mouth comments, therefore damaging the firm's reputation.

Previous literature has identified numerous factors that influence customer loyalty. The first of these is satisfaction, which has been defined by Kotler et al. (1995) as the positive result of comparing "perceived value" with "expected value." A considerable cohort of scholars consider that affective elements must also be included in this definition (Sanzo et al., 2003; Bigné and Andreu, 2004; Bigné et al., 2010). Mägi (2003), Fandós et al. (2009), Liang and Wang (2008), as well as Ashley and Varki (2009) find a direct link between satisfaction and loyalty. Furthermore, quality is a determinant element of satisfaction and therefore influences customer loyalty (Del Río et al. 2001; Cronin et al. 2000). Other factors are trust in the company's ability to fulfill expectations (Garbarino and Johnson, 1999; Gommans et al., 2001) and physical or mental switching costs (Jones et al., 2000; Apaolaza et al., 2004). Finally, a sense of commitment –for rational or emotional reasons– also has a positive effect on loyalty (Demoulin and Zidda, 2009; Lacey and Morgan, 2009).

3 Loyalty Development Strategies

Loyalty programs may be categorized according to their principal strategic goals. Previous research identifies two broad groups: a) costumer heterogeneity management, in which specific marketing activities are designed for different customer purchase behavior profiles (Lacey and Sneath, 2006; Nunes and Dréze, 2009); and b) customer relationship management, in which efforts are made to modify consumer behavior in order to increase the value of the relationship over its entire life cycle (Meyer-Waarden and Benavent, 2003; Uncles et al., 2003; Reinares et al., 2010).

The first step in customer heterogeneity management is to segment the customer base in order to identify consumer groups with different needs and buying behaviors. Each segment can then be offered products and services that better suit its needs. In the financial sector, some authors consider that it can be outright harmful to design a unified marketing strategy for the entire client base.

While this type of personalization is already being put into practice for a small group of clients (particularly in private banking), a future challenge for the financial industry is to introduce a reasonable level of personalization to larger segments in what Mas (2003) defines as "mass customization strategies." In essence, combining a standardized approach with personalization elements can be a cost-effective way of increasing the client's perceived value, satisfaction, and therefore loyalty to the financial institution (Huete et al., 2003). For customer segmentation to be effective, two other strategies are necessary: differentiated marketing strategies for each segment and the creation of client portfolios for each segment in order to manage marketing efforts through individualized channels.

Relationship marketing processes aim to develop customer loyalty. Customer loyalty programs therefore have customer relationship management as a second goal. If a company has strong relationships with its customers –in other words, it knows them and can offer products and services that they want– then they will show loyalty (Reinares and Ponzoa, 2002). This leads to the last loyalty strategy or multichannel strategy, which consists of reaching the client through a multitude of communication channels (virtual and/or physical).

4 Empirical Analysis

4.1 Goals

The primary goal of our empirical analysis is to ascertain the pervasiveness and importance of loyalty development strategies (as described in section 3) among Spanish financial institutions.

The secondary goal is to verify whether the use of certain tools is helpful in obtaining positive results from these strategies. The successful implementation of loyalty strategies requires the collection and management of large amounts of client data. Information on current and future behavioral and consumption habits is essential if an individualized marketing approach is to replace the traditional mass marketing approach[1].

In order to optimize client relationships and obtain positive results from the loyalty-development process, it is key to have information in sufficient quantity and quality, as well as appropriate systems to internally disseminate this information (Day, 2003).

The supporting tools that this paper studies are classified into scientific marketing tools such as datawarehouse, data mining, neural networks, among others; and customer relationship management (CRM) tools. All of these categories of tools collect, analyze, and incorporate information on all interactions with the client.

4.2 Methodology

This paper studies the two main groups of financial institutions in Spain –banks and savings banks– holding 95% of the sector's assets. Banco de España provides a list of all financial institutions in its 2008 Statistical Bulletin. From this list, the authors excluded the smaller and non-retail banks. The final group of 16 banks and 37 savings banks holds 92% of the sector's assets. Primary data was collected from mail and online surveys sent out to the marketing directors of participating institutions. 40 institutions replied, resulting in a response rate of 72.7% that represents 88.6% of total assets in the Spanish banking sector.

Table 1 Technical specifications

Universe	Spanish financial institutions (banks and savings banks) with retail presence
Sampling method	Random: survey sent to financial institutions in the universe
Data collection method	Online and mail questionnaire
Surveyed position	Marketing Directors
Population	53
Sample size	40
Confidence level	95% (z=1,96; p=q=0,5)
Sampling error	7,7%
Information collection period	November 2, 2009 to February 28, 2010

[1] According to research by Unysis Corporation and The American Banker, 46% of US banks that utilize a customer information system have a client retention ratio of over 90% versus 17% of banks that do not ("Customer Focus in Retail Banking: Opinions & Practices at 150 Top Banks," 2000).

5 Results

Results on the pervasiveness of the different loyalty-development strategies among Spanish banks and savings banks are shown in Table 2.

Table 2 Pervasiveness of loyalty-development strategies

% of Institutions	Not in use	Future implementation	Implemented 1-2 years ago	Implemented 3 or more years ago
Client segmentation	0%	0%	0%	100%
Differentiated marketing strategies for each segment	0%	0%	20%	80%
Client portfolios	0%	0%	27,5%	72,5%
Multichannel strategy	0%	0%	22,5%	77,5%

The results show that all loyalty-development strategies are fully implemented across the sample. Particularly, client segmentation was implemented 3 or more years ago in all institutions and 75% implemented the other strategies in the same time period. Results concerning the importance attached to customer loyalty strategies are shown in Table 3.

Table 3 Importance attached to customer loyalty strategies

% of Institutions	Not important	Somewhat important	Quite important	Very important
Client segmentation	0%	0%	47,5%	52,5%
Differentiated marketing strategies for each segment	0%	0%	47,5%	52,5%
Client portfolios	10%	20%	22,5%	47,5%
Multichannel strategy	0%	15%	67,5%	17,5%

All participating institutions consider client segmentation and differentiated marketing strategies for each segment as quite or very important in their loyalty-development efforts. Conversely, client portfolios and multichannel strategies are not at all or only somewhat important for 30% and 15% of the sample, respectively.

In order to evaluate whether the implementation of scientific marketing and CRM tools contributes to obtaining positive results from customer loyalty strategies, the survey first asked about the levels of implementation among financial institutions, as shown in Table 4.

Table 4 Implementation of loyalty program support tools

% of Institutions	Not in use	Future implementation	Implemented 1-2 years ago	Implemented 3 or more years ago
Scientific marketing tools	0%	32,5%	52,5%	15%
CRM tools	0%	25%	32,5%	42,5%

The implementation of customer loyalty support tools is relatively widespread (67.5% and 75% of institutions); as well as recent. Only 15% implemented scientific marketing tools 3 or more years ago, while 42.5% implemented CRM tools 3 or more years ago.

In order to evaluate the possible benefits of these support tools for loyalty-development programs, the authors studied client desertion rates among participating institutions, in accordance with previous literature that identifies this as one of the best factors to measure customer loyalty. By cross-listing the evolution of desertion rates with the implementation of scientific marketing or CRM tools, we obtain the results shown in Tables 5 and 6.

Table 5 Development of client desertion rates as a function of implementation of scientific marketing tools

% of Institutions	Change in client desertion rates				
Implementation of scientific marketing tools	Significantly reduced	Slightly reduced	No change	Slightly Increased	Significantly increased
Not in use or future implementation	0,0%	46,2%	53,8%	0,0%	0,0%
Implemented 1-2 years ago	0,0%	71,4%	28,6%	0,0%	0,0%
Implemented 3 or more years ago	0,0%	66,7%	33,3%	0,0%	0,0%

Most financial institutions that implemented scientific marketing techniques one or two years ago saw a slight decrease in their client desertion rates in comparison with those that did not (71.4% versus 46.2%). There is not a large difference between institutions that implemented these tools 1 or 2 years ago and those that implemented them 3 or more years ago (71.4% vs. 66.7%).

Table 6 Development of client desertion rates as a function of implementation of CRM tools

% of institutions	Change in client desertion rates				
Implementation of CRM tools	Significantly reduced	Slightly reduced	No change	Slightly increased	Significantly increased
Not in use or Future implementation	0,0%	30%	70%	0,0%	0,0%
Implemented 1-2 years ago	0,0%	61,5%	38,5%	0,0%	0,0%
Implemented 3 or more years ago	0,0%	64,7%	35,3%	0,0%	0,0%

Client desertion rates are less likely to have diminished in financial institutions that implemented CRM tools 1 or 2 years ago, in comparison to those that did not (61.5% vs. 30%). Nevertheless, results are similar for those that implemented CRM tools 1 or 2 years and up to 5 years ago (61.5% vs. 64.7%).

6 Conclusions and Implications for Future Research

As shown in Table 2, loyalty-development strategies have been implemented in 100% of the financial institutions in the sample. More than 70% of them did so over three years ago, particularly client segmentation (100%). As a consequence, Spanish banks and savings banks have had ample time to evaluate the results of these decisions and the importance they attach to each one of them (Table 3). All the financial institutions that participated in the survey consider client segmentation and differentiated marketing strategies for each segment to be important. On the other hand, the effects of client portfolios and multichannel strategy are not unanimously seen as relevant by the marketing directors of these institutions (30% and 15%, respectively, considered that these strategies where only somewhat or not at all important to develop loyalty). This lack of agreement could be symptomatic of the need for other elements in a successful loyalty-development program, such as support tools. Evaluating the use of these tools within the context of loyalty development constitutes the second goal of this paper.

Customer loyalty support tools are widespread within this sub-sector of the financial industry, but not so much as the loyalty-development strategies previously studied. As seen in Table 4, 32.5% of the participating financial institutions have not yet implemented scientific marketing tools, and 25% have not implemented CRM tools. As a way of evaluating the possible influence of these tools on loyalty-development strategies, we analyzed how the client desertion rate has evolved during the last 5 years according to the level of implementation of these tools. Tables 5 and 6 show that client desertion rates have evolved more favorably both for financial institutions that have adopted scientific marketing tools (71% saw slightly reduced desertion rates, in comparison to slight decreases in 46% of those that have not implemented these tools); as well as for those that are using CRM tools (61% saw slightly reduced desertion rates, in comparison to slight decreases in 30% of those that have not implemented these tools). The effects of these improvements on desertion rates appear within the first two years of implementation and then reach a plateau.

Given these results, it would be valuable to further investigate the influence on customer loyalty of other characteristics of financial institutions. Among these, cultural elements such as orientation to the client or the market and leadership commitment with loyalty programs could shed more light on the practices of Spanish financial institutions. Another promising venue would be to study the implementation and effectiveness of customer-centric collaborative marketing, a strategy that is starting to be implemented by certain financial institutions.

References

Apaolaza, V., Hartmann, P., Zorrilla, P.: Antecedentes de la lealtad del cliente de energía doméstica: calidad del servicio, satisfacción, confianza y costes de cambio. In: Proceedings of 16th Encuentro de Profesores Universitarios de Marketing, pp. 160–173. Editorial ESIC, Madrid (2004)

Ashley, C., Varki, S.: Loyalty and its influence on complaining behavior and service recovery satisfaction. Journal of Consumer Satisfaction, Dissatisfaction and Complaining Behavior 22, 21–35 (2009)

Bigne, E., Andreu, L.: Emociones, satisfacción y lealtad del consumidor en entornos comerciales. Distribución y Consumo 76, 77–87 (2004)

Bigne, E., Currás, R., Sanchéz, I.: Consecuencias de la insatisfacción del consumidor: un estudio en servicios hoteleros y de restauración. Universia Business Review 28, 78–100 (2010)

Cortiñas, M., Elorz, M., Múgica, J.M.: Loyalty cards: Are retailers ignoring non-cardholder behaviour? European Retail Digest. 45, 18–20 (2005)

Cronin Jr., J.J., Brady, M.K., Hult, G.T.: Assessing the effects of quality, value, and customer satisfaction on consumer behavioral intentions in service environments. Journal of Retailing 76(2), 193–216 (2000)

Quest, D.: Clubes y tarjetas de fidelización. The Marketing Intelligence Review 5, 4–9 (2005)

Day, G.: Creating a superior customer-relating capability. MIT Sloan Management Review 44(3), 77–82 (2003)

Del Río, A., Vázquez, R., Iglesias, V.: La influencia de la marca en la evaluación del producto: consideraciones sobre el efecto halo. Revista Europea de Dirección y Economía de la Empresa 12(3), 25–40 (2003)

Demoulin, N.T.M., Zidda, P.: Drivers of customers´ adoption and adoption timing of a new loyalty card in the grocery retail market. Journal of Retailing 85(3), 391–405 (2009)

Faine, I., Tornabell, R.: Pasión por la Banca. Ediciones Deusto, Bilbao (2001)
Fandos, J., Sánchez, J., Moliner, M.: Perceived value and customer loyalty in financial services. Services Industries Journal 29(6), 775–789 (2009)
Garbarino, E., Johnson, M.S.: The different roles of satisfaction, trust, and commitment in customer relationships. Journal of Marketing 63(2), 70–84 (1999)
Gommans, M., Krishnan, K.S., Scheffold, K.B.: From brand loyalty to E-loyalty: a conceptual framework. Journal of Economic and Social Research 3(1), 43–58 (2001)
Heskett, J.L., Sasser, W.E., Schlesinger, L.A.: The service profit chain. How leading companies link profit and growth to loyalty, satisfaction and value. The Free Press, New York (1997)
Huete, L.M., Serrano, J., Soler, I.: Servicios y Beneficios. Ediciones Deusto, Bilbao (2003)
Jones, M.A., Mothersbaugh, D., Beatty, S.: Switching barriers and repurchase inten-tions in services. Journal of Retailing 76(2), 259–274 (2000)
Kotler, P., Cámara, D., Grande, I.: Dirección de Marketing, 8th edn. Editorial Pearson, Madrid (1995)
Lacey, R., Sneath, J.: Customer loyalty programs: are they fair to consumers? Journal of Consumer Marketing 23(7), 458–464 (2006)
Lacey, R., Morgan, R.: Customer advocacy and the impact of B2B loyalty programs. Journal of Business & Industrial Marketing 24(1), 3–13 (2009)
Liang, C.-J., Wang, W.-H.: Do loyal and more involved customers reciprocate retailers relationship efforts? Journal of Services Research 8(1), 63–90 (2008)
Mägi, A.W.: Share of wallet in retailing: the effects of customer satisfaction, loyalty cards and shopper characteristics. Journal of Retailing 79(2), 97–106 (2003)
Mas, S.: Servicios Financieros: La era del cliente. Editorial Gestión, Barcelona (2003)
Meyer-Waarden, L., Benavent, C.: Les cartes de fidelité comme outils de segmentation et de ciblage. Le cas dúne enseigne de distribution. Décisions Marketing 32, 19–30 (2003)
Porter, M.: Competitive Advantage. The Free Press, New York (1985)
Reichheld, F., Sasser, W.: Zero defections: quality comes to services. Harvard Business Review 68, 105–111 (1990)
Reinares, P., Ponzoa, J.M.: Marketing Relacional. Un nuevo enfoque para la seducción y fidelización del cliente. Editorial Financial Times- Prentice Hall, Madrid (2002)
Reinares, P., Reinares, E., Mercado, C.: Gestión de la heterogeneidad de los consumidores mediante programas de fidelización. Revista Europea de Dirección y Economía de la Empresa 19(3), 143–160 (2010)
Sanzo, M., Santos, M., Vázquez, R., Álvarez, L.: The effect of market orientation on buyer-seller relationship satisfaction. Industrial Marketing Management 32, 327–345 (2003)

[illegible]

The Dimensions of the Financial Condition in Spanish Municipalities: An Empirical Analysis

Roberto Cabaleiro Casal[1], Enrique J. Buch Gómez[1], and Antonio Vaamonde Liste[2]

[1] Department of Financial Economics and Accounting, University of Vigo, Lagoas Marcosende s/n, 36310 Vigo, Spain
rcab@uvigo.es, ebuch@uvigo.es
[2] Department of Statistics and Operations Research, University of Vigo, Torrecedeira, 105, 36208 Vigo, Spain
vaamonde@uvigo.es

Abstract. The current financial crisis highlights the relevance to have frameworks and analytical tools adapted to the financial specificities of the local public sector. The complex nature of the concept of financial condition has been studied from both the doctrinal and institutional perspective. The aim of this study was to analyze whether the multiple indicators of the financial realities of local entities in Spain reflect the dimensions of analysis of financial condition developed by the ICMA or the dimensions developed by the CICA. For the analysis, we used a large number of indicators and a large sample of Spanish municipalities. Our methodology combined statistical techniques of factorial analysis, canonical correlation, and cluster analysis. The results of the analysis in local governments in Spain showed a combination of the dimensions espoused by both theoretical frameworks.

Keywords: Financial Condition, Financial Analysis, Financial Management, Municipal Finance, Municipal Governments.

1 Introduction

The international financial crisis that has shaken the different economies has had important effects on the finances of public institutions. The global financial and economic crisis has finally impacted all urban communities and investment financing systems around the world and the landscape of this financing sector is currently one of devastation (Paulais, 2009). In general terms, the slowdown in global economic activity is impacting local budgets in both the emerging and the most developed countries. The impact on revenues of local government results from six direct causes: the reduction in revenues obtained from own taxes, fees and charges, the reduction in yields of specific tax shares, the reduction in

A.M. Gil-Lafuente et al. (Eds.): Soft Comput. in Manag. and Bus. Econ., STUDFUZZ 287, pp. 137–152.
springerlink.com

transfers from the national budget, the constraints on local government access to their own revenues, the reduction in revenues from capital holdings, and the loss of assets/deposits (Martínez et. al., 2009).

Although the current crisis is more pronounced, these institutions are usually subject to financial difficulties (Honadle, 2003; 2004). These circumstances have led to the development of concepts and tools aimed at specific knowledge of the financial situations of local governments. Among the many concepts is the financial condition.

The complex concept of financial condition and its evaluation has led to several studies embodied in concrete approaches, both from the literature and by various institutions. Two of the most important institutional documents on the subject are provided by the International City / County Management Association (ICMA, 2003) in USA and the Canadian Institute of Chartered Accountants (CICA, 1997; 2009).

The contribution of this work to the economic literature is to provide knowledge, through various techniques of multivariate analysis, whether the multiple indicators of the financial realities of local entities in Spain reflect the framework of analysis of financial condition developed by the ICMA (2003) (based on the dimensions of cash solvency, long-run solvency, budgetary solvency and service-level solvency) or the framework developed by the CICA (1997, 2009) (based on the dimensions of sustainability, flexibility and vulnerability). The results of this work could be useful to focus the work aimed to the knowledge of the financial situation in the Spanish municipal sector. For this, we have taken a large sample that represents over 60% of the municipalities in Spain.

2 Literature Review

The financial analysis of public institutions has been supported (Hendrick, 2004) in various concepts, such as fiscal strain (Stonecash and McAfee, 1981, Clark and Ferguson, 1983; Clark, 1994; ...), fiscal distress (Cabill and James, 1992; Kleine et al., 2003; Kloha et al., 2005a; 2005b; Jones and Walker, 2007; ...), fiscal crises (Hirsch and Rufolo, 1990; Inman, 1995; Chernick and Reschovsky, 2001; Carmeli, 2003; Coe, 2008; ...), financial condition (Berne, 1992; Mercer and Gilbert, 1996; Chase and Phillips, 2004; Wang et al., 2007), etc. This shows the existence of alternative names that correspond to different methodological approaches on a common reality.

The term financial condition has been used frequently in the literature to refer to this problem, although with different connotations (Groves et al., 1981; Mead, 2001). Honadle et al. (2004) noted that financial condition of local institutions is a term closely linked to the concept of fiscal health.

Financial condition (Berne 1992) is likely to meet the financial obligations due to creditors, employees, taxpayers, and other stakeholders, as well as obligations to serve their constituents in both the present and future. According Lin and Raman (1998), a weak financial condition means that a government has a relatively low probability of being able to sustain the current level of services at acceptable levels of taxation.

Wang et al. (2007) pointed out that this concept represents the ability of an organization to meet its financial obligations on time. Mead (2001) defined financial condition as the ability of an institution to meet its obligations as they come due and to finance the services its constituency requires.

As financial condition is a concept that is not directly observable, the problem that arises is what are the most appropriate instruments to measure it. There are several proposals to assess financial condition and they are influenced by both the conceptual approach as the information available in each particular environment. The literature suggests a variety of dimensions to consider, and some indicators to use. There has not been consensus on what dimensions and specific indicators represent their status or value (Wang et al., 2007).

Diverse ratios and benchmarks have been used to evaluate governments' financial condition, but there is no consistency in their selection, use, and application (Clark, 1976; 1994; Groves et al., 1981; Alter et. al., 1984; Zehms, 1991; Berne, 1992; Brown, 1993; Petro, 1998; Wolff and Huges, 1998; Kleine et al., 2003; Hendrick, 2004; Ammar, 2005; Kloha et al., 2005a; 2005b; Wang et al., 2007; Zafra et al., 2009a; 2009b).

Hendrick (2004), following Berne and Schramm (1986), developed an approach to fiscal health assessment according to different dimensions. In her study to assess the financial condition and fiscal health of local governments of suburban municipalities in the Chicago metropolitan region, Hendrick (2004) summarized the dimensions of this approach in properties of government`s environment (revenue wealth – spending needs – socioeconomic; political; and demographic features – changes in the previous), balance of fiscal structure with environment, and properties of the government`s fiscal structure (fiscal slack – relativity of components within major structural areas – current operating conditions – future financial obligations – changes in fiscal slack; relativity of components; and current operating conditions).

In the study carried out by Ammar et al. (2005), four main aspects were considered to evaluate the financial condition: short-run financial condition, long-run financial condition, economic conditions, and performance.

Groves et al. (1981) considered that the term financial condition has many meanings. In a narrow accounting sense, as a government`s capacity to generate enough cash or liquidity to pay its bills (cash solvency). It can also refer to city`s ability to generate sufficient revenues over its normal budgetary period to meet its expenditure obligations and not incur deficits (budgetary solvency). In a broader sense, financial condition refers to the long-run ability of a government to pay all the costs of doing business, including expenditure obligations that normally appear in each annual budget, as well as those that show up only in the years in which they must be paid (long-run solvency). Furthermore, it also refers to whether a government can provide the level and quality of services required for the general health and welfare of a community (service-level solvency). The approach based on these four dimensions has been used frequently in the financial literature (Wang et al., 2007; Zafra et al., 2009a; 2009b; …).

Clark (1976) developed 29 indicators to construct measures of municipal fiscal strain. Using factor analysis, he reduced these to four dimensions: long-rung debts, short-run debts, expenditures, and tax efforts. Following this approach, Morgan and England (1983) identified three basic components of urban fiscal decline. These tree stress variables were long term debt per capita, per capita expenditures for nine common functions (police, fire, sanitation, …), and own source revenue per capita.

Furthermore, using the statistical technique of factorial analysis on a set of more than 100 measures of financial condition applied to 55 municipalities in the province of Nova Scotia, based on ratios of revenues and expenses, taxation, debt burden, and economic, Mercer and Gilbert (1996) reduced the number of indicators to 17. These 17 indicators were able to explain most of the variance of the data and were reduced to three factors: fiscal condition, debt burden, and revenue base.

Owing to the importance of the issue, normalizing institutions have considered both the definition of financial condition as the development of appropriate mechanisms for measurement.

Research Report published by the GASB (Berne, 1992) highlighted that the financial condition is multidimensional or multiconstituency with complex interdependencies among the various parts, and it is rooted in a government´s economic environment and involves implicit as well as explicit obligations that are not necessarily reflected in cash flows or financial contracts.

Given the confusion generated by using the terms financial position and financial condition interchangeably, the GASB (2004) has adopted the term "economic condition" in its Statement No. 44. Economic condition is a composite of a government's financial health and its ability and willingness to meet its financial obligations and commitments to provide services. Economic condition includes financial position (the status of a government's asset, liability, and net asset accounts, as displayed in its basic financial statements), fiscal capacity (a government's ongoing ability and willingness to raise revenues, incur debt, and meet its financial obligations as they come due), and service capacity (a government's ongoing physical ability and willingness to supply the capital and human resources needed to meet its commitments to provide services) (Mead, 2001b; Chaney et al., 2002).

Recently, the CICA, to understand that, to date, governments do not have a common methodology, issued the Statements of Recommended Practice (SORP) 4. The aim is to establish a common framework of indicators for assessing the financial condition (CICA, 2009). This statement has taken a Research Report published by the institution previously as a precedent (CICA, 1997).

For the CICA (2009), government financial condition is a government's financial health as assessed by its ability to meet its existing financial obligations both in respect of its service commitments to the public and financial commitments to creditors, employees, and others.

Although local governments were not included in the Research Report (CICA, 1997), these institutions have been considered in the recently published SORP 4 (CICA, 2009).

The government financial condition is measured by sustainability (the degree to which a government can maintain its existing financial obligations without increasing the relative debt or tax burden on the economy within which it operates), flexibility (the degree to which a government can change its debt or tax burden on the economy within which it operates to meet its existing financial obligations), and vulnerability (the degree to which a government is dependent on sources of funding outside its control or influence, or is exposed to risks that could impair its ability to meet its existing financial obligations) (CICA, 2009). Both documents (CICA, 1997; 2009) propose a set of indicators that follow the same line for the measurement of these dimensions. This same analysis approach has also been followed by Greenberg and Hillier (1995).

Similar problems are analyzed by other normalizing institutions. The Federal Accounting Standards Advisory Board (FASAB) of the United States uses the term fiscal sustainability (FASAB, 2009). The Local Government Association of South Australia (LGAs) uses the concept of financial sustainability and similar indicators for its valuation (LGAs, 2006).

One of the most important contributions, for its extensive application in local governments over several decades, is the Financial Trend Monitoring System (FTMS). This contribution is intended as an internal monitoring system and has been released by the ICMA. Current version of FTMS (ICMA, 2003) is adapted to the reporting requirements of Statement No. 34 (GASB, 1999) to local governments. This current edition of FTMS is based on the definition and dimensions that they are reflected in the concept of financial condition developed by Groves et al. (1981) (cash solvency, long-run solvency, budgetary solvency and service-level solvency). This mechanism is based on 11 factors. They are classified in environmental, organizational, and financial factors. This scheme uses a set of 42 indicators.

3 The Financial Condition Dimensions of the CICA vs. ICMA: An Empirical Application to the Spanish Municipalities

The review of economic literature has shown that there are several approaches from an institutional standpoint to determine the financial condition of public institutions. Among them is the proposal of the ICMA (2003), which follows the scheme developed by Groves et al. (1981) and moves in line with GASB No. 34. Another approach has been the formal framework developed by the CICA (1997; 2009) for assessing the financial health aspects that constitute the financial condition of public institutions. Therefore, two institutional frameworks of analysis of financial condition directly applicable to local governments are taken as reference in this work.

For the purpose of the work, to determine if the financial realities of local entities in Spain reflect the dimensions of the approach of ICMA (2003) and / or dimensions of the framework of CICA (1997; 2009), we have used various techniques of multivariate analysis.

3.1 Methodology

To examine the dimensions proposed by the CICA and ICMA, we used a wide range of data concerning Spanish municipalities. For the analysis, the data were obtained from various institutional sources. The settlements of the budgets of the local administrations and the amount of borrowing for the year 2009 were extracted from two databases of the Ministry of Economy and Finance (MEH). Data of the local population for 2009 were obtained from the National Statistics Institute (INE).

Following the existing literature and taking into account the data actually available, we developed a battery of 33 indicators to collect all the theoretical dimensions proposed by both the approaches (CICA, ICMA). This has been done using as reference the indicators proposed in the documents of both institutions. Some ratios had to be adapted to the information that follows the accounting information system in Spain, without losing sight of the philosophical approach of these indicators by CICA and ICMA.

A full set of indicators used, with its description and its corresponding allocation approach, is shown in Table 1.

Table 1 Indicators of Financial Condition

	Indicator	Components
I1	$\frac{\text{CCE+AR-DAR-EF}}{\text{OO}}$	Refined short-term solvency: Cash and Cash Equivalents (CCE) plus accounts receivable (AR), less doubtful accounts receivable (DAR) and excess funds to finance expenditures earmarked funding (EF), divided by the outstanding obligations (OO) at year end.
I2	$\frac{\text{CCE+AR- DAR}}{\text{OO}}$	Gross short-term solvency: Cash and Cash Equivalents (CCE) plus accounts receivable (AR) and less doubtful accounts receivable (DAR), divided by the outstanding obligations (OO) at year end.
I3	$\frac{\text{CCE}}{\text{OO}}$	Quick Ratio: Cash and Cash Equivalents (CCE) divided by outstanding obligations (OO) at year end.
I4	$\frac{\text{Long-term debt}}{\text{TNBR}}$	Long-term debt in relation to the total net budgetary revenues (TNBR)
I5	$\frac{\text{Long-term debt}}{\text{NBR Ch.* 1 to 8}}$	Long-term debt divided by net budgetary revenues (NBR) from non-financial operations.
I6	$\frac{\text{Long-term debt}}{\text{NBR Ch. 1 to 5}}$	Ratio between the long-term debt and net budgetary revenues from current operations.
I7	$\frac{\text{Long-term debt}}{\text{Pop.}}$	Long-term debt per inhabitant (Pop)
I8	$\frac{\text{NBR Ch. 1 to 5}}{\text{NBO Ch.1 to 4}}$	Net current budgetary revenues divided by net budget obligations (NBO) from current expenditures
I9	$\frac{\text{NBR Ch. 1 a 5}}{\text{NBO Ch.1 to 4 and 9}}$	Net current budgetary revenues divided by budget obligations from non-financial current expenditures, minus debt service.
I10	$\frac{\text{Net savings}}{\text{Pop.}}$	Difference between the receivables from current budget resources and the budget obligations from non-financial current expenditures, minus debt service per inhabitant.

Table 1 *(continued)*

I11	NBO Ch. 3 and 9 / NBR Ch.1 to 5	Debt service (interest and principal) divided by net current budgetary revenues
I12	NBO Ch.. 3 and 9 / Pop.	Debt service per inhabitant.
I13	NBO Ch.. 3 / Pop	Debt interest per inhabitant.
I14	RBS / Pop.	Result of the budget settlement (RBS) per inhabitant.
I15	Total NBR / Pop.	Total net budgetary revenues per inhabitant.
I16	NBR Ch 1 to 5 / NBR Ch. 4	Ratio between net current budgetary revenues and current grants received.
I17	NBR Ch 1 to 3 / NBO Ch. 1 to 3	Direct and indirect taxes and fees divided by obligations from net expenditure of personnel, services and debt interest.
I18	NBR Ch 1 to 3 / NBO Ch 1 to 4	Direct and indirect taxes and fees divided by net budget obligations from current expenditures.
I19	NBR Ch 1 to 3 / Pop.	Direct and indirect taxes and fees per inhabitant.
I20	NBR Ch 1 a 3 to 5 / NBO Ch. 1 to 4	Net current budgetary revenues less current grants received, divided by net budget obligations (NBO) from current expenditures.
I21	NBR Ch 1 to 3, 5, 6, 8, 9 / Total NBO	Difference between total net budgetary revenues and budgetary current and capital transfers received divided by total net budget obligations.
I22	NBR Ch. 7 / Pop.	Capital transfers received per inhabitant
I23	NBR Ch. 4 and 7 / Pop.	Current and capital transfers received per inhabitant
I24	NBO Ch. 6 and 7 / Pop.	Investments per inhabitant: Net budget obligations from capital expenditures, capital transfers and capital grants per inhabitant.
I25	NBO Ch.. 6 and 7 / Total NBO	Investments effort: Net budget obligations from capital expenditures, capital transfers and capital grants divided by total net budget obligations.
I26	Expend. CP and PS / Pop.	Expenditures (Expend.) on civil protection and public safety (CP and PS) per inhabitant.
I27	Expend. SS, SP and SPR / Pop.	Expenditures on social security, social protection and social promotion (SS, SP an SPR) per inhabitant.
I28	Expend. E / Pop.	Expenditures on education (E) per inhabitant.
I29	Expend. H and UD / Pop.	Expenditures on housing and urban development (H and UD) per inhabitant.
I30	Expend. CW / Pop.	Community welfare spending (CW) per inhabitant.
I31	Expend. C / Pop.	Expenditure on culture (C) per inhabitant.
I32	Expend. OCS and OSS / Pop.	Expenditure on other community and social services (OCS and OSS) per inhabitant.
I33	Expend. BI and T / Pop.	Expenditure on basic infrastructure and transport (BI and T) per inhabitant.

*Ch.: Budgetary chapter of the economic classification in Spain.

All indicators of the table 1 are present in both approaches, with the exception of those related to the analysis of cash solvency of the institution (I1, I2, and I3), which are not explicitly covered by the approach for evaluating the financial condition of the CICA.

Due to this circumstance, the first step was to perform two sets of ratios (I1 to I3; I4 to I33) that they were subjected to canonical correlation analysis. The objective was to determine whether there is correlation between the group of variables that make up the cash solvency (I1 to I3) and the group composed of the ratios that form the theoretical aspects covered simultaneously by both the approaches (CICA, ICMA) (I4 to I33).

If the canonical correlation analysis performed shows the existence of high correlation between the set of variables not common (I1 to I3) and the group being common (I4 to I33), it would mean that the analysis of the indicators that comprise the common group would be sufficient to obtain a founded vision of the financial condition of the local entity.

Subsequently, we proceeded to identify the dimensions that may exist in common group, for which we used the cluster analysis technique applied to the benchmarks provided by both the approaches (CICA, ICMA). The aim was to obtain a comprehensive and disaggregated view of all the dimensions that could configure the financial condition of the local authorities in Spain through a joint approach. Among the different alternatives of the cluster analysis, we chose a hierarchical method because, starting from the reading of each individual indicator, it allows vertical groups to be increasingly broad and heterogeneous.

3.2 Data

After crossing the debt database with the settlement of the last municipal budgets prepared annually by the MEH for the year 2009, we found information about 5823 municipalities. Following a purification process that involved the elimination of those municipalities that did not have complete information or was flawed, we were left with a sample of 5158 municipalities. Accordingly, we rejected less than 11.5% of all the available data. The universe of local governments in Spain is composed of 8112 municipalities. This means that the sample used represents 63.58% of all the local governments in the country. For all these municipalities, we obtained the information regarding the location and population of the INE database.

The sample disaggregated by geographical areas (Autonomous Communities (CCAA)) and segments of the population (Pop.) can be seen in Table 2.

Table 2 Distribution of the sample of municipalities by population size and CCAA

	Pop. < 5000		5000 < Pop. < 20000		20000 < Pop. < 50000		50000 < Pop. < 200000		200000 < Pop.	
CCAA	S	S/U %	S	S/U %	S	S/U %	S	S/U %	S	S/U %
Andalucía	273	53.74	126	69.61	43	82.69	21	87.50	4	80.00
Aragón	278	39.32	12	60.00	2	100	1	100.00		
Canarias	14	66.67	30	75.00	14	73.68	6	100.00	2	100.00
Cantabria	60	73.17	9	60.00	3	100.00	1	50.00		

Table 2 *(continued)*

Castilla y León	1192	54.45	29	65.91	4	66.67	7	87.50	1	100.00
Castilla-La Mancha	398	47.16	42	70.00	7	87.50	6	85.71		
Cataluña	573	77.64	129	89.58	39	95.12	14	77.78	5	100.00
Cdad. Foral de Navarra	192	76.49	14	77.78	1	50.00	1	100.00		
Comunidad de Madrid	58	57.43	39	84.78	10	83.33	16	94.12	3	100.00
Comunitat Valenciana	328	85.42	87	92.55	48	97.96	12	100.00	2	66.67
Extremadura	244	70.72	32	96.97	3	75.00	3	100.00		
Galicia	164	82.41	80	85.11	11	73.33	5	100.00	2	100.00
Illes Balears	18	66.67	27	96.43	10	100	1	100.00	1	100.00
La Rioja	128	77.58	7	100	1	100	1	100.00		
País Vasco	124	68.13	42	82.35	8	66.67	3	75.00	2	100.00
Principado de Asturias	32	68.09	16	66.67	2	66.67	1	50.00	1	50.00
Región de Murcia	2	22.22	17	89.47	10	76.92	2	100.00	2	100.00
Total	4078	59.98	738	80.39	216	85.71	101	88.60	25	86.21

S: Sample; U: Total number of municipalities in the CCAA.

3.3 Analysis of the Results

We carried out a calculation of all the indicators shown in Table 1. The descriptive analysis of the sample is summarized in Table 3.

Table 3 Descriptive Statistics.

Indicator	Minimum	Maximum	Mean	Std. Deviation
I1	-47.6119	279.2265	3.0322	10.4794
I2	-47.6119	279.2265	3.4404	10.6035
I3	-215.3450	270.9052	2.0982	9.8859
I4	0.0000	2.7624	0.1831	0.2315
I5	0.0000	3.5033	0.1991	0.2573
I6	0.0000	8.0124	0.2980	0.3996
I7	0.0000	9,250.0000	279.8486	422.9547
I8	0.4106	38.6805	1.2034	0.7410
I9	0.1299	38.6805	1.1623	0.7383
I10	-3,531.6322	17,348.2217	130.0467	588.8943
I11	0.0000	5.5975	0.0476	0.1157
I12	0.0000	2,618.0817	44.9031	99.9258
I13	0.0000	433.2755	10.7749	18.2515
I14	-2,279.0228	17,215.1485	54.4503	488.3428
I15	66.3647	20,267.8768	1,646.7433	1,214.9197
I16	1.0275	80.9780	2.8414	2.5492
I17	0.0162	44.9581	0.6757	0.7316
I18	0.0161	38.0132	0.6086	0.6189

Table 3 *(continued)*

I19	0.9350	17,501.4033	511.3774	611.0050
I20	0.0305	38.0230	0.6962	0.6468
I21	0.0200	18.1341	0.4309	0.3249
I22	0.0000	12,940.3914	569.1930	705.6636
I23	12.9603	14,790.8385	978.6584	863.7673
I24	1.8732	16,325.9618	720.8649	817.8102
I25	0.0099	0.9313	0.4093	0.1518
I26	0.0000	787,562.9783	1,674.5382	15,693.2326
I27	0.0000	604,975.7177	3,043.7052	19,550.7891
I28	0.0000	311,503.1133	1,040.5805	7,551.5409
I29	0.0000	1,468,021.0385	4,173.3603	30,266.8607
I30	0.0000	1,075,930.1892	3,241.3566	22,981.1056
I31	0.0000	652,659.9770	3,004.2931	19,158.6866
I32	0.0000	122,752.9533	462.8842	3,703.5887
I33	0.0000	841,136.5449	2,959.8383	20,759.1642

Valid N (listwise): 5158.

Subsequently, we explored the correlation between the variables that represent the cash solvency (I1, I2 and I3) and the set of indicators common to both the approaches (I4 to I33), using a canonical correlation analysis.

Following the process of canonical correlation analysis, we concluded that there is no relationship between the set of variables representing the cash solvency dimension and the set of common indicators of the two approaches. The highest correlation of the canonical functions that represent the cash solvency only reached a correlation of 0.097 and the test carried out indicated that we cannot exclude the hypothesis that the canonical correlations between the indicators of cash solvency and common indicators is zero (Table 4).

Table 4 Canonical Correlation Analysis

	Canonical Correlations	Test that remaining correlations are zero: Wilk's	Chi-SQ	DF	Sig.
1	0.097	0.983	88.996	90.000	0.510
2	0.066	0.992	40.184	58.000	0.964
3	0.059	0.997	18.019	28.000	0.926

After determining that the cash solvency represents a differentiated dimension, i.e., uncorrelated with the other aspects contained in the proposals of the ICMA and CICA, the next step was to try to identify the dimensions of the group of common indicators.

For this, the procedure was to try to understand the indicators that were having a similar variability in the common group to both the approaches (I4 to I33). Thus, we could obtain a common reading of the groups of indicators that were based on the theoretical concepts considered by the CICA and/or ICMA.

We considered it appropriate to use a cluster analysis technique. Among the different methods of cluster analysis, the method that we presumed to provide an adequate answer to our objective is the sequential, agglomerative, hierarchical, and exclusive method based on the minimization of the variance of the dissimilarity measure, known as the Ward Method. The dissimilarity measure used was the squared Euclidean distance.

Consequently, we proceeded to typify the data for the set of common variables and the application of the technique mentioned. Figure 1 shows the dendrogram of clustering where four clusters of indicators have been displayed, which can be identified with the theoretical dimensions of both the approaches.

The first set of indicators that showed similar variability (I8, I9, I14, I17, I18, I20, I21, I10, I19, and I16) was focused essentially on issues related to differentials among echelons of income and expenses, either blocks analyzed by income and economic costs, or through differences that represent savings or results. The issues covered by these indicators were essentially of a budgetary nature. These were linked to the feasibility with which the entity has to meet the new commitments that may arise as a result of changes in the demands of citizens. Consequently, this block of variables can be identified with the concept of flexibility (CICA), or with a substantial part of the concept of budgetary solvency (ICMA).

The second set was formed by the indicators I22, I23, I15, I24, and I25. In this group, the indicators I22 and I23 represented the level of current or capital transfers received by the local institutions from other institutions. This reflected the degree of financial dependence on the institution, that is, the dimension of vulnerability (CICA). Similar behavior that these two indicators have with the ratios I24 and I25, which represent the investment effort, could be justified in that the transfers received by the municipalities have a role that is essentially finalist in the Spanish context. The presence in this set of I15 ratio, which shows the per capita volume of financial resources available to the institution, confirms the situation of dependence shown by the finances of local public institutions in Spain. Owing to the budgetary nature of the issues that are covered in this set of ratios, it would be a part of budgetary solvency (ICMA).

The ratios I11, I12, I4, I5, I6, I7, and I13 formed the third cluster of variables. The type of information collected in this block referred to two aspects clearly associated, the volume of debt (I4, I5, I6, and I7) and debt service (I11, I12, and I13) that results from it. Similar variability shown by these two aspects seemed to indicate a similar behavior in terms of maturity debt and homogeneity of interest rates that the financial market has in this sector. Therefore, we considered that the dominant character in the behavior of this group of variables is linked to the capacity of local institutions to maintain or increase the existing expenditure programs, regardless of the level of debt they have assumed. This represents the concept of sustainability, because the volume of debt variables has a primary character on the set. This group would approximate to the dimension of long-run solvency of the ICMA. The last variables (I11, I12, and I13) would collect any aspect related to flexibility in accordance with the CICA and to the budgetary solvency in the approach ICMA.

The last block of indicators that showed the statistical analysis was composed of a set of ratios (I26, I30, I28, I29, I27, I31, I33, and I32) that are referred to the

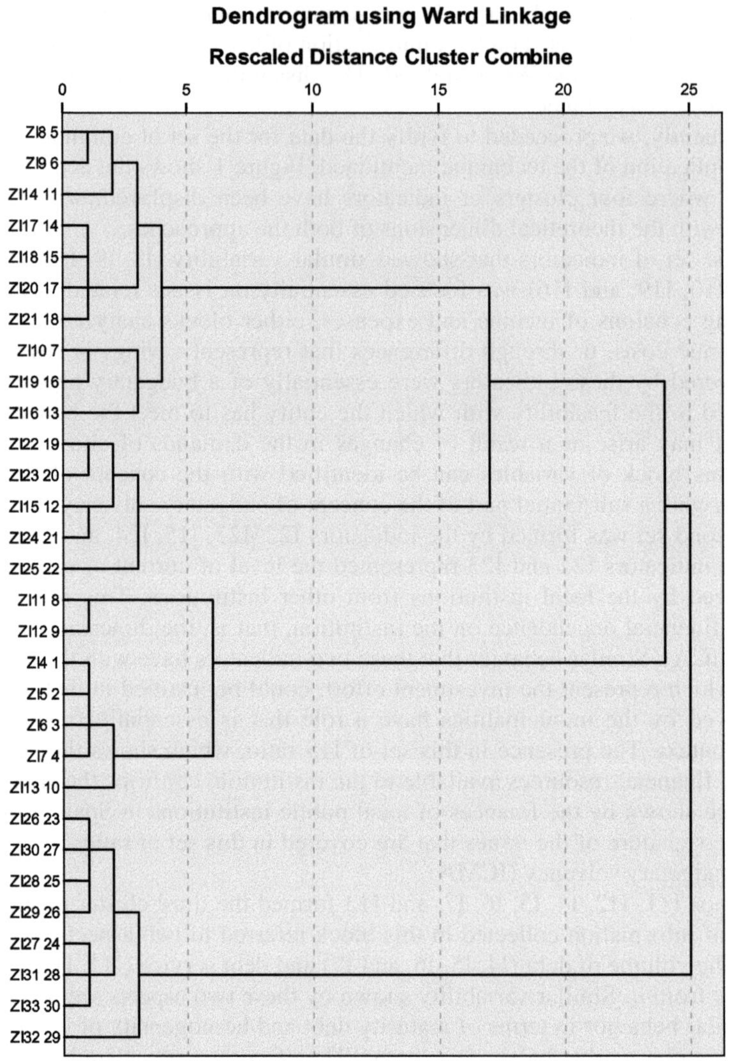

Fig. 1 Cluster process

levels of expenditure by objectives, which the institution incurred. The grouping of expenditures by different functions or programs seemed to indicate a certain consistency in the provision of different services in each institution. In the approach to the assessment of the financial condition developed by the CICA, this group of indicators would be framed within the concept of flexibility. However, on analyzing the results, we observed that this block has a differentiated variability. This would come close to the views of the ICMA, which postulates the existence of a differentiated dimension of level-service solvency.

Because the result of canonical correlation analysis showed no correlation between cash solvency indicators (following the approach ICMA) and the common indicators (ICMA, CICA), we make a check of whether the evidence reveals a single dimension of cash solvency. The choice of the 3 indicators (I1, I2 and I3) is a consequence of legal and economic complexities that local governments have in Spain. The nuances that distinguish these three indicators may reflect different behaviors.

Consequently, we proceeded to conduct a factorial analysis of the cash solvency. The analysis of the correlation matrix of the indicators showed a high and significant level of correlation among all the variables. This aspect was highlighted by both the high and significant value of correlation coefficients, as low coefficient showed by the determinant (0.004). The existence of correlation between the variables was formally contrasted with the Bartlett test of sphericity, which had a χ^2 statistic of 27,931.733 and a p value of less than 0.001. Kaiser-Meyer-Olkin measure of sampling adequacy (KMO) was favorable because it showed a high value (0.729). The model demonstrated that a single factor explained the 94.724 of the variability of the data. The analysis of the communality of each variable after extraction of the factor showed high values (Table 5).

Table 5 Total Variance Explained / Communalities

	Total Variance Explained						Communalities	
Factor	Initial Eigenvalues			Extraction Sums of Squared Loadings			Variables	Extract.
	Total	% Variance	% Cumulative	Total	% Variance	% Cumulative		
1	2.842	94.724	94.724	2.842	94.724	94.724	I1	0.970
2	0.148	4.923	99.648				I2	0.972
3	0.011	0.352	100.000				I3	0.899

Extraction Method: Principal Component Analysis.

These data show that the three indicators used seem to make up a single representative dimension of the concept of cash solvency. Consequently, none of the differentiated aspects incorporated to quantify each of these ratios was, in fact, a distinguishing element in the joint behavior.

4 Conclusions

The concern to know the financial condition of public institutions to fulfill their objectives has led to various institutions to carry out diverse research projects that have been materialized in documents both of theoretical and applied character. Two fundamental exponents for local governments are the frameworks developed by the CICA and the ICMA. Although the two frameworks are based on the concept of financial condition, in each of them the concept is articulated in different dimensions.

Using a large sample of Spanish local institutions, through canonical correlation analysis between the indicators of cash solvency and the universe of indicators representative of the dimensions covered by the approach of the CICA, which can also be seen as representative of the rest of the dimensions of the ICMA, we have verified that the concept of cash solvency is not implicitly included in the set of indicators common to both approaches. Moreover, using factor analysis, we concluded that the different nuances of cash solvency, which arise from legal and economic reality in Spain, do not imply distinct sub-dimensions, so cash solvency is a single-dimension concept.

Through a cluster analysis of variables of the set of indicators common to both approaches, we found the existence of four groups of variables that have a homogeneous behavior. The study revealed the existence of all dimensions theorized by the approaches of the CICA and ICMA in this universe, with some qualifications. Aspects of the dimensions of sustainability and flexibility are included in the long-run solvency. Budgetary solvency, which forms a unique dimension in the approach of ICMA, contains the vulnerability, along with other aspects that focus on flexibility. The service-level solvency forms a part of the flexibility.

Through the various statistical analysis techniques used, we found that for a more extensive analysis of the financial condition of local public institutions in Spain, it would be advisable to use a combination of the dimensions espoused by both the theoretical frameworks.

Finally, the methodology used in this paper is fully applicable to center the analysis of the dimensions that comprise the financial condition of local institutions in accordance with the legal and economic realities elsewhere.

References

Alter, T.R., McLaughlin, D.K., Melniker, N.E.: Analysing Local Government Fiscal Capacity. Pennsylvania State University, Cooperative Extension Service, University Park, PA (1984)

Ammar, S., Duncombe, W., Jump, B., Wright, R.: A Financial Condition Indicator System for School Districts: A Case Study of New York. Journal of Education Finance 30(3), 231–258 (2005)

Berne, R., Schramm, R.: The financial analysis of governments. Prentice Hall, Englewood Cliffs (1986)

Berne, R.: Relationships between Financial Reporting and the Measurement of Financial Condition. GASB Research Report 18. GASB, Norwalk (1992)

Brown, K.W.: The 10-Point Test of Financial Condition: Toward an Easy-to-Use Assessment Tool for Smaller Cities. Government Finance Review 9(6), 21–26 (1993)

Cabill, A.G., James, J.A.: Responding to Municipal Distress: An Emerging Issue for State Governments in the 1990s. Public Administration Review 52(1), 88–94 (1992)

Carmeli, A.: Introduction: Fiscal and Financial Crises of Local Governments. International Journal of Public Administration 26(13), 1423–1430 (2003)

Chaney, B., Mead, D., Schermann, K.: New governmental Financial Reporting Model: What It Means for Analyzing Government Financial Condition. Journal of Government Financial Management 51(1), 26–31 (2002)

Chase, B.W., Phillips, R.H.: GASB 34 and government financial condition: an analytical toolbox. Government Finance Review 20(2), 26–31 (2004)

Chernick, H., Reschovsky, A.: Discussion Paper. A Lost in the Balance: How State Policies Affect the Fiscal Health of Cities. The Brookings Institution Center on Urban and Metropolitan Policy (2001)

CICA (Canadian Institute of Chartered Accountants): Research Report. Indicators of Government Financial Condition. CICA, Toronto, Canada (1997)

CICA (Canadian Institute of Chartered Accountants): Public Sector Statements of Recommended Practice (SORP) 4. Indicators of Financial Condition. CICA, Toronto, Canada (2009)

Clark, T.N.: How many more New Yorks? New York Affair 3, 18–27 (1976)

Clark, T.N.: Municipal Fiscal Strain: Indicators and Causes. Government Finance Review 10(3), 27–29 (1994)

Clark, T.N., Ferguson, L.C.: City money: Political processes, fiscal strain, and retrenchment. Columbia University Press, New York (1983)

Coe, C.K.: Preventing Local Government Fiscal Crises: Emerging Best Practices. Public Administration Review 68(4), 759–767 (2008)

FASAB (Federal Accounting Standards Advisory Board): Reporting Comprehensive Long-Term Fiscal Projections for the U.S. Government. FASAB, Washington, DC (2009)

GASB (Governmental Accounting Standards Board): Statement nº 34. Basic Financial State-ments—and Managements Discussion and Analysis—for State and Local Governments. GASB, Norwalk, CT (1999)

GASB (Governmental Accounting Standards Board): Statement nº 44. Economic Condition Re-porting: The Statistical Section an amendment of NCGA Statement. GASB, Norwalk, CT (2004)

Greenberg, J., Hillier, D.: Indicators of financial condition for governments. In: 5th Conference of Comparative International Governmental Accounting Research (CIGAR), CIGAR, Paris-Amy (1995)

Groves, S.M., Godsey, W.M., Shulman, M.A.: Financial Indicators for Local Government. Public Budgeting and Finance 1(2), 5–19 (1981)

Hendrick, R.: Assessing and Measuring the Fiscal Heath of Local Governments. Urban Affaires Review 40(1), 78–114 (2004)

Hirsch, W.Z., Rufolo, A.N.: Public Finance and Expenditure in a Federal Systems. Harcourt Brace Jovanovich, San Diego (1990)

Honadle, B.W.: The States Role in U.S. Local Government Fiscal Crises: A Theoretical Model and Results of a National Survey. International Journal of Public Administration 26(13), 1431–1472 (2003)

Honadle, B.W., Costa, J.M., Cigler, B.A.: Fiscal Heath for Local Governments. Elsevier Academic Press, San Diego (2004)

Inman, R.P.: How to have a fiscal crisis: lessons from Philadelphia. Fiscal problems of cities. In: AEA Papers and Proceedings, vol. 85(2), p. 378 (1995)

ICMA (International City/County Management Association): Evaluating financial condition. A Handbook for Local Government. ICMA, Washington, DC (2003)

Jones, S., Walker, G.: Explanators of Local Government Distress. Abacus 63(3), 396–418 (2007)

Kleine, R., Kloha, P., Weissert, C.S.: Monitoring local government fiscal health: Michigan´s new 10 points scale of fiscal distress. Government Finance Review 19(3), 18–23 (2003)

Kloha, P., Weissert, C.S., Kleine, R.: Developing and Testing a Composite Model to Predict Local Fiscal Distress. Public Administration Review 65(3), 313–323 (2005a)

Kloha, P., Weissert, C.S., Kleine, R.: Someone to Watch Over me: State Monitoring of Local Fiscal Conditions. The American Review of Public Administration 35(3), 236–255 (2005b)

Lin, W., Raman, K.K.: The housing value-relevance of governmental accounting information. Journal of Accounting and Public Policy 17(2), 91–118 (1998)

LGAs (Local Government Association of South Australia): Information Paper 9. Local Government Financial Indicators (2006), http://www.lga.sa.gov.au/webdata/resources/files/FinancialSustainabilityInfo_Paper_9_Local_Government_Financial_Indicators.pdf (accessed August 2011)

Martinez, J., Smoke, P., Vaillancourt, F.: The Impact of the 2008-2009 Global Economic Slowdown on Local Governments. In: The Impact of the Global Crises on Local Governments. United Cities and Local Governments, Barcelona (2009)

Mead, D.M.: Assessing the Financial Condition of Public School Districts: Some Tools of the Trade. In: Fowler Jr., W.J. (ed.) Selecting Papers in School Finance 2000-2001, pp. 2000–2001. National Center for Educational Statistics, Washington, DC (2001)

Mercer, T., Gilbert, M.: A Financial Condition Index for Nova Scotia Municipalities. Government Finance Review 12(5), 36–38 (1996)

Morgan, D.R., England, R.E.: Explaining Fiscal Stress Among Large U.S. Cities: Toward an Integrative Model. Policy Studies Review 3(1), 73–78 (1983)

Paulais, T.: Local Governments and the Financial Crisis: An Analysis, The Cities Alliance, Washington, DC (2009), http://www.citiesalliance.org/ca/sites/citiesalliance.org/files/Paulais_LocalGovernmentsandtheFinancialCrisis_Eng.pdf (accessed October 2011)

Petro, J.: Fiscal Indicators Reports and Ratio Analysis: Benchmarking Ohio Municipalities and School Districts. Government Finance Review 14(5), 17–21 (1998)

Stonecash, J., McAfee, P.: The ambiguities and limits of fiscal strain indicators. Policy Studies Journal 10, 379–395 (1981)

Wang, X., Dennis, L., Tu, Y.S.: Measuring Financial Condition: A Study of U.S. States. Public Budgeting and Finance 27(2), 1–21 (2007)

Wolff, L.W., Hughes, J.: Net Available Assets as a Proxy for Financial Condition: A Model for Measuring and Reporting Resources Available to a Local Government. Government Finance Review 14(3), 29–33 (1998)

Zafra Gómez, J.L., López Hernández, A.M., Hernández Bastida, A.: Developing a Model to Measure Financial Condition in Local Government. The American Review of Public Administration 39(4), 425–449 (2009a)

Zafra Gómez, J.L., López Hernández, A.M., Hernández Bastida, A.: Developing an alert system for local governments in financial crisis. Public Money & Management 9(3), 175–181 (2009b)

Zehms, K.M.: Proposed Financial Ratios for Use in Analysis of Municipal Annual Financial Reports. Government Accountants Journal Fall, 79–85 (1991)

Empirical Evidence of Spanish Banking Efficiency: The Stakeholder Theory Perspective

Leire San-José[1], José Luis Retolaza[2], and José Torres Pruñonosa[3]

[1] University of the Basque Country (UPV/EHU), Bilbao, Spain
University of Huddersfield, United Kingdom
[2] Institute of Applied Business Economics (IEAE), (UPV/EHU)
AURKILAN Institute for Business Ethics Research, Spain
[3] Euncet Business School
Ctra. de Terrassa a Talamanca, km 3, 08225 Terrassa (Barcelona), Spain
leire.sanjose@ehu.es, joseluis.retolaza@ehu.es,
jtorres@euncet.es

Abstract. Over recent years the structure of Spanish banking has been changing rapidly and a primary motivation, at least theoretically, has been the search for greater efficiency. This study aims to contribute to the established literature by using the frontier methodology to estimate and compare the economic, social and overall efficiency for the population of Spanish savings banks and banks in 2005 and 2009. The perspective used is based on the Stakeholder Theory because of the differentiation in terms of stakeholder participation between savings banks and banks. The results obtained indicate significant differentiation between these two groups of financial institutions in relation to social and economic efficiency, even though Spanish savings banks are not less efficient globally than banks.

Keywords: Economic Efficiency, Social Efficiency, Banks, Savings Banks, Data Envelopment Analysis.

1 Introduction

Savings banks are an important part of the Spanish banking system accounting for more than 30% its assets. In terms of customer deposits, savings banks are even stronger holding over 40% of market share (Asociación Española de Banca, 2009). Currently, the economic viability of savings banks is been questioned (Beltran et al., 2009) and public institutions have provided incentives to transform them into traditional banking institutions which are governed as capitalist companies by shareholders. One feature that still has differentiated the Spanish savings banks from traditional banks is their governance form because savings banks' organisation is based on a multi-fiduciary model (García-Cestona and Surroca, 2008) instead of a property right model.

The importance of analysing the efficiency of savings banks is highlighted by Altunbaş (2001) that concludes that this type of financial institutions should be

A.M. Gil-Lafuente et al. (Eds.): Soft Comput. in Manag. and Bus. Econ., STUDFUZZ 287, pp. 153–165.
springerlink.com

managed efficiently with earning capacity to be competitive. The aim of this study is to advance in the established literature by using a frontier analysis and mean comparison in order to show if the multi-fiduciary governance model in banking is confirmed by a significant differentiation in terms of efficiency in comparison with banks. Therefore, we intend to deal with the existence of significant differences between the efficiency of banks and savings banks that may be attributed to their governance model.

Literature review about the efficiency of banking concludes that the study of the efficiency of financial institutions is measured in terms of costs and benefits. For instance, in their extensive review of the bank efficiency literature, Berger and Humphrey (1997) analysed 130 studies from 21 countries applying frontier analysis based on costs. However, there is still a lack of conclusive findings mostly because of the definition of efficiency only based on costs. Therefore, this paper is focused, firstly, on this perspective because of the effort to define not only the economic efficiency but also the social and overall ones; and secondly, on the differentiation between the most important types of financial institutions in Spain: banks and savings banks.

The methodology used in the present research is based on statistical hypothesis testing and Data Envelopment Analysis (DEA). Studying efficiency has made it possible to give each institution a score from 0 to 100 that corresponds to its relative efficiency among the entire population. The empirical evidence to compare efficiency between banks and savings banks comes out both from an ANOVA parametric technique and non-parametrical techniques in order to strengthen the results of the hypothesis contrasts.

The findings suggest that the current process to restructure savings banks into banks is not founded on reasons of efficiency, but rather on economic and political interests, and possibly on the legal weakness of multi-stakeholder governance compared with governance based on property rights.

2 Literature Review

Stakeholder theory (Freeman, 1984) states that organisations should not only prioritise their shareholders' interest, but also respond to the interests of the stakeholders on the whole, with executives assuming the responsibility of striving to satisfy all involved parties in a balanced way. For the purposes of this paper, we are going to avoid the debate over whether stakeholders should be defined in terms of their power or their condition within the company, as well as various points of view regarding the theory itself. The multi-stakeholder approach (Goodpaster, 1991) raises a well-known paradox based on agency theory: if executives are fiduciary agents of the shareholders (the principal), they are not entitled (either morally or legally) to improve the contractual conditions that the shareholders of the company would be willing to provide them. The multi-fiduciary theory (Boatright, 1999) was hatched in response to this problem, which seeks to solve the problem of agency by considering that the principal does not just include the shareholders, but all the stakeholders as well. Thus, the agent is the fiduciary of all the stakeholders; as such, he would be not only entitled, but actually obligated to respond to their interests. Jensen (2002, 2008) stressed that multi-fiduciary theory makes businesses ungovernable because there is nobody with the right to control the agent. As part of this debate, Boatright (2008) argues that in shareholder-oriented businesses, the conflicts of interest among them and the ineffectiveness of control mechanisms hinder the principal's control of the agent; thus, the multi-fiduciary approach would not add to the problem.

Differences in ownership structure, and hence in the governance of savings banks, result in differences in priorities compared with other financial institutions structured differently. Therefore, multi-objective and multi-stakeholder organisations, and particularly Spanish savings banks, should not be evaluated solely in terms of profits (or costs), because they pursue goals that differ from those of institutions that focus on shareholders (García-Cestona and Surroca, 2008).

In this context, there are studies that have analysed the situation of savings banks. Financial literature has studied these issues with a narrow view towards making a profit. Thus, Kumbhakar et al. (2008) studied the technical efficiency of Spanish savings banks during the years 1986-1995 and concluded that it diminished over the period, even if they also found evidence of an increase in productivity in savings banks in Spain. Other authors like Tortosa-Ausina et al. (2002) have also studied the efficiency of Spanish savings banks, but for the years 1992-1998. They used the (frontier) DEA efficiency analysis technique, as we do in this paper. With regard to productivity rates, the conclusions that emerged from this study state that there is an increase in productivity due to improved production possibilities. As for efficiency, they concluded that the technical efficiency mean was very high and did not vary much throughout the period studied. However, it seems that there were significant differences among the banks. Their findings coincide with those obtained by Pastor (1995), but differ from those obtained by Grifell-Tatjé and Lovell (1996); this is mainly due to the choice of different outputs and to studying a period different from the same sample.

3 Methodology, Hypothesis and Sample

The problem that concerns us is analysing whether there is evidence that multi-stakeholder governance adversely affects the efficiency of a financial institution. To solve it, we have resorted to statistical hypothesis testing using the hypothetical-deductive method. Prior to this, we employed the synthetic analytical method to identify the components of the problem and move them to a system of inputs and outputs.

The fundamental hypothesis (H1) can be stated as follows: "There is a significant difference between savings banks and banks in relation with their overall efficiency". To conduct a more exhaustive analysis, this hypothesis is broken into two sub-hypotheses:

1. (H_1a) *"There is a significant difference between savings banks and banks in relation with their economic efficiency"*.
2. (H_1b) *"There is a significant difference between savings banks and banks in relation with their social efficiency"*.

To pursue the hypotheses above, we must first be able to identify the population of banks and savings banks in Spain and then be able to differentiate them using these two categories. Nowadays, a foreseeable impact variable exists: the current financial crisis. That is the reason why a comparative analysis of efficiency before and during the crisis has been done. To do so, we have just used two cutoff points: 2005, before the financial crisis; and 2009. To obtain this information we used the Anuario Estadístico de la Banca en España from the year 2005 and 2009, published by the Spanish Banking Association (Asociación Española de Banca, AEB); and the Anuario Estadístico de las Cajas de Ahorros, also from 2005 and 2009 and published by the Spanish Confederation of Savings Banks (Confederación Española de Cajas de Ahorros, CECA). It should be noted that credit cooperatives have been excluded from the study; while they make for a highly interesting financial model, they represent an intermediate (pluri-fiduciary) approach in terms of multi-fiduciary theory, so their possible relationship with efficiency is not so clear when addressing Jensen's "problem of governance".

Secondly, it is essential to obtain a measurement of the efficiency of each financial institution. For this, we opted to use Data Envelopment Analysis (DEA) methodology, based on measuring relative efficiency, which traces back to Farrell (1957), building on the previous work of Debreu (1951) and Koopmans (1951), who define business efficiency considering multiple inputs. Specifically, it proposes to measure efficiency based on two basic components: technical efficiency, which reflects a company's ability to get maximum output from a group of inputs; and allocative efficiency, which reflects a company's ability to use inputs in optimal proportions and according to their respective prices. Combining these two measurements allows us to measure economic efficiency. The concept of efficiency is defined as a decision-making unit's (DMU) position relative to the frontier of best execution, mathematically established by the ratio composed of the sum of outputs compared to inputs and their corresponding adjustments (Charnes et al., 1978).

DEA methodology is a deterministic, non-parametric technique proposed by Charnes et al. (1978) and developed by Banker et al. (1984) and Banker (1984). This technique is particularly suitable for calculating the efficiency of productive units with multiple outputs and inputs, and its non-parametric nature prevents the imposition of a particular functional form. The methodology estimates the production frontier using linear programming techniques, and the production frontier is determined by some envelopment functions of the combinations of outputs-inputs provided by empirical data, which require information from various decision-making units [for a more thorough review of this position, see Charnes et al., (1995); for a discussion of the methodologies used in the 1970s and 1980s, when it really gained in importance, see Seiford & Thrall (1990) and Seiford (1996)].

The DMUs can be companies, strategic business units, departments, and even specific processes within the companies themselves.

In our case, we will consider each financial institution (bank or savings bank) as a DMU. The value of maximum efficiency is equal to 1, or 100%, depending on the scale used by the DEA model; here we will use "Frontier Analyst 4", which employs a scale of 100. To the extent that a company is far from the frontier (which is determined by the group of decision-making units that obtain maximum efficiency), the value will fall to between 100% and 0%. This method allows us to obtain relative, but not absolute, efficiency. In this way we obtain the most efficient DMUs compared with the selection under consideration, meaning that the units that are more efficient when compared with the others are identified. This analysis works best when, as in our case, we can perform it over the entire population and not just a sample of it.

Given that the DEA is based on a relationship between inputs and outputs, selection and quantification of the same will be key to obtaining the results. Table 1 shows the inputs and outputs considered in our case.

Table 1 Inputs and outputs

INPUTS	OUTPUTS
Equity (E)	
Total Assets (TA)	Customer credit (CC)
Results (RS)	Jobs (J)
Risk (R)	
Economic Efficiency (EE)	***Social Efficiency (SE)***
Overall Efficiency (OE)	

Equity and total assets are the inputs to the system that have been considered; both variables are seen as pertinent since they are part of the indicators of financial and economic profitability, respectively; and we did not use the Foundation Endowment Fund 1 as inputs because it is irrelevant for savings banks with respect to equity and assets, so using it would lead to disproportionate efficiencies.

Four outputs have been considered, partly linked with economic efficiency, and the result of the exploitation activity2 obtained and the risk borne were taken into account. Given that some institutions had obtained negative results in 2009 and that efficiency analysis only works with positive results, we created a consistent transformation algorithm: y = x + (min. x +1)3. With regard to risk, it was

[1] The equivalent of social capital for a bank.

[2] We chose this parameter because it reflects annual efficiency better than profits before taxes, as extraordinary gains and losses can distort the results if one only considers the period of one year, as in the present case.

[3] *Y* is the resulting score; *x* is the raw score obtained by each DMU; *min. x* is the lowest score obtained by a DMU.

obtained from the summation of the contingent risks and commitments recognised by the different institutions.

As for social efficiency, it was more difficult to identify possible outputs since there is no standardised system of indicators that measures social profitability or the profitability provided to other groups of stakeholders other than shareholders. However, from the data provided by the institutions we have chosen three: 1) the number of jobs created, since job creation has clear social value with an obvious positive impact on the economy and tax revenue; 2) customer credit, which we consider an incomplete indicator of the institutions' support for the real economy (it would have been more appropriate to analyse the quality of asset investment, but most institutions are opaque in this regard) (San-Jose et al., 2011); and 3) the mathematical reverse of the risk, also considered as part of economic efficiency, since it is directly related with the moral risk assumed by the institutions and its negative social impact has been established on the international level since the very beginnings of the crisis. As for the risk variable, we must point out that there are some institutions with a value of 0, and in order to include them in this study we changed this score to 0.01 so that we could work with them without creating a mismatch between the relative efficiencies.

With regard to the DMUs, we have considered the entire population composed of 109 institutions and discarded 10 units for reasons of efficiency analysis4. As for the DEA, six efficiency analyses were performed: three only considering total assets as input and the outputs of economic efficiency, social efficiency and overall efficiency, respectively; and three others using the same outputs and total assets and equity together as inputs (this second group for analysis included one for which only the output result was considered). This last analysis was not conducted using a sole input since at least three variables are necessary to be able to perform a Data Envelopment Analysis.

For the data return, we used the variable returns to scale (BBC) proposed by Banker et al. (1984), which is an improvement to the commonly used constant returns to scale (CCR) of Charnes et al. (1978).

4 Empirical Results

Using de Data Envelopment analysis with the sample of Spanish banks institutions we have obtained the efficiency scores. The efficiency scores for banks and savings banks on each of the three analyses (economic, social and overall) are shown in Table 2. As expected, savings banks are less efficient economically but more efficient socially. The overall efficiency shows in absolute terms that in 2005 savings banks are more efficient but not in 2009.

---

[4] Caja de Castilla la Mancha, for being the only unit with negative equity, which detracted from any analysis of efficiency; Banesto Banco de Emisiones, BBV Banco de Financiación, Banco Industrial de Bilbao, Banco de Promoción de Negocios, Banco Occidental, Banco de Albacete and Banco Alicantino de Comercio for not having employees, which suggests that they are not engaged in regular banking activities; and Banco Liberta because its equity and assets coincided.

Table 2 Economic, Social and Overall Efficiency Mean and Standard Deviation

	Entities	2005 Mean (Standard Deviation)	2009 Mean (Standard Deviation)
Economic Efficiency	Savings Bank	33.67 (18.17)	75.88 (10.40)
	Banks	40.90 (31.15)	87.92 (11.98)
Social Efficiency	Savings Bank	83.73 (12.37)	88.33 (8.12)
	Banks	67.35 (33.75)	71.53 (32.04)
Overall Efficiency	Savings Bank	84.22 (12.51)	94.49 (5.75)
	Banks	74.65 (28.01)	96.15 (5.75)

Once the efficiency scores for each of the DMUs was obtained, a comparison of measurements was made with the ANOVA technique, using Snedecor's F test statistic with a significance level of 95%. The results obtained appear in Table 3.

Table 3 Economic, Social and Overall Efficiency Tests

2005						
Variables	Entities	N	F-Test	Welch	Levene[5]	Brown-Forsythe
Economic Efficiency	Savings Bank	44	1,940 (0,166)	2,402 (0,000)***	21,511 (0,000)***	2,402 (0,000)***
	Banks	68				
	Total	112				
Social Efficiency	Savings Bank	44	9,517 (0,003)***	13,273 (0,001)***	79,52 (0,000)***	13,273 (0,001)***
	Banks	68				
	Total	112				

[5] The Levene test is based on the Analysis of Variance (ANOVA).

Table 3 *(continued)*

Overall Efficiency	Savings Bank	44	4,538 (0,035)**	6,066 (0,153)	32,56 (0,000)***	6,066 (0,153)
	Banks	68				
	Total	112				
			2009			
Variables	Entities	N	F-Test	Welch	Levene[5]	Brown-Forsythe
Economic Efficiency	Savings Bank	43	23,484 (0,000)***	29,036 (0,000)***	3,56 (0,062)*	29,036 (0,000)***
	Banks	58				
	Total	101				
Social Efficiency	Savings Bank	43	8,809 (0,004)***	12,846 (0,001)***	51,32 (0,000)***	12,846 (0,001)***
	Banks	58				
	Total	101				
Overall Efficiency	Savings Bank	43	0,390 (0,534)	2,077 (0,153)	0,290 (0,591)	2,077 (0,153)
	Banks	58				
	Total	101				

*** Significant at 1%. ** Significant at 5%. *Significant at 10%.

As can be seen, there are significant differences, both in 2005 and 2009, favourable to savings banks in relation with their social efficiency. On the other hand, there is a significant mean difference favourable to banks in relation with their economic efficiency, but only in 2009. Finally, overall efficiency is favourable to savings banks, and it is significant in 2005 but not in 2009. However, we must bear in mind that ANOVA compares the measurements of groups similar in size and assumes similar variances between each group. To verify that our sample complies with the second premise, we conducted a Levene test (Levene (1960)) whose results are integrated in Table 3.

There is equality among variances in all cases, except for overall efficiency in 2009 (in this case, test Snedecor's F-distribution would not be conclusive). To solve the problem, we have used two complementary statistics tests: those of Welch and Brown-Forsythe (see Table 3), which provide greater robustness to the analyses of equality of means (Brown-Forsythe (1974)). Nevertheless, as can be noticed, concerning economic efficiency in 2005 and overall efficiency in 2009, although the outcomes point at the same direction of the results obtained by means of Snedecor's F, they are not conclusive.

As there is no equality among the variances, test Snedecor's F-distribution would not be conclusive. To solve the problem, we have used two complementary statistics tests: those of Welch and Brown-Forsythe (see Table 3), which provide greater robustness to the analyses of equality of means (Brown-Forsythe (1974)). As can be noted, the results from both tests mutually coincide. They also coincide with the results obtained by test Snedecor's F-distribution.

So that there would be no doubt about the normality or homoscedacity of the data, we also used the non-parametric Kolmogorov-Smirnov technique (see Table 4). As can be seen in the table, its results fully coincide with those obtained from the ANOVA analysis and Welch and Brown-Forsythe statistics.

Table 4 Non-parametric Test: Z Kolmogorov-Smirnov

	2005	2009
Variables	Z Kolmogorov-Smirnov	Z Kolmogorov-Smirnov
Economic Efficiency	1,244 (0,091)*	2,632 (0,000)***
Social Efficiency	1,935 (0,001)***	1,585 (0,013)**
Overall Efficiency	1,555 (0,016)**	1,008 (0,261)

*** Significant at 1%. ** Significant at 5%. *Significant at 10%.

In summary, the results obtained allow us to draw the following conclusions regarding our established hypotheses:

3. **Hypothesis (H_1a) is not refuted:** *"There is a significant difference between savings banks and banks in relation with their economic efficiency"*.
4. **Hypothesis (H_1b) is not refuted:** *"There is a significant difference between savings banks and banks in relation with their social efficiency"*.

So, **Hypothesis (H_1) is refuted:** *"There is a significant difference between savings banks and banks in relation with their overall efficiency"*. For overall efficiency, understood as the sum of economic and social efficiency, we must uphold the null hypothesis, since no significant differences were noted between banks and savings banks.

5 Discussion

The comparison made with data from 2005 (before crisis) and 2009 (fully immerse in crisis) in terms of efficiency among banks and savings banks using the Spanish population and Data Envelopment Analysis leads us to determine that savings banks are as efficient as banks or even more efficient than them in both periods. Nevertheless, the financial crisis has affected in a different way the efficiency of banks and savings banks. Hence, in 2005, economic efficiency of both types of institutions was similar, although overall efficiency of saving banks was significantly more positive. On the contrary, in 2009, economic efficiency of savings banks was inferior to the one obtained by banks, although overall efficiency was similar.

With regard to the problem of governance that caused the analysis, we can say that before crisis saving banks were as efficient economically as banks, and that their social and overall efficiency was similar. As a consequence, multi-fiduciary governance of this type of institutions does not reduce its efficiency, on the contrary, increases it in relation with its mission. In fact, there is no evidence that this risk is higher in multi-stakeholder based entities. The efficient scores take into account the established risk by banks and savings banks. Hence, the assumed high-risk decisions with political motivation should decrease the efficiency of the entity, but the mean efficiency of entities with participated governance has not decreased. Thus, the fact that overall efficiency of multi-fiduciary governance institutions was not inferior in any of the two cases leads us to reject Jensen's governance hypothesis (Jensen (2002, 2008); Retolaza et al. (2010)) because both types of institutions have, in the worst of the scenarios, similar levels of efficiency; and globally, favourable to those with a multi-stakeholder corporative governance.

Savings banks are one of the few types of organisations in which governance is not linked with property rights, which are indisputably of the greatest economic magnitude in Spain. The fact that multi-fiduciary governance reaches levels of overall efficiency similar to those in banking prompts us to corroborate the possibility of this type of governance in economically significant institutions, and this opens a new perspective for the applicability of stakeholder theory from its most ontological and constituent standpoint, in which stakeholder rights are not mediated via executive management (agent), but rather find their voice through participation in the institution's governing bodies (principal).

The results obtained lead us to question the premises underpinning recent government decisions on transforming savings banks into banks, because while it is true that the economic performance of these types of institutions has decreased more than banks, the fact remains that the social outcome is higher and that overall efficiency is similar, even during crisis. Arguments have been made, in general, that savings banks are less solvent than banks. Yet that difference is not significant. Thus, one may wonder why measures are generally applied to one type of institution when analysis shows that a lack of efficiency is distributed randomly between banks and savings banks. Perhaps the answer lies once again in property rights theory, since it may be much easier to regulate a subsector (savings banks) in which there are no adversely affected "owners" and debate is limited to policy areas. Respecting property rights and non-interference in banking, in addition to dismantling a financial system that could be put forward as an alternative to the current global financial system focused on shareholders' profits that contributed so much to the present crisis may be well received by the "financial markets", especially if a significant part of the savings banks' business goes to generate private profit, which so far has not been the case. What is clear is that it is not inefficiency in managing savings banks (although it obviously can be improved, like some banks can also do) that leads to the impoverishment of the model by public authorities, but the pressure of some markets that don't understand the model (in the best of cases) or even see it as a threat to their speculative interests. The famous recasting of capitalism has actually led to the consolidation of capitalism, eliminating a model that specifically would have been good to use for such recasting.

6 Conclusion

This paper advances in the established literature by means of analyzing the efficiency of banking by using the frontier methodologies to estimate and compare the economic, social and overall efficiency for the population of Spanish savings banks and banks in 2005 and 2009. The aim of the paper is to prove if there are differences in terms of efficiency before and during the beginning of the current financial crisis in different types of financial entities; banks and savings banks, because of their governance demarcation.

The results of this paper are similar to those obtained by former European studies, and they suggest that European savings banks can be efficient –at least, as much as banks are- because of their strategies around costs and economic scales. Moreover, before crisis global, economic, social and overall efficiency of savings banks was similar to the one obtained by banks. Furthermore, saving banks are linked properly with their mission. The results of the period of crisis are also similar; then the lack of efficiency is not a reason to determine the necessity of establishing similar rules to these two types of financial entities.

Several important and interesting findings are reported in this study. It appears that the different stakeholder models followed by Spanish savings banks do not convey any obvious difference in the level of inefficiency.

We would like to point out that the research's main limitations come from the use of data from just two fiscal years and from the lack of standardised indicators related with social efficiency (hence, the selected indicators used in this research can be discussed).

Finally, future lines of research could be the three following: 1) replicating the study to cover a wider period of years in order to control variability of results; 2) advancing in the establishment of a system of indicators that would enable the objectivity, with a higher inter-subjective consensus, of the social value generated; and 3) since the aim of the study is not referred to financial entities, but to multi-fiduciary governance efficiency, replicating the research in other sectors.

References

AEB: Statistical Yearbook of Banking in Spain: Asociación Española de Banca, Madrid (December 2005) (in Spanish)

AEB: Statistical Yearbook of Banking in Spain. Asociación Española de Banca, Madrid (December 2009) (in Spanish)

Altunbaş, Y., Gardener, E.P.M., Molyneux, P., Moore, B.: Efficiency in European banking. European Economic Review 45(10), 1931–1955 (2001)

Banker, R.: Estimating most productive scale size in data envelopment analysis. European Journal of Operational Research 17, 35–44 (1984)

Banker, R., Charnes, A., Coopers, W.: Some models for estimating technical and scale inefficiencies in data envelopment analysis. Management Science 30(9), 1078–1092 (1984)

Beltran, J., Torres, J., Prado, M.: The current process of savings banks mergers: an empirical analysis of their possible economical and financial causes. In: Redondo, J.A. (ed.) Creating Clients in Global Markets, AEDEM, Santiago de Compostela (2009) (in Spanish)

Berger, A., Humphrey, D.: Efficiency of financial institutions: International survey and directions for future research. European Journal of Operational Research 98, 175–212 (1997)

Boatright, J.: Ethics in Finance. Blackwell, Malden (2008)

Brown, M.B., Forsythe, A.: Robust tests for the equality of variances. Journal of the American Statistical Association 69, 364–367 (1974)

CECA: Statistical Yearbook of Savings Banks in Spain: Confederación Española de Cajas de Ahorros, Madrid (2005) (in Spanish)

CECA: Statistical Yearbook of Savings Banks in Spain: Confederación Española de Cajas de Ahorros, Madrid (2009) (in Spanish)

Charnes, A., Cooper, W., Rhodes, E.: Measuring the efficiency of decision making units. European Journal of Operational Research, 429–444 (1978)

Debreu, G.: The coefficient of resource utilisation. Econometrica 19, 273–292 (1951)

Farrell, J.: The measurement of productive efficiency. Journal of the Royal Statistical Society 120, 253–281 (1957)

Freeman, R.: Strategic Management: A Stakeholder Approach. Pitman, Boston (1984)

García-Cestona, M., Surroca, J.: Multiple goals and ownership structure: Effects on the performance of Spanish savings banks. European Journal of Operational Research 187(2), 582–599 (2008)

Goodpaster, K.: Business ethic and stakeholder analysis. Business Ethics Quarterly 1(1), 53–73 (1991)

Grifell-Tatjé, E., Lovell, C.: Deregulation and productivity decline: The case of Spanish savings banks. European Economic Review 40(6), 1281–1303 (1996)

Jensen, M.: Value maximization, stakeholder theory, and the corporate objective function. Business Ethics Quarterly 12(3), 235–256 (2002)

Jensen, M.: Non-rational behaviour, value conflicts, stakeholder theory, and firm behaviour. Business Ethics Quarterly 18(2), 167–171 (2008)

Koopmans, E.: Activity Analysis of Production and Allocation, Cowles Commission for Research in Economics. Monograph No. 13. Wiley, New York (1951)

Kumbhakar, S., Lozano-Vivas, A., Knox, C., Hasan, I.: The effects of deregulation on the performance of financial institutions: The case of Spanish savings banks. Journal of Money, Credit and Banking 33(1), 101–120 (2001)

Levene, H.: In: Olkin, I., et al. (eds.) Contributions to Probability and Statistics: Essays in Honor of Harold Hotelling. Stanford University Press (1960)

Pastor, J.: Efficency, productive change and technological change banks and savings banks: a non parametrical frontier analysis. Revista Española de Economía 12(1), 35–73 (1995) (in Spanish)

Retolaza, J.L., San-Jose, J., Ruiz, M.: Toward a new approach to the stakeholder theory: four innovative proposals. In: Maximsev, I., Krasnoproshin, V., Prado, C. (eds.) Global Financial & Business Networks and Information Management Systems, AEDEM, Minsk and San Petersburg (2010) (in Spanish)

San-Jose, L., Retolaza, J.L., Gutierrez-Goiria, J.: Are ethical banks different? A comparative analysis using the radical affinity index. Journal of Business Ethics 100(1), 151–173 (2011)

Seiford, L.: Data Envelopment Analysis: The evolution of the state of the art (1978-1995). Journal of Productivity Analysis 7, 99–138 (1996)

Seiford, L., Thrall, R.: Recent developments in DEA: The mathematical programming approach to frontier analysis. Journal of Econometrics 46, 7–38 (1990)

Tortosa-Ausina, E., Grifell-Tatjé, E., Armero, C., Cones, D.: Sensitivity analysis of efficiency and Malmquist productivity indices: An application to Spanish savings banks. European Journal of Operational Research 184(3), 1062–1084 (2002)

Disability Caused by Occupational Accidents in the Spanish Long-Term Care System

Ramón Alemany, Catalina Bolancé, and Montserrat Guillén

Department of Econometrics, Riskcenter and IREA, University of Barcelona
Av. Diagonal, 690, 08034 Barcelona, Spain
ralemany@ub.edu, bolance@ub.edu, mguillen@ub.edu

Abstract. This study assumes that long-term care may be caused by occupational accidents. We quantify the number of disabled individuals that qualify to receive public funds in the context of the Spanish Long-Term Care System, for those individuals whose need of support is caused by an injury related to working activities. Using data from a Spanish Survey we also estimate the number of workers by age and severity level.

Keywords: Degree of dependence, Work related injuries, Welfare.

1 Introduction

Hazardous working conditions affect the labour market in many respects. Low safety creates job vacancies and enhance the difficulty to recruit workers unless a risk premium supplements the salary, but occupational accidents put pressure on the health system a, particularly, increase the cost of long-term care when disabled workers need the support of a third person to perform daily life activities, a condition known as dependency. The reason why dependency caused by occupational accidents is extremely worrisome is because it affects adults who may require care for a very long period of time, i.e. decades. It is a phenomenon similar to disability caused by traffic accidents. The costs of care associated to occupational and traffic accidents create concern in modern societies. There is a long tradition in Spain about collecting exhaustive information on the number of occupational accidents. These data are available at the Ministry of Labour (*Ministerio de Empleo y Seguridad Social*) and have been presented yearly also in combination with the occupational sector, age of the worker and sex. However from these data there is no way to know the effect of those accidents on dependence system.. The data on the evolution of the workers who suffered those accidents is not available from that source, so even if an accidents is serious it is not possible to infer from the analysis of the statistics on occupational accidents the impact on long-term care (LTC) needed by those affected workers.

A.M. Gil-Lafuente et al. (Eds.): Soft Comput. in Manag. and Bus. Econ., STUDFUZZ 287, pp. 167–176.
springerlink.com

2 Background

Occupational diseases have deserved a lot of research in the past few decades. Grondtrom et al. (1980) were mainly concerned about accident prevention, Greenwood (1984) recommended a shift from a medical to a socioeconomic frame of reference for work-related disability that would overtly recognize regional variations in the static and dynamic factors of the economy and that would promote return to work and rehabilitation. Kettle (1984) found evidence that disabled workers as compared with the accident rates of other workers. Persson and Larsson (1991) studied occupational accidents resulting in a permanent medical disability in Sweden and found that the Swedish woodworking industry had the gravest problem when severe work injuries and young people were concerned. Laflamme and Menckel (1995) found a relationship between aging and occupational accidents and suggested to emphasize greater precision with regard to the type of accident in focus. Evidence about the relationship between age and occupational injuries was again found by Pransky et al. (2005).

Miller and Galbraith (1995) were the first to estimate the aggregate cost of occupational accidents in the US, and they included medical and emergency services, lost productivity, insurance costs, and lost quality of life. More specific studies are those by Bylund and Björnstig (1998) who studied mechanics and construction metal workers, Sheu et al. (2000) analyzed human capital loss of occupational injury including cumulative injury rate, proportion of potential workdays lost, and potential salary lost for a steel company in Taiwan. Larsson and Field (2002) analyzed workers' compensation data from the public fund in Australia. Wang et al. (2005) presented a method that can be applied to assess the effectiveness of intervention trials on other populations at high risk of occupational injury. Macedo and Silva (2005) studied occupational accidents in Portugal and Camino López et al. (2008) analyzed construction industry accidents in Spain. Shalini (2009) estimated the number of occupational accidents have occurred amongst the working population in Mauritius. Karlsson et al. (2008) concluded that besides socio-demographic risk factors, the sick-leave diagnoses constitute an important both medium and long-term predictor of disability pension among both men and women on long-term sickness absence in Sweden.

Dembe (2001) found complex and multi-factorial relationships due to occupational injuries and illnesses. Social consequences involve filing and administration of workers' compensation insurance claims, medical care experiences, domestic function and activities of daily living, psychological and behavioral responses, stress, vocational function, rehabilitation and return to work, and equity and social justice. Weil (2001) suggested that estimated costs of injuries and fatalities tend to understate the true economic costs from a social welfare perspective, particularly in how they account for occupational fatalities and losses arising from work disabilities. Most of those articles contributed much to understand that substantial burdens are caused by workplace illnesses and injuries (see Boden et al., 2001; Reville et al., 2001; Keller, 2001; Reville and Schoeni, 2004; Ho et al., 2006; Carlos-Rivera, 2009) and indicated that money spent on prevention is worth a lot more as it reduces the number of accidents (see Rauner et al., 2005; Benavides et al., 2006; Sousa Santana et al., 2006; Lin et al., 2008;

Tüchsen et al., 2009; Sanmiquel et al., 2010). Regarding determinants and consequences of occupational safety and health Pouliakas and Theodossiou (2011) have made a very recent attempt to formalised a very complete economic theory model to analyse affordability and efficiency in social security funds and workers' compensation schemes.

In Spain, efforts have been made in the past decade to have an official registry of occupational diseases (see Benavides and Benach, 2001). Benavides et al. (2004) showed the existence of substantial regional differences in the incidence of non-fatal occupational accident injuries. More information was provided by Lilian and Serra (2009) for Catalonia and a longitudinal analysis was presented by García Mainar and Montuenga Gómez (2009) where they claimed that an excessive number of hours worked is a significant risk factor that leads to more work-related accidents. Benavides et al. (2010) observed substantial differences in the incidences of permanent disability by demographic, social and geographical characteristics using a cohort of Spanish workers. In Europe, an adequate insight into the costs of accidents and the potential benefits of accident prevention is given by the report of the European Agency for Safety and Health at Work (2002) and later it was regularly updated (European Commission, 2004 and 2008).

In the US, Yee (2002) argued whether the he public sector should be the first or last source of payment for long-term care; and the extent to which choice is afforded privately with regard to the types, settings, and amount of long-term care desired to complement family care.

Social inequalities in injury occurrence and in disability retirement attributable to injuries were studied by Hannerz et al. (2007a and 2007b), whereas differences due to urban-rural differences were outlined by Young et al. (2008). To our knowledge, the first to investigate the longer-term health consequences of work-related injuries among young workers were Koehoorn et al. (2008). They noted that as a result of musculoskeletal injuries in particular among males, increased odds of belonging to the higher health care trajectories defined by general practitioner with an intensive number of visits.

The legal framework in Spain is presented in a very concise and understandable contribution of Moretón Sanz (2007). Carnero and Pedregal (2010) showed and predicted the evolution of Spanish occupational accidents of different levels of severity. This paper was the first contribution that forecasted occupational accidents for different levels of severity. Gené et al. (2011) studied the cost of nursing home care. They found that there exists a positive correlation of workload with variables related to disease severity and a negative correlation with variables related to cognitive impairment. They also showed that home care nursing in Catalonia is basically demand-oriented. Just recently, Remo Diez (2011) shows that legislative efforts in the prevention of occupational hazards in recent years, show that social costs of occupational accidents and occupational diseases in 2007 in Spain has reached 2% of GDP.

3 The Spanish Long-Term Care System

The Spanish system recognizes the right to receive support in case of dependence. Emphasis is out on need of support, not on disability or handicap, while budged allocated to social protection in Spain is among the very small in Europe. In the Spanish law, the concept of dependence is defined as "a permanent state of a person, as a result of ageing-related processes, illness o disability and linked to the absence or loss of physical, intellectual or sensorial autonomy that requires the assistance of a third person or support of any kind to perform basic daily life activities, or, in the case of people with intellectual disorders or mental illness, of other types of help for personal autonomy". Namely, a person is considered dependent situation if three circumstances occur:

(i) The person suffers physical, psychological or intellectual limitations that permanently reduce his or her capabilities.
(ii) Basic daily life activities (BDLA) cannot be performed autonomously.
(iii) The person requires assistance or care by a third party.

We note that BDLA are the most basic tasks that allow someone to cope with a minimum of autonomy and independence, these tasks are: personal care, basic domestic activities, essential mobility, to recognise persons and objects, to keep orientation, to understand, to execute orders and simple tasks.

One of the key issues in any LTC protection system in Spain is how the level of needs requiring support is defined, that is, how to classify any individual applying for care into one group. This is usually done by means of a scale. Once the applicant's situation has been assessed by experts, he or she is either assigned to one of the three possible degrees of dependence or denied eligibility.

A scale called *Baremo de Valoración de la Dependencia* (BVD) measures an individual's inability to perform daily life activities by means of a scale from 0 to 100 points. The scale used in Spain puts a lot of emphasis on the intensity of support needed and the tasks for which assistance is required.

The number of points given by the BVD scale is obtained by adding the severity coefficient times the weight assigned on the scale to that particular task for every existent daily life activities (DLA) or instrumental daily life activities (IDLA). Using the BVD, there are three severity degrees in the Spanish LTC evaluation system (see Esparza, 2010) and two levels in each degree.

A person is eligible in:

- Degree 1, if support is needed once a day (Level I: 25-40 points, Level II: 40-49 points)
- Degree 2, if assistance is to be provided two or three times per day (Level I: 50-64 points, Level II: 65-74 points)
- Degree 3, if assistance is demanded several times during the day (Level I: 75-89 points, Level II: 90+ points)

Someone who is eligible to receive LTC support from the public system can obtain up to 833.96 euros in cash as a monthly payment for the services received or 520.69 euros monthly for family care, if he prefers that relatives take care of him. In that case, an extra sum of 162.49 euros is given monthly to cover the social security taxes, and the training and education of the person that is employed as the family care-giver.

Information from the survey EDAD 2008 which stands for Survey of Disabilities, Personal Autonomy and Situations of Dependence (Instituto Nacional de Estadística, INE) is analyzed. This survey was conducted in 2008 and it was a huge statistical instrument that involved more than 220,000 respondents. The statistical accuracy and ability to represent the Spanish population is guaranteed. We identified 26 variables in the survey questionnaire that inquire about each DLA that are then used in the BVD scale to measure dependence severity. When a respondent indicates that he or she has a difficulty to perform a specific DLA, then he or she must indicate whether the difficulty is moderate, severe or total. According to the law, we assigned a coefficient value of 0.90 for moderate difficulty, 0.95 for severe difficulty and 1.00 for total inability to perform that particular DLA. In practice, a medical team decides the level of difficulty.

We also identified individuals with disorders when performing IDLAs (instrumental daily life activities), because the BVD scale has some particular coefficients for those individuals who have a difficulty in performing tasks due to some cognitive or intellectual challenges. Inability to perform any of the following four activities is considered sufficient to have a disorder in IDLA: difficulty to pay attention when listening or looking, significant difficulty to learn to read, to write, to count (or calculate), to copy or to learn to handle devices, major difficulty to perform simple tasks without help or supervision and major difficulty to perform complex tasks without assistance and without supervision.

4 Results

With data from the survey we estimated the population in each degree of dependence at 2008, the results are presented in Table 1. We emphasize that the estimated number of women in need of care is more than double the number of men. In total there are nearly one million individuals 65 or older with some degree of dependence (data not shown).

Table 1 Population in each degree of *dependence and sex*(all ages)

	Men	Women	Total
Degree 1	213,426	407,304	620,730
Degree 2	113,321	239,087	352,407
Degree 3	112,562	232,696	345,257
Total	439,308	879,087	1,318,394

Source: EDAD (2008) and own analysis.

For the previous estimates, we also estimated the number of individuals that are dependent due to a work-related injury or an occupational accident. The figures are shown in Table 2.

Table 2 Dependent population due to work-related injury or occupational accident in each degree of *dependence and sex* (row percentage in parenthesis)

	Men	Women	Total
Degree 1	24,044 (59.7%)	16,240 (40.3%)	40,284
Degree 2	8,557 (47.2%)	9,555 (52.8%)	18,112
Degree 3	4,568 (51.1%)	4,375 (48.9%)	8,943
Total	37,169 (55.2%)	30,170 (44.8%)	67,339

Source: EDAD (2008) and own analysis.

Table 3 Percentage of dependent population whose condition is due to work-related injury or occupational accident with respect to the population in each degree of *dependence*

	Men	Women	Total
Degree 1	11.3%	4.0%	6.5%
Degree 2	7.6%	4.0%	5.1%
Degree 3	4.1%	1.9%	2.6%
Total	8.5%	3.4%	5.1%

Source: EDAD (2008) and own analysis.

We observed that work-related injuries and occupational accident is much more prominent for men than for women in the moderate level of severity (see Table 2). In Table 3, we see that the prevalence of dependence due to work-related injury or occupational accident is much higher in men than in women, but it is more prevalent in moderate levels than in the highest severity degree of dependence.

We also confirm that there is a large age effect. The population of work-related and occupational accident dependents increases with age, as shown in Table 4. We believe that this is due to the fact that the older a worker is the longer has he been exposed to risk. However, dependence due to occupational injuries or work-related accidents is much more prevalent among younger individuals than for the elderly. This is due to the ageing effect, which implies that elderly people have more health related impairments than younger people.

Table 4 Dependent population due to work-related injury or occupational accident in each degree of dependence by age

	18 to 34	35 to 64	+65
Degree 1	1,322	17,846	21,117
Degree 2	410	5,835	11,867
Degree 3	330	2,178	6,435

Source: EDAD (2008) and own analysis.

Table 5 Percentage of dependent population due to work-related injury or occupational accident in each degree of *dependence* by age with respect to total number of dependent people in that age group

	18 to 34	35 to 64	+65
Degree 1	5.0%	12.3%	5.3%
Degree 2	2.6%	9.3%	4.7%
Degree 3	1.7%	5.6%	2.4%

Source: EDAD (2008) and own analysis.

Table 6 Severity distribution of dependent population due to work-related injury or occupational accident

	Men	Women	Total
Degree 1	64.7%	53.8%	59.8%
Degree 2	23.0%	31.7%	26.9%
Degree 3	12.3%	14.5%	13.3%

Source: EDAD (2008) and own analysis.

Tables 5 and 6 indicate that injuries and accidents related to professional activities lead to dependence situations that are concentrated on moderate levels of severity. However, the distribution for women shows a slight tendency to higher degrees.

5 Conclusions

In this work we show a set of results attempting to show the effect of the work-related injuries and the occupational accidents in the needs of long term care of the population.

Our results show as percentage of dependent population due to work related injuries and occupational accident is much higher for men than for women, especially in the moderate level of severity.

In summary, the percentage of dependent individuals that need support of a third person due to an occupational injury or a work-related injury is between 1.7%, for the younger and the higher severity level, and 12.3%, for the adults in the moderate level. The cost of these accidents implies that all these people which probably would not have needed support, have a worse quality of life than they would have had if they had not suffered the work related accident. At least for the population of 65 years or more, the reduction of work-related misfortunes would lead to a considerable reduction in long-term care costs.

Highest lifetime LTC costs are associated to a sustained dependency over time and work-related injuries affect young adults, therefore a significant percentage of these highest lifetime LTC costs is due to hazardous professional activities or work related accidents. The lifetime cost of lost quality of life is not studied here but it should certainly affect the societal interest for prevention.

Acknowledgments. We thank the Spanish Ministry of Economy and Knowledge grant number ECO2011-21787-C03-01.

References

Benavides, F.G., Benach, J.: La prevención de riesgos laborales, las estadísticas de accidentes de trabajo y el "Informe Durán". Arch. Prev. Riesgos. Labor. 4(2), 53–54 (2001)

Benavides, F.G., Castejón, E., Giráldez, M.T., Catot, N., Delclós, J.: Lesiones por accidente de trabajo en España: Comparación entre las comunidades autónomas en los años 1989, 1993 y 2000. Revista Española de Salud Pública 78, 583–591 (2004)

Benavides, F.G., Delclos, J., Benach, J., Serra, C.: Lesiones por accidentes de trabajo, una prioridad en salud pública. Revista Española de Salud Pública 80, 553–565 (2006)

Benavies, F.G., Durán, X., Martínez, J.M., Jódar, P., Boix, P., Amable, M.: Incidencias de incapacidad permanente en una cohort de trabajadores afiliados a la Seguridad Social, 2004-2007. Gacena Sanitaria 24(5), 385–390 (2010)

Boden, L.I., Biddle, E.A., Spieler, E.A.: Social and economic impacts of workplace illness and injury: Current a future directions for research. American Journal of Industrial Medicine 40, 398–402 (2001)

Bylund, P.O., Björnstig, U.: Occupational injuries and their long term consequences among mechanics and construction metal workers. Safety Science 28(1), 49–58 (1998)

Camino López, M.A., Ritzel, D.O., Fontaneda, I., González Alcantara, O.J.: Construction industry accidents in Spain. Journal of Safety Research 39, 497–507 (2008)

Carlos-Rivera, F., Aguilar-Madrid, G., Gómez-Montenegro, P.A., Juárez-Pérez, C.A., Sánchez-Román, F.R., Durcudoy Montandon, J.E.A., Borja-Aburto, V.H.: Estimation of health-care cost for work-related injuries in the Mexican institute of social security. American Journal of Industrial Medicine 52, 195–201 (2009)

Carnero, M.C., Pedregal, D.J.: Modelling and forecasting occupational accidents of different severity levels in Spain. Reliability Engineering and System Safety 95, 1134–1141 (2010)

D'Amico, G., Guillén, M., Manca, R.: Full backward non-homogeneous semi-Markov processes for disability insurance models: a Catalunya real data application. Insurance: Mathematics and Economics 45(2), 173–179 (2009)

Dembe: The social consequences of occupational injuries and illnesses. American Journal of Industrial Medicine 40, 403–417 (2001)

Esparza Catalán, C.: Métodos de cálculo de la gravedad de la discapacidad, Informes Portal Mayores, vol. 103, Madrid (2010) (published on: June 25, 2010)

García Mainar, I., Montuenga Gómez, V.: Causas de los accidentes de trabajo en España: análisis longitudinal con datos de panel. Gaceta Sanitaria 23(3), 174–178 (2009)

Gené, J., Borrás, A., Contel, J.C., Camprubí, M.D., Cegri, F., Heras, A., et al.: Nursing workload predictors in Catalonia (Spain): a home care cohort study. Gaceta Sanitaria 25(4), 308–313 (2011)

Greenwood, J.G.: Invention in work-related disability: the need for an integrated approach. Social Science and Medicine 19(6), 595–601 (1984)

Grondstrom, R., Jarl, T., Thorson, J.: Serious occupational accidents-An investigation of causes. Journal of Occupational Accident 2, 283–289 (1980)

Hannerz, H., Mikkelsen, K.L., Nielsen, M.L., Tüchsen, F., Spangenberg, S.: Social inequalities in injury occurrence and in disability retirement attributable to injuries: a 5 year follow-up study of a 2.1 million gainfully employed people. BMC Public Health 7(215) (2007a), doi:10.1186/1471-2458-7-215

Hannerz, H., Spangenberg, S., Tüchsen, F., Nielsen, M.L., Mikkelsen, K.L.: Prospective analysis of disability retirement as a consequence of injuries in a labour force population. Journal of Occupational Rehabilitation 17, 11–18 (2007b)

Ho, J.J., Hwang, J.S., Wang, J.D.: Estimation of reduced life expectancy from serious occupational injuries in Taiwan. Accident Analysis and Prevention 38, 961–968 (2006)

Karlsson, N.E., Carstensen, J.M., Gjestal, S., Alexanderson, K.A.E.: Risk factors for disability pension in a population-based cohort of men and women on long-term sick leave in Sweden. European Journal of Public Health 18(3), 224–231 (2008)

Keller, S.D.: Quantifying social consequences of occupational injuries and illnesses: State of the art and research agenda. American Journal of Industrial Medicine 40, 452–463 (2001)

Kettle, M.: Disabled people and accidents at work. Journal of Occupational Accidents 6, 277–293 (1984)

Koehoorn, M., Breslin, F.C., Xu, F.: Investigation the longer-term health consequences of work-related injuries among youth. Journal of Adolescent Health 43, 466–473 (2008)

Larsson, Y.J., Field, B.: The distribution of occupational injury risks in the State of Victoria. Safety Science 40, 419–437 (2002)

Laflamme, L., Menckel, E.: Aging and occupational accidents. A review of the literature of the last three decades. Safety Science 21, 145–161 (1995)

Lilian, F., Serra, D.: An analysis of the costs of work-related accidents and illnesses in Catalonia. A methodological proposal and figures for the years, and, Generalitat de Catalunya. Departament de Treball (2009)

Lin, Y.H., Chen, C.Y., Luo, J.L.: Gender and age distribution of occupational fatalities in Taiwan. Accident Analysis and Prevention 40, 1604–1610 (2008)

Macedo, A.C., Silva, I.L.: Analysis of occupational accidents in Portugal between 1992 and 2001. Safety Science 43, 269–286 (2005)

Miller, T.R., Galbraith, M.: Estimating the cost of occupational injury in the United States. Accident Analysis and Prevention 27(6), 741–747 (1995)

Moreton Sanz, M.F.: De la minusvalía a la discapacidad y dependencia: pasarelas y asimilaciones legales. UNED, Madrid (2007)

Persson, I., Larsson, T.: Accident-related permanent disabilities of young workers in Sweden 1984-1985. Safety Science 14, 187–198 (1991)

Pinquet, J., Guillén, M., Ayuso, M.: Commitment and lapse behavior in long-term Insurance: a case study. Journal of Risk and Insurance 78(4), 983–1002 (2011)

Pouliakas, K., Theodossiou, I.: The economics of health and safety at work: an interdiciplinary review of the theory and policy. Journal of Economic Surveys (2011), doi:10.1111/j.1467-6419.2011.00699.x

Pransky, G.S., Benjamin, K.L., Savageau, J.A.: Early retirement due to occupational injury: Who is a risk? American Journal of Industrial Medicine 47, 285–295 (2005)

Rauner, M.S., Harper, P., Shahani, P., Schwarz, B.: Economic impact of occupational accidents: Resource allocation for auva's prevention programs. Safety Science Monitor 9(1), Article 3 (2005)

Remo Díez, M.N.: Costes socials de siniestralidad laboral (2000-2007). Pecunia Monográfico, 213–231 (2011)

Reville, R.T., Schoeni, R.F.: The fraction of disability caused at work. Social Security Bulletin 65(4), 31–37 (2004)

Reville, R.T., Bhattacharya, J., Sager Weinstein, L.R.: New methods and data sources fro measuring economic consequences of workplace injuries. American Journal of Industrial Medicine 40, 452–463 (2001)

Sanmiquel, L., Freijo, M., Edo, J., Rossell, J.M.: Analysis of work related accidents in the Spanish mining sector from 1982-2006. Journal of Safety Research 41, 1–7 (2010)

Shalini, R.T.: Economic cost of occupational accidents: Evidence from a small island economy. Safety Science 47, 973–979 (2009)

Sheu, J.J., Hwang, J.S., Wang, J.D.: Diagnosis and monetary quantification of occupational injuries by indices related to human capital loss: analysis of a steel company as an illustration. Accident Analysis and Prevention 32, 435–443 (2000)

Sousa Santana, V., Araújo-Filho, J.B., Alburquerque-Oliveira, P.R., Barbosa-Branco, A.: Occupational accidents: social insurance costs and work days lost. Rev. Saude Publica. 40(6), 1–8 (2006)

Tüchsen, F., Christensen, K.B., Feveile, H., Dyreborg, J.: Work injuries and disability. Journal of Safety Research 40, 21–24 (2009)

Wang, K., Lee, A.H., Yau, K.K.W., Carrivick, P.J.W.: A bivariate zero-inflated Poisson regression model to analyze occupational injuries. Accident Analysis and Prevention 35, 625–629 (2003)

Weil: The Valuing the economic consequences of work injury and illness: A comparison of methods and finding. American Journal of Industrial Medicine 40, 418–437 (2001)

Yee, D.L.: Long term care policy and financing as a public of private matter in the United States. Journal of Aging and Social Policy 13(2-3), 35–51 (2002)

Young, A.E., Wasiak, R., Webster, B.S., Shayne, R.G.F.: Urban-rural differences in work disability after an occupational injury. Scandinavian Journal of Work, Environment and Health 34(2), 158–164 (2008)

A Paradigm Shift in Business Valuation Process Using Fuzzy Logic

Anna M. Gil-Lafuente, César Castillo-López, and Fabio Raúl Blanco-Mesa

Department of Business Administration, University of Barcelona,
Av. Diagonal 690, 08034 Barcelona, Spain
amgil@ub.edu, cclopez@economistes.com, frblamco@yahoo.com

Abstract. The aim of this paper is proposing an overview of valuation of companies' process, associated to uncertainty modelization. This exposition represents a paradigm shift in the significance of what today we understand as valuation of a company. Through previous investigation published about finance literature, we prove there's still a long way to investigate on valuation methodology in terms of fuzzy methodology. The result given in this paper represents a new working line for researchers in the valuation field, and in general, for the scientific community, who wants to study further on value of business organizations, taking into account drivers, elements and variables, than common valuation models don't consider in the calculations. The exposition, which we propose by construction, is set like a fundamental axis to reach the higher efficiency levels in the valuation process.

Keywords: Uncertainty, Valuation of companies, estimation, subjective variables.

1 Introduction

Decision making problems are very common in the literature (Figueira et al., 2005; Gil-Aluja, 1999; Kaufmann and Gil-Aluja, 1986; Merigó, 2008) and can be applied in a lot of fields. For example, we can use them for the selection of policies in a government, in a company, etc. Selecting the best policy in a government is one of the key problems to be solved in order for a good development of a country. There are a lot of different types of policies such as fiscal policies, monetary policies and commercial policies. In order to select the optimal policy, the government has to develop a selection process because they have to choose the best policy in each moment. Among the great variety of studies existing in selection, this work will follow those models that develop the decision process using ideals (Gil-Aluja, 1998; 1999; A.M. Gil-Lafuente, 2005; A.M. Gil-Lafuente and Merigó, 2006; J. Gil-Lafuente, 2001; 2002; Merigó, 2008; Merigó and A.M. Gil-Lafuente, 2006; 2007a; 2008a; 2008b; 2008c; 2008d; 2010).

The company valuation has been, and remains today, a fundamental and complex process, both in economic and financial systems, in which firms operate. The

A.M. Gil-Lafuente et al. (Eds.): Soft Comput. in Manag. and Bus. Econ., STUDFUZZ 287, pp. 177–189.
springerlink.com

evolution of the economic context and the appearance of different theoretical and methodological instruments over time have led to the existence of different value systems, each of them raised and aimed at solving specific issues during the time.

As the context has evolved towards more complex and uncertain notes the urgent need to introduce new business valuation models imbued with a more general than those used so far.

The basic objective of this present paper lies in performing a gap analysis underlying the models that have been used so far and establish more basic approaches on which to build new dynamic models for global forecasting techniques based on the treatment and management of uncertainty and decision-making. Through studied scientific contributions we have found that traditional models of business valuation, hasn't reached to date to include different tools from the models which treat uncertainty. Those models could be of great help when it comes to evaluate subjective elements, as well as those drivers that are not easily quantifiable, but which are transcendental to adjust the calculated value to real market value of assets.

This research starts trying to avoid ambiguity and confusion identifying what, when, how and where is the value creation in the company (Froud et al. 2000, p. 81).

The approach described below, provides new methods for valuing companies based on new techniques for the treatment of management problems, under uncertainty contexts. In our work we will proceed to highlight the techniques based on fuzzy numbers, linguistic variables, multi-skilled techniques and aggregation operators.

In order to full fill the requirements for this paper, we are going to carry out an analysis of the most important contributions and the newest ones, allowing investigating how to improve business valuation process through techniques and methods, including non-numerical mathematics of uncertainty and multivalent logic. By combining hybrid models will be achieved traditional techniques with contributions arising from the mathematics of uncertainty. The results allow estimating the value of organizations in a global sense, considering the cumulative interaction of all its components that affect market value.

2 On the Previous Work

The main problem identified in the literature on valuation of companies, is on such subjective factors, involved in determining the value of the company. You could drill down to try to explain, not only the financial data involved in the final value, but also experimental evidence produced by speculative risks (Baberis, et al. 2001, p. 48). Since the beginning, it has worked hard to differentiate the concepts of price and value. Price is the result of the agreed amount upon between buyer and seller in the sale of a business. It is obvious that price contains valuable information for calculating the value, but is not the best intrinsic value index (Bhojraj and Lee, 2002, p. 435). The value is the estimated price that could have a company based on multiple factors. This value depends on the seller/buyer's expectations.

In fact, different buyers have different expectations, so the process for the valuation will be different.

Although there is extensive academic literature on methods and models used in the valuation of companies, most studies rely on a common scientific basis: the company's value is determined by the present value of company's expected cash flows (González Jiménez, et al. 2010, p. 67). Hrvol'ova, et al. (2011, p. 148) also concludes that the currently used models to value companies are based on discounted cash flows variations. Studies also proof that based models on the earnings per share (EPS), adjust the calculated value substantially better than the discounted cash flows (Liu, et al. 2007, p. 66). The same study concludes the relative superiority of EPS as a method in most of analyzed sectors, but not in all of them. So there are not taken into account drivers that could add a further adjustment in the found value. The work of Kaplan and Ruback (1995, p. 1059), provides the evidence that discounted cash flows differ about 10% of market value.

In general, the main methods and business valuation models, accepted commonly, are grouped into four types: methods based on the balance, based on the profit, mixed and those based on discounted cash flows (DCF). In practice, as mentioned earlier, the DCF model has become the most popular, as it may appear to be more consistent with the objective of value creation, it includes the majority of drivers with influence in the company value. Moreover, DCF is also used in different sectors such as might be project management, insurance and financial management. Even it provides a more appropriate result than the balance based methods (Copeland, et al. 2000, p. 177-180). Also the profit-based methods are much used, because they are easier to handle and understand.

Among the value drivers that affect the valuation of a company or its assets, we can distinguish three groups: those that affect future cash flows, those that affect the required return of shares (Fernandez, 2005, p. 56) and those that influence the relationship with the market. The issue could be the difficult to identify and quantify some drivers among the three groups, company valuation models fail to include them properly in its formulation.

There are studies (Yao, et al. 2005, p. 222) with numerical examples showing that the inclusion of fuzzy philosophy in a valuation model as popular as the discounted cash flow, get a more accurate assessment and help investors to be more precise with the measured value of their assets. Also in the DCF model, it has been applied a fuzzy binomial approach to estimate the associated uncertainty with these cash flows when facing the decision-making (Ho and Liao, 2011, 15 301). Smith and Trigeorgis (2006, p.110), concerned about the importance of creating shareholder value, apply a combination of real options and analysis of set theory to make value creation to flow from an strictly financial point of view to the strategic one. Thus, the real options make it the preferred methodology of academic literature at a time to minimize the effects of uncertainty, over the traditional methods of valuation (Ucal and Kahraman, 2009, p. 666) or probabilistic approaches (Carlsson and Fuller, 2003, p. 310). In some studies it has been used "Subtle Sets" to measure the value of goodwill, leaving the accounting methods that do not properly handle uncertainty, and thus determine a more adjusted company's value (Ionita and Stoica, 2009, p. 122).

3 Research Methodology

At present there are many studies on valuation of companies and assets, but always from a deterministic perspective. The use of modern management techniques based on the uncertainty treatment should help to improve the classical models for valuation of companies, because it includes a focus on techniques and methods based on numerical mathematics of uncertainty and multivalent logic.

The valuation of companies is vital to solve problems arising from corporate governance. The creation of value is the benchmark for measuring performance and the management results. Those results also measured by its effect on the stakeholders' organization environment. Stakeholders are individuals or entities groups, which can influence the objectives of the company, and therefore, that these goals affect them also as individuals (Gil-Lafuente and Barcellos, 2010; Freeman and Reed, 1983, p. 89). Managers have to identify these groups in their environment. Groups that influence their management style through legitimate claims, such a, moral, legal or property issues (Mitchell, et al. 1997, p. 882), and exercise power and legitimacy over decisions taken in the organization. Therefore, value creation is no longer just for shareholders, also affects the stakeholders.

This steering behaviour affects businesses, individuals and the entire organization. More and more studies collect the interest on corporate responsibility over ethics codes and social responsibility. Those are ethical and moral principles that arise in the corporate environment, committed to non-economic values, based not only on legal aspects (Clegg, 2007, p. 118).

Business ethics also affects the company valuation. Business ethics is reflected in the stock market (Choi, 2007, p. 451), as there is a positive correlation between corporate social responsibility and the valuation made of the company.

A valuation could be used for different purposes in business management. It is essential in the buying and selling of companies, assets or shares. The valuation could also be used for the quantification of value creation by managers. It identifies and ranks value drivers responsible for the creation or destruction of company value. From a strategic and planning perspective, it helps to identify the products, processes, lines of business, customers and / or countries which the organization should maintain, enhance or leave. It helps, to make decisions about staying in business and about sell, merge, grow, or buy other companies. With the valuation we measure the impact of the company's strategic decisions. Changes in company structure also affect the value of the shares (Damodaran, et al. 2005, p. 24).

Some discussions between academic and business consultants subsist about the appropriateness of valuation methods that should be applied. Perhaps the best known index facing the academic community with managers is the Economic Value Added (EVATM) of Stern Steward (Froud et al. 2000, p. 82). In general, despite being an indicator of company value widely used, it also provides no relevant information beyond that we get the financial statements of the company (Biddle, et al. 1997, p. 331-332) neither a method to cover uncertainty.

Traditional methods of valuation of companies couldn't explain why diversification destroys shareholder value. The added value of the separate parts in most of the cases is greater than the market value of the entire diversified firm (Martin and

Sayrak, 2003, p. 52-53). In the literature there isn't explanation about why some companies succeed with diversification and its value increases more than the sum of the parts separately.

The academic contributions, with examples of industries, countries, or applications to specific methods, conclude that the empirical evidence is not complete, and that in any case, the model isn't explained in its fullness. There are always things to keep in mind that could be covered with fuzzy mathematics and provide a beneficial perspective to create value for the organization.

The main limitations found in the methodology of valuation of companies are that those valuations are based on calculations between quantitative indicators. But, since it has been explained until now, they do exist infinity of qualitative drivers that influence in the real value, and which they are not included in the valuation methods. Fuzzy mathematics would allow us to be able to include these qualitative drivers in the calculations, thus to try to determine a value of the company more fit to its real value.

4 Fuzzy Methodology in the Assessment of Organizations

In order to see how fuzzy methodology could influence in improving the value of a company calculation, it is necessary to update the contributions of this methodology to the management of organizations, from its beginnings.

Research in the field of uncertainty has been an issue for many decades. It has attracted growing interest because it is a fundamental and recurrent concept in science. Since the scope of the fuzziness which is part of this project can be highlighted as a key date 1965 with the first publication of Lotfi A. Zadeh, "Fuzzy sets". The initial idea was a new logic that allowed the binary logic (which can only take value 0 or 1) switch into a multivalent logic (in which the variables can take any result between 0 and 1 of the characteristic function of permanence). Against the Aristotelian principle of excluded third, prevailing in the modelazing process for over 2000 years, won in 1996 the so-called "Principle of Simultaneity Gradual" (Gil Aluja, 1996), in which a proposal can be both true or false, condition for assigning a value to a truth and a value to a falsehood. Thus, mathematics was achieved more complete information that would represent a more appropriate way to the complex reality in which we live. From a business context and taking as support the concept of confidence interval (Gil-Lafuente, 2005), is possible represent inaccurate information as a more general form allowing to consider all possible scenarios, from optimistic position to more pessimistic one, in each of the considered variables. This has been very useful, especially to manage expectations, in particular ex-ante estimations, because events are usually influenced by subjective and uncertain elements.

In the sixties, these ideas were not readily accepted by the scientific community and not until the late seventies that these tools were gaining weight in the scientific community. Among the pioneering works, it includes the book "Introduction to the theory of fuzzy subsets" of Arnold Kaufmann. This book, published in 1975, is the first written on issues of uncertainty-fuzzy in the world (cited over 1,000 times in the ISI). Throughout the eighties, these mathematical approaches experienced a great development extending to almost all known major disciplines such as engineering, biology, physics, medicine and economics. And in the nineties, this growth continued in geometrical progression to be able to be seen as a revolution. At present, it is almost incalculable the number of annual publications that are published on these topics. For orientation, we can say that in the ISI Web of Science in recent years have been published over 3,000 articles per year with the word fuzzy. The total number of articles in which the word fuzzy appears is more than 30,000.

According to the exposed until now, may be interesting to develop a review of the most important scientific contributions that have addressed the issue of market value of organizations with fuzzy numbers. The study is a contribution to the scientific community for its novelty. Not work has been found that relates company valuation and fuzzy mathematics.

To prepare the review we proceed to discard those papers that analyze the uncertainty as a variable to determine the risk facing the company. We're focused on studies published in the ISI Web of Science that relate the market value of the company with the fuzzy methodology. The goal is to know the depth that techniques for the treatment of uncertainty has been discussed or applied to the valuation of companies.

5 Analysis of the Research

The selection of the sources consulted for this work became cradled in existing publications in the ISI Web of Knowledge, without considering other possible publications that may exist outside their surroundings. It was considered that we had to focus the study on publications of recognized academic quality, reason why we excluded other data bases. We know that this decision may forget information from any other source valuable for the study. We recommend in further studies to increase the number of data sources outside the ISI Web of Knowledge. Nevertheless, we think that the journals including in this research, are recognized by their greater index of quality, reason why the work that we present, is based on literature with greater academic recognition.

In order to prepare the selection of articles object of the study, we have considered only articles that study valuation of companies through fuzzy mathematics. Also we have included those papers that included some associated variables to the valuation process. It even features those who determine the stock market value, but were discarded those methods or models who only were looking to predict a stock index value, since they moved away of the objective of this present work.

The searching criteria were done associating the concepts of the valuation of companies with the fuzzy methodology. We search for papers that related "company value", "market value", "valuation process", "valuation methods" and "shares value", with "fuzzy methodology" or "fuzzy logic".

The total number of papers selected, that fulfilled the characteristics and objectives of the present work, were 65. Those were the articles that we used to analyze with more detail.

The analysis process was developed under four parameters. First we created a chronological evolution of the past studies, in order to find if the interest of the aim of this paper is increasing. Secondly we found which Journals had published more articles relating the valuation of companies and the fuzzy methodology. Thirdly, it was important to know the most important authors in the filed. And finally, which were the most interesting subjects in the existing literature.

The scientific community's interest in studies about companies valuation combined with fuzzy methodology is increasing in recent years, as seen in Figure 1, which analyzes trends in the number of scientific publications between the publication years (1997-2011) of each of the papers. The two years with more papers published are 2009 with fifteen papers and 2011 with fourteen papers.

During the last four year (2008-2011) of the analyzed were published 68% of the selected papers, as shows table 1. It is a significant percentage of the growing importance that the business valuation has acquired among academics.

Table 1 Evolution of the number of articles published per year

Year of publication	Number of papers
2011	14
2010	7
2009	15
2008	8
2007	2
2006	3
2005	1
2004	3
2003	4
2002	3
2001	2
2000	-
1999	2
1998	-
1997	1

Source: own elaboration based on selected papers.

The Journals selected, in which there was more papers on company valuation related with fuzzy models are ordered as follow, according to the number of papers published (we have only included in the list those Journals with two or more papers):

1. Expert systems with applications (Nº papers: 17; Impact factor: 1,926; 5-years impact factor: 2,195; Total cites: 6.615).
2. International journal of innovative computing information and control (Nº papers: 5; Impact factor: 1,667; 5-years impact factor: 7,797; Total cites: 1.936).
3. European journal of operational research (Nº papers: 2; Impact factor: 2,159; 5-years impact factor: 2,513; Total cites: 21.307).
4. Fuzzy sets and systems (Nº papers: 2; Impact factor: 1,875; 5-years impact factor: 2,250; Total cites: 9.879).
5. IEEE Transactions on systems man and cybernetics part C applications and reviews (Nº papers: 2; Impact factor: 2,105; 5-years impact factor: 2,132; Total cites: 1.390).
6. Information sciences (Nº papers: 2; Impact factor: 2,836; 5-years impact factor: 3,009; Total cites: 6.719).
7. International journal of intelligent systems (Nº papers: 2; Impact factor: 1,331; 5-years impact factor: 1,256; Total cites: 1.087).
8. Physica a-statistical mechanics and its applications (Nº papers: 2; Impact factor: 1,522; 5-years impact factor: 1,467; Total cites: 13.244).

The most important journals of the ISI Web of Scince for our selected papers are listed above. The journal Expert Systems with Applications highlights with seventeen published articles. The International Journal of Innovative Computing Information and Control, has five published articles. Not listed in the table, not being journals, but have been taken into account in the study: Lecture notes in artificial intelligence with six items and Lecture Notes in Computer Science with two articles. In the table we include impact factor in ISI Web of Science for each journal, the 5-years impact factor and total journal citations.

By impact factor journal highlights the Information Sciences. By number of citations the most prominent is the European Journal of Operational Research, with over 21,000 citations.

The authors who have contributed to research on value of the company through a fuzzy methodology are shown in Table 2, which includes authors with two or more published papers. There are only three authors with four or more papers: Chang Pei Chann, Chen Tai Liang y Cheng Ching Hsue, the latter with five publications is the first in the list.

Most authors focused their research on the value of shares in combination with fuzzy mathematical techniques, only assessing price trends, so we can't properly say that was valuation methods. Although through value of shares we could determine the market value of the company.

The results in Table 2 are only a guideline. The significance of the author may also come from other perspectives than the number of articles published in Journals of the ISI Web of Science.

For further analyses, we have refined the list of studied papers. Two thirds of the papers (44 papers), are held only on the value of the shares. Excluding these items, we have twenty-one papers on other issues related with valuation of organizations listed in Table 3.

Table 2 Relevant Authors - value of the company through a fuzzy methodology

Author	Number of papers
Cheng Ching Hsue	5
Chang Pei Chann	4
Chen Tai Liang	4
Fan Chin Yuan	3
Wang Yf	3
Wei Liang Ying	3
Zmeskal Z	3
Abraham A	2
Atsalakis George S	2
Hung Jui Chung	2
Jalili Kharaajoo M	2
Liu Chen Hao	2
Tolga A Cagri	2

Source: own elaboration based on selected papers.

Table 3 Selected articles - combine fuzzy mathematics and business valuation

Topics	Number of papers
Real options	6
Company value	3
Investment and assets	3
Forecast	2
Acquisitions	1
Comparison with probability	1
Earnings per share	1
Equity value	1
Set theory	1
Shareholders value	1
Transaction value	1
Total	21

Source: own elaboration based on selected papers.

Among those twenty-one papers, the ones with real option topic are not a proper business valuation model, but nevertheless, have been included in the study because give us a kind of valuation of the organization. This has been the most studied topic. Of the remainder, only three papers relate fuzzy mathematics as a tool to evaluate the company value. The other papers, study a particular part of the business valuation process, some of the involved drivers and others implications of the valuation in itself. So we can conclude this is an area yet to be explored, so scientific community could contribute much to explore.

The objective of this paper was to find in the existing literature, the application of fuzzy methodology in the companies valuation process, to determine the best valuation method to be used. However, from this work we could appreciate that not in-deep study has been done that developed a complete method of valuation of companies through the fuzzy methodology. The three papers which have approached more to the analysis of company value do not developed any concrete methodology. The three papers are:

1. Magni CA; Malagoli S; Mastroleo G, "An alternative approach to firms' evaluation: Expert systems and fuzzy logic". Source: International journal of information technology & decision making. Volume: 5; Issue: 1; Pages: 195-225; DOI: 10.1142/S0219622006001812. Published: Mar 2006 (Times Cited: 3 «from Web of Science»).
2. Zmeskal Z, "Application of the fuzzy-stochastic methodology to appraising the firm value as a European call option". Source: European journal of operational research. Volume: 135; Issue: 2; Pages: 303-310; DOI: 10.1016/S0377-2217(01)00042-X. Published: Dec 1 2001 (Times Cited: 30 «from Web of Science»).
3. Zmeskal Z, "Fuzzy-stochastic estimation of a firm value as a call option". Source: Finance a uver". Volume: 49; Issue: 3; Pages: 168-175. Published: 1999 (Times Cited: 0 «from Web of Science»).

We remark that Zmeskal, Z., with two of the three papers, becomes the author with more articles on valuation of companies related with fuzzy methodology. And he is one of the first to develop this new line of study in the academic community. Unfortunately, his work was based also on call options as an important part of his research. That is why we could not use it as a seminal author for future researchers in the field.

6 Final Thoughts: Conclusions and Future Research

From the previous analysis that has been explained in this work, we have not found clear and developed academic evidences of the application of fuzzy methodology to the study of valuation of companies. According to the review made, there are sufficient missing qualitative variables and drivers that should be included in the value of a company calculation. It could be necessary to develop a future line of study in a branch as important as the determination of the value of a company.

The interest of the scientific community in the object of this study has been increasing in the last years. The application of fuzzy mathematics to business valuation, we believe, enable a breakthrough for the scientific community and society in general, as new methodologies are working to improve the approaches that address the value of an organization. The fuzzy mathematics amplified the variables and fields to research on it so far, because the nature and characteristics of those studies on valuation of a company.

The current research is directed towards determining the most appropriate tools for the treatment of various issues affecting the determination of the value of the company, from models for the treatment of non-numerical mathematics of uncertainty, through related algorithms, mapping, grouping and sorting (Gil-Aluja, 1999) for the treatment of subjective content variables without the possibility of having a deterministic quantification and scope stochastic-probabilistic form (that is to say, in which it is not possible to fulfil the axiomatic of Barel- Kolmogorov), to the instruments that allow including objective data susceptible to be given a numerical assignation or hybridization process.

References

Adimando, C., Butler, R., Malley, S.: Stern Steward EVATM Round Table. Journal of Applied Corporate Finance 7(2), 46–70 (1994), doi:10.1111/j.1745-6622.1994.tb00405.x

Barberis, N., Huang, M., Santos, T.: Prospect theory and asset prices. The Quarterly Journal of Economics 116(1), 1–53 (2001), doi:10.1162/003355301556310

Beaver, W., Ryan, S.: Biases and lags in book value and their effects on the ability of the book-to-market ratio to predict book return on equity. Journal of Accounting Research 38(1), 127–148 (2000), doi:10.2307/2672925

Bhojraj, S., Lee, C.: Who is my peer? A valuation-based approach to the selection of comparable firms. Journal of Accounting Research 40(2), 407–439 (2002), doi:10.1111/1475-679X.00054

Bhojraj, S., Lee, C.: Who is my peer? A valuation-based approach to the selection of comparable firms. Journal of Accounting Research 40(2), 407–439 (2002), doi:10.1111/1475-679X.00054

Biddle, G., Bowen, R., Wallace, J.: Does EVA (R) beat earnings? evidence on associations with stock returns and firm values. Journal of Accounting Economics 24(3), 301–336 (1997), doi:10.1016/S0165-4101(98)00010-X

Carlsson, C., Fuller, R.: A fuzzy approach to real option valuation. Fuzzy Sets and Systems 139(2), 297–312 (2003), doi:10.1016/S0165-0114(02)00591-2

Choi, T., Jung, J.: Ethical commitment, financial performance, and valuation: An empirical investigation of korean companies. Journal of Business Ethics 81(2), 447–463 (2008), doi:10.1007/s10551-007-9506-1

Clegg, S., Kornberger, M., Rhodes, C.: Business ethics as practice. British Journal of Management 18(2), 107–122 (2007), doi:10.1111/j.1467-8551.2006.00493.x

Copeland, T., Koller, T., Murrin, J.: Valuation: Measuring and managing the value of companies, 2nd edn., Makinsey and Company, United States of America (2000)

Damodaran, A., John, K., Liu, C.: What motivates managers? evidence from organizational form changes. Journal of Corporate Finance 12(1), 1–26 (2005), doi:10.1016/j.jcorpfin.2004.03.001

Fernández, P.: Valoración de empresas. In: como medir y gestionar la creación de valor (Tercera edición ed.), Gestión, España (2005)

Frederickson, Miller, J.: The effects of pro forma earnings disclosures on analysts' and nonprofessional investors' equity valuation judgments. The Accounting Review 79(3), 667–686 (2004), doi:10.2308/accr.2004.79.3.667

Eduard, F.R., Reed David, L.: Stockholders and stakeholders: A new perspective on corporate governance. California Management Review XXV(3), 88 (1983)

Froud, J., Haslam, C., Johal, S., Williams, K.: Shareholder value and financialization: Consultancy promises, management moves. Economy and Society 29(1), 80–110 (2000)

Gil Aluja, J.: Lances y desventuras del nuevo paradigma de la teoría de la decisión. In: Proceedings Del III Congreso SIGEF, Buenos Aires (Noviembre 1996)

Gil Aluja, J.: Elements for a theory of decision in uncertainty. Kluwer Academic Publishers, The Netherlands (1999)

Gil-Lafuente, A.M.: Fuzzy logic in financial analysis. Springer (2005)

Gil-Lafuente, A.M., Barcellos, L.: The expertons method applied in the dialogue with stakeholders. In: The 2nd International Conference on Computer Supported Education (CSEDU), Valencia-Spain (April 7, 2010)

Gonzalez Jimenez, L., Blanco Pascual, L.: Enterprise valuation with track-record ratios and rates of change. European Journal of Finance 16(1), 57–78 (2010), doi:10.1080/13518470902853343

Hillman, A., Keim, G.: Shareholder value, stakeholder management, and social issues: What's the bottom line? Strategic Management Journal 22(2), 125–139 (2001), doi:10.1002/1097-0266(200101)22:2<125::AID-SMJ150>3.0.CO 2-H

Ho, S., Liao, S.: A fuzzy real option approach for investment project valuation. Expert Systems with Applications 38(12), 15296–15302 (2011), doi:10.1016/j.eswa.2011.06.010

Hrvol'ova, B., Markova, J., Nincak, L.: Modern methods of valuation of shares. Ekonomický časopis 59(2), 148–162 (2011)

Ionita, I., Stoica, M.: A new approach method of company valuation. Romanian Journal of Economic Forecasting 10(1), 115–122 (2009)

Johnson, L., Neave, E., Pazderka, B.: Knowledge, innovation and share value. International Journal of Management Reviews 4(2), 101–134 (2002), doi:10.1111/1468-2370.00080

Joos, P., Plesko, G.: Valuing loss firms. The Accounting Review 80(3), 847–870 (2005), doi:10.2308/accr.2005.80.3.847

Kaplan, S., Ruback, R.: The valuation of cash flow forecast – An empirical analysis. The Journal of Finance 50(4), 1059–1093 (1995), doi:10.2307/2329344

Kaufmann, A.: Introduction to the theory of fuzzy subsets (D. L. Swanson Trans.). Academic Press, New York (1975)

Laeven, L., Levine, R.: Complex ownership structures and corporate valuations. The Review of Financial Studies 21(2), 579–604 (2008), doi:10.1093/rfs

Liu, J., Nissim, D., Thomas, J.: Is cash flow king in valuations? Financial Analysts Journal 63(2), 56–68 (2007), doi:10.2469/faj.v63.n2.4522

Magni, C., Malagoli, S., Mastroleo, G.: An alternative approach to firms' evaluation: Expert systems and fuzzy logic. International Journal of Information Technology Decision Making 5(1), 195–225 (2006), doi:10.1142/S0219622006001812

Magoc, T., Modave, F.: The optimality of non-additive approaches for portfolio selection. Expert Systems with Applications 38(10), 12967–12973 (2011), doi:10.1016/j.eswa.2011.04.093

Martin, J., Sayrak, A.: Corporate diversification and shareholder value: A survey of recent literature. Journal of Corporate Finance 9(1), 37–57 (2003), doi:10.1016/S0929-1199(01)00053-0

Maury, B., Pajuste, A.: Multiple large shareholders and firm value. Journal of Banking Finance 29(7), 1813–1834 (2005), doi:10.1016/j.jbankfin.2004.07.002

Miller, D., Lester, R., Cannella, A.: Are family firms really superior performers? Journal of Corporate Finance 13(5), 829–858 (2007), doi:10.1016/j.jcorpfin.2007.03.004

Mitchell, R., Agle, B., Wood, D.: Toward a theory of stakeholder identification and salience: Defining the principle of who and what really counts. The Academy of Management Review 22(4), 853–886 (1997), doi:10.2307/259247

Rhodes Kropf, M., Viswanathan, S.: Market valuation and merger waves. The Journal of Finance 59(6), 2685–2718 (2004), doi:10.1111/j.1540-6261.2004.00713.x

Smit, H.T.J., Trigeorgis, L.: Real options and games: Competition, alliances and other applications of valuation and strategy. Review of Financial Economics 15(2), 95–112 (2006), doi:10.1016/j.rfe.2005.12.001

Ucal, I., Kahraman, C.: Fuzzy real option valuation for oil investments. Ūkio Technologinis Ir Ekonominis Vystymas 15(4), 646–669 (2009), doi:10.3846/1392-8619.2009.15.646-669

Wang, Y.: Predicting stock price using fuzzy grey prediction system. Expert Systems with Applications 22(1), 33–38 (2002), doi:10.1016/S0957-4174(01)00047-1

Wyatt, A.: What financial and non-financial information on intangibles is value-relevant? A review of the evidence. Accounting and Business Research 38(3), 217–256 (2008), doi:10.1080/00014788.2008.9663336

Yao, J., Chen, M., Lin, H.: Valuation by using a fuzzy discounted cash flow model. Expert Systems with Applications 28(2), 209–222 (2005), doi:10.1016/j.eswa.2004.10.003

Yermack, D.: Higher market valuation of companies with a small board of directors. Journal of Financial Economics 40(2), 185–211 (1996), doi:10.1016/0304-405X(95)00844-5

Zadeh, L.: Fuzzy sets. Information and Control 8(3), 338 (1965), doi:10.1016/S0019-9958(65)90241-X

Description of a Straightforward Strategy to Invest: An Experiment in the Spanish Stock Market

Eugenio M. Fedriani, Jesús López, Ignacio Moreno, and Jesús Trujillo

Universidad Pablo de Olavide, de Sevilla,
Ctra. de Utrera, km 1, 41013 Sevilla, Spain
{efedmar,imorgab}@upo.es, jesusferreras@mba-camara.com,
jesus@nebrimatica.com

Abstract. In this paper we reintroduce the topic of arbitrage, explaining an easy method to obtain reasonable results for every possible behavior in the stock market. It is based on purchases at low prices and sales at higher prices, and it is closely related to volatility. Obviously, our technique has its limitations, but we study the hypotheses needed to assure a good performance. Our aim is to contribute to the better understanding of the topic by the researchers, giving an idea of the way how algorithmic strategies work in practice. Finally, we exemplify and test the method through the use of a real data set from some shares of Madrid Stock Exchange, for the time period 1994-2011, and express some concluding remarks.

Keywords: Algorithmic trading, Efficient Market Hypothesis, arbitrage, Madrid Stock Exchange, IBEX 35.

1 Introduction

Several phenomena in the real word are governed by specific, well-known physical laws; these are the cause-and-effect systems, where causal determinism establishes that any effect is determined by prior states of the system. However, others are elusive and seem to be playing with researchers along the centuries. According to the Efficient Market Hypothesis (EMH), stock market fluctuations are considered a priori as stochastic processes. In these stochastic (or random) processes, there is always some indeterminacy, possibly given by probability distributions. The statistical models used to analyze or predict the behavior of the stock market assume that it is impossible to know the situation of the variables in the future. But determinism does not require the perfect prediction either, since the system can be too complex (or poorly known) to be modeled. The degree to which a prediction can be made is usually called predictability, and this concept depends on the theoretical state-of-the-art, on the complexity of the system, on the amount and accuracy of the available information, etc.

A.M. Gil-Lafuente et al. (Eds.): Soft Comput. in Manag. and Bus. Econ., STUDFUZZ 287, pp. 191–200.
springerlink.com

Sometimes, there is no need in a perfect determination of the process to take some advantage against competitors. In this sense, if any agent has access to partial but reliable predictions, this fact may generate the possibility of obtaining profits without risk. This situation can be referred to as arbitrage, and we will use this concept later.

This paper has the following parts, after this brief introduction. Firstly, we present different possibilities of going against the EMH. Secondly, we show the theoretical explanation of the proposed method. Then we apply the technique to a real data set from Spanish stock market (five shares from IBEX 35). Finally, we provide the reader with some discussion and conclusions.

2 Some Studies on Behavioral Finance

Although it is a fascinating topic, we are not discussing the causes for the possible patterns or trends detected in the stock market. We only assume here that stock market behavior is difficult to be predicted. We cannot say "impossible", because lots of practitioners hold that there exists the possibility of finding any kind of pattern. For instance, the "technical analysis" uses graphical and statistical techniques, applied to information reflected on past stock prices, when trying to forecast price trends, and lots of authors support the benefits derived from the use of related techniques; some examples are Grech and Pamula (2008), Green and Fielitz (1997), Mandelbrot and Hudson (2004), and Peters (1991). Others think that any correct prediction corresponds to a singularity, a coincidence due to some unknown reason included in the market situation.

In opposition to the possibility of predicting the stock market behavior, we find the EMH, which has counted on the support of a large proportion of the academia for the past thirty years. An efficient capital market is one in which share prices always fully reflect available information (Fama, 1970). And subsequently stock prices are fairly priced, considering this available information. Thus, given that stock prices are fairly priced, a trading strategy based on selling a stock whenever its price rises and opposite, buying this stock whenever its price falls, would not generate positive profits because a recent price rise does not imply that the stock is overvalued, the same way that a price fall does not imply that the stock is undervalued.

Moreover, the EMH implies that, as long as stock prices fully incorporate all available information, the future price variations are random and caused by new unexpected information which is incorporated to the price. Consequently, according to the EMH, there is no point in developing trading strategies based on past information in order to get higher returns than the return associated to the same risk.

There are three conditions, any of which will lead to market efficiency: (a) rationality; (b) independent deviations from rationality, and (c) arbitrage (Shleifer, 2000). For a market to be efficient there is only needed to hold one of the three conditions. Much of the criticism to the FMH points to the irrationality of the investors, but efficient market economists defend themselves arguing that even though there may be some irrational investors who misprice the assets, this will led the possibility of arbitrage by rational investors who will correct the situation

(Friedman, 1953). On the other hand, behavioral economists argue that none of the three conditions, including arbitrage, hold in capital markets (Hillier *et al.*, 2010). This is what they call the limits to arbitrage. Indeed, behavioral finance rests on limits to arbitrage and investor psychology (Schlleifer and Summers, 1990). According to Brealey, Myers, and Allen (2009, p. 369), "arbitrage means an investment strategy that guarantees superior returns without any risk"; therefore, according to the theory, arbitrage is possible whenever the stock prices are above or below their efficient prices. If prices are higher than their fundamental value, then, the arbitrageur would sell the overpriced securities pushing their price down; while if the securities are underpriced, the arbitrageur would buy these securities pushing the price up. Thus, it seems that arbitrage is a strong tool to keep markets efficient, even though there may be irrational investors, as long as there are no limits to such arbitrage. However, according to behavioral economists (see Barberis and Thaler, 2003, for a comprehensive survey), one cannot disregard these limits to arbitrage, as it is clearly noted by the following statement attributed to J.M. Keynes: "Markets can remain irrational far longer than you or I can remain solvent". Therefore, arbitrage, when possible, only works whenever the period of time needed to take advantage of the market irrationality is not very long and near-term risk can be avoided. However, if rational arbitrageurs are few compared to irrational investors, their arbitrage strategy might not be enough to move prices back to their fair value, causing them a large loss. Moreover, given that possibility, in order to avoid risk, the positions taken by arbitrageurs would be too small to bring prices back to its efficient price. After all, these arbitrage opportunities, which by definition should be riskless, can be in fact very risky and costly, even when the asset is highly mispriced.

If markets are not efficient, the next logical step is the design of an appropriate procedure to invest in order to get positive abnormal returns. Indeed, the existence of a successful algorithm has a crucial importance. One of the main consequences of this idea is the possibility of using computers to invest in the stock market. In fact, the algorithmic trading (also called "algo", "automated", "black-box" or "robo" trading) may use electronic platforms to enter trade orderings without human intervention. In these cases, the algorithm can decide about timing, price, or quantity of each order. Although the literature on this subject comprises the results reached for different algorithms (Aite Group, 2006), there is no a proper description of the basis of these methods (as this paper does), at least to the authors' knowledge. Besides, according to some experts, the involvement of automatic programs to drive stock trades is spreading very fast: in 2006 they meant approximately one third of the total equity trading volume in Europe and North America (Aite Group, 2009); nowadays, these numbers have probably doubled up.

3 Trading Proposal

Sellers and buyers in the stock market can be small individuals but also large hedge fund traders. Each one has its own strategy to invest. Some of them have a more scientific appearance, and others seem to be more exotic (or even exoteric). As we mentioned earlier, computers play an increasingly important role, especially for the so-called “algorithmic trading”. Some well-known algorithmic strategies are: delta neutral, mean reversion, pair trading, scalping trading, transaction cost reduction techniques (as “iceberging” or “stealth”), trend following, and other arbitrage. Our simple method is closer to the mean reversion strategy. This theory holds that too high or too low prices are temporary, and every stock price will tend to an average over time. Once detected this average price, if the current market price is less than the average, the stock is interesting for purchase (because its price should rise); on the contrary, when the current market price is above the average, it is expected to fall, hence it is attractive to be sold. Summarizing, deviations from the average price are expected to revert to such average and this fact could be usable.

We can introduce our strategy even from another point of view. If the price of the stocks rises, then our proposal is inspired in something known as “covering a short position”. In this situation, speculators borrow stock that is sold hoping for its price to fall. On the contrary, if the stocks fall, our idea is more similar to “margin buying”, when the traders borrow money to buy a stock hoping that it will rise. The aim of our procedure is not a fast enrichment, but a way to get reasonable results no matter what the performance of the market is.

Now we are introducing the technique by means of an example. First, let us suppose that the value of shares (understood as the unit of account for stocks and other assets) only take discrete values that can be labeled by positive integers (1, 2, 3… 98, 99, 100, 101, 102, 103…), but we can divide infinitely many times each share to buy or sell it. This simplification will allow an easy description of the basis, but it is not as coarse as its appearance makes the impression to be. We introduce this first example when we have, let us say, 1,000 shares of a stock. If the price goes down, we buy a fixed amount; if it goes up, we sell the same amount. We also suppose that trading commissions and, later, fees are inexistent. We will see the complete model in a second step.

Under the preliminary assumptions of the previous paragraph, let us call 100 the initial price, p_0, of our specific share. If this value goes down, then $p_1 = 99$ and we buy one monetary unit of the share. If the share reverts to the previous value, $p_2 = 100$, we consequently sell one unit. Now we have the same initial price (and value, $p_0 = 100$) for the shares, but we received a dividend from these two stock market operations: the purchase price was 99 and the sale gave us 100 monetary units per share. On the other hand, if the share first goes up, $p_1 = 101$ and we sell one unit; in this case, if the share price returns to $p_2 = 100$, we buy one unit, and this process also provides us with a benefit of 1 monetary unit per share exchanged. Obviously, these are only two possibilities. Positive or up trends (usually referred to as ‘bull markets’) or negative or down trends (the so-called bear markets) may last for long time periods.

As another example, let us suppose $p_1 = 99$, $p_2 = 98$, $p_3 = 99$, and $p_4 = 100$. Here we buy in p_1 and p_2, and sell in p_3 and p_4. Orders in p_1 and p_3 cancel each other, so we gain the difference between p_2 and p_4. Now, let us consider the following three sequences of prices: $p_1 = 101$, $p_2 = 102$, $p_3 = 101$, $p_4 = 100$; $p_1 = 99$, $p_2 = 98$, $p_3 = 99$, $p_4 = 98$, $p_5 = 99$, $p_6 = 100$; $p_1 = 101$, $p_2 = 102$, $p_3 = 101$, $p_4 = 102$, $p_5 = 101$, $p_6 = 100$. Each price change implies a new trade ordering. The interested reader can check the profit given by each sequence, greater in the last two cases. These results suggest that price volatility of stocks can help our system to obtain additional benefits.

We have shown how different movements of the stocks prices provide us with some gain. However, we still need an explanation of the complete technique. The first step is the description of the flowchart that can be consulted in Fig. 1. From this algorithm, we designed specific software to perform the corresponding analysis in the real situations.

The starting parameters for the software are the following: a list of daily trading of the share value on the stock market (opening, minimum, maximum, and last); the amount of initial investment (this number is the balance on the first day); the percentage of balance that will be used in *first buy*; the amount of each operation *buy shares* and *sell shares*; the minimum percentage of increase or decrease from the value of the share in the last operation to meet the *condition of buy* or the *condition of sell*; the percentage commission or fee on each operation *buy shares* and *sell shares*.

We now proceed to the description of the *first buy* step (Fig. 1). Here, the program makes a first purchase of shares using the default percentage on the balance, calculating the number of shares as follows:

$$number\_shares = \left\lfloor \frac{\dfrac{balance \cdot pct\_first\_buy}{1 + pct\_commission}}{opening\_value} \right\rfloor, \tag{1}$$

where $\lfloor\cdot\rfloor$ stands for the floor function.

Once the investment is started, the automatic system should proceed, step by step, according to the market behavior and the ideas stated above. But sometimes the user has to make a decision, and *invest again* stands for a first dilemma. When all the shares have been sold, it queries the user if the process is restarted from the balance at that time. Afterwards, we find the *condition of buy*, where the system has to decide if a purchase has to be done. When the minimum value of daily trading falls below a predetermined percentage to the value of the share in the last operation (*first buy*, *buy shares* or *sell shares*), then the condition is satisfied.

On the other hand, *increase investment* decides if one agrees with the idea of adding more money: when the predetermined amount to *buy shares* is greater than the balance, it prompts the user whether to increase investment to complete the amount.

With respect to the purchase of shares (*buy shares*), the system calculates the *buy price* as follows:

$$share\_value = \left[(1 - pct\_decrease\_value) \cdot last\_operation\_value\right], \tag{2}$$

where the square brackets stand for the round function with two decimal digits, that is, the nearest integer function applied to 100 times the result and finally divided by 100.

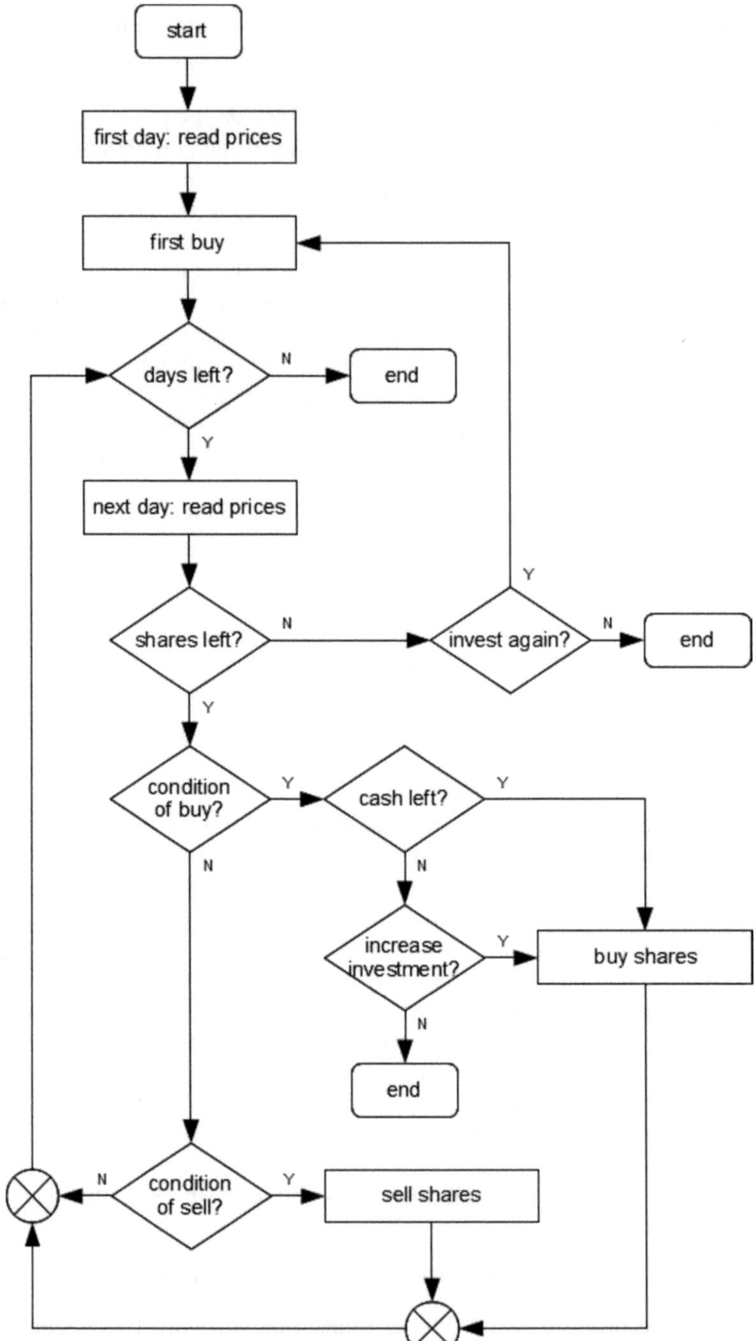

Fig. 1 A simple flowchart to represent the algorithm used in the trading proposal

Eq. (2) adds to the total the number of shares calculated as follows:

$$number\_shares = \left\lfloor \frac{buy\_value}{share\_value} \right\rfloor \tag{3}$$

This is complemented by the *condition of buy*. When the maximum value of daily trading exceeds a predetermined percentage to the value of the share in the last operation (*first buy*, *buy shares* or *sell shares*), then the condition is satisfied.

Finally, we have to explain something about the sell shares. It calculates the sell price as follows:

$$share\_value = \lceil (1 - pct\_increase\_value) \cdot last\_operation\_value \rceil \tag{4}$$

If it is big enough, it subtracts from the total the number of shares calculated as follows:

$$number\_shares = \left\lfloor \frac{sell\_value}{share\_value} \right\rfloor \tag{5}$$

On the contrary, if the total number of shares is less than that number, it sells all the shares available at that moment.

In this general algorithm we have faced several extra difficulties with respect to the preliminary example. The first one could be the existence of too high or low trends, not having enough shares which guarantee the payments to the seller or enough money to buy the planned shares. Under the assumption of considering stocks which are not always below the initial level, this can be easily solved. All we have to do is fixing a maximum amount of money to be invested or a maximum amount of shares to be sold.

The second problem is the different scales for each stock. Although this makes the description more complicated, we have used a percentage of the value of the share to determine the orderings. Hence, we buy when the gain is, say, 5% of the share. As a consequence, we usually do not have the same prices for purchases and sales. Let us see: if the initial price is 100 monetary units, and it goes down 5% and then it goes up 5%, we first buy at 95 units and later we sell at 99.75 units. Note that we obtain the same price (99.75) if the share first goes up and then goes down. As the time series is longer, the prices may differ a lot to the described above. However, the reasoning remains the same.

The third difficulty, quite related to the previous one, is the consideration of fees or commissions, which can lower the benefits (too many orderings will increase the costs). Its solution lies in the consideration of percentages that are big enough to compensate the fees. The exact determination of the minimum percentage depends on the value of each share. But one have also to bear in mind that too high percentages will decrease the number of orderings, what goes against our financial objectives. Unfortunately, the optimum percentage can only be calculated *a posteriori*, so the investor has to choose the percentage according to personal preferences or past behavior observed for the stock.

Finally, another issue that deserves a comment is the nature of units considered in the orderings. And can we divide each share infinitely many times? Our proposal consists of using approximations, so we should warn that if the percentage or the amount per sale are too low, our approximations may corrupt the results. We confronted the discretization of the problem; although this made the algorithm more complex, we think the results are adequate.

So we have seen that all the troubles (even the presence of a long bear market) can be overcome by a determined waiting time. In the following section, we introduce the experiment carried out in the Spanish Stock Market.

4 Illustration with Specific Data from IBEX 35

We have used our specific software to carry out multiple comparative profitability analysis for 46 time periods (of at least one year each) selected at random and running between January, 1st, 1994 and March, 21st, 2011. The length of 16 of them was less than 2 years, and the largest one lasted almost 17 years. We have developed only these few experiments, to be able to analyze each one independently.

In the different scenarios, we have taken different companies' shares, all of them belonging to the IBEX 35. The companies chosen for the aforementioned analysis were the ones with the most liquid stocks traded in Madrid Stock Exchange: BBVA, Banco Popular, Banco Santander, Mapfre, and Telefónica. They represent more than a half of the total volume of the IBEX 35. We have focused on periods of different length in order to compare, on the one hand, market profits in such periods and, on the other, profits arisen from our automatic system.

We supposed that the surplus taxes have been paid at the end of each analysis period. As benefits and losses compensate through the period, this calculation is similar to the cumulative annual computation of taxes, so it does not affect the final comparison between trading strategies. On the other side, the trading costs considered are the same faced by one of the authors of the paper in his real trading operations.

For each period of time, we have performed two types of analysis which directly depended on the amount invested at the beginning of the period. In the first case that amount reached €100,000. In the second, it went up to €50,000. In both situations €10,000 were bought and subsequently sold every time the price increased or decreased a 5%.

With an initial investment of €100,000, the behavior of the market for the 46 periods rises up to 14% average annual profit (for the 5 selected shares and the considered scenarios). When dealing with the automatic system and the second type of investment, the average annual profit ascended to 13.95%. It looks similar to the general one, but its standard deviation was higher here. However, in the first type of automatic trading, with a smaller initial investment, the results obtained for the 46 periods rise up to 21.69%, and the market had 11.13% average annual profit. This means that profitability doubles up when dealing with the automatic system and initial investments of €50,000.

When analyzing each case independently, we observe that our automatic trading system beats the market most of the times (see Table 1). With respect to the two types of investments analyzed (modifying the initial investment), the second type is more profitable in 71.7% of scenarios, worse in 10.9%, and similar in the rest of cases.

Table 1 Comparison between our trading proposal (automatic) and the market; note that "Draw" means that the profits obtained with our method are similar to the market behavior

	More profitable trading proposal		
Initial investment	*Automatic*	*Draw*	*Market*
€50,000	76.1%	15.2%	8.7%
€100,000	65.2%	10.9%	23.9%

We must note that profitability obtained by the system highly differs in the first and second cases due to the fact that all the money has already been used to buy more stock. This fact originates a higher profitability in the second case since investors may interrupt a too long bear market. Besides, we cannot forget that as the amount invested is smaller than in the first case, profitability is higher in the market lateral movements.

We would like to point out the interest which lies in our system for researchers, because it helps us see that an automatic trading system can generate higher profitability than the usual given by the market. It depends, though, on the specific variables used for that investment, either the initial amount, the fluctuation on the shares' values for each purchase and sale, or the amount spent in each one of them.

5 Conclusions

The first idea we want to emphasize is that it is possible to obtain reasonable profits through the use of very simple automatic trading techniques. In our simplified proposal, the solvency of the investor is a key factor in the process, since we cannot assure a quickly enrichment. Moreover, the highest profits in long bull or long bear markets are usually obtained when one can maintain his/her investment despite the market tendency. Of course, the solvency and patience of investors in practice are finite.

With respect to the risk assumed by the investor in our proposal, we argue that it would be at most the same as the supported by another strategy when choosing the same shares. However, to evaluate our technique, it is necessary the comparison of our results with others derived from the application of different strategies. But we have to bear in mind that the assessment of our ideas against an *ad hoc* technique or the comparison of our results with the profits of decisions taken a posteriori is not fair at all. That is, the best strategy is always unknown before the market activity.

To conclude this paper, we would like to say something about two future research lines. On the one hand, we would like to determine the influence and relevance of each variable involved in the analyzed phenomenon (initial investment, percentage to decide purchases and sales, length of the time period, trading costs, etc.); this can be performed by taking into account many more scenarios, and by using different statistical methods to detect the significance of each factor (specifically, we recommend the use of an artificial neural network). On the other hand, it may be interesting the consideration of volatility as a cause-effect variable to explain the validity of our technique. In this same sense, we have to admit that there are some papers which may be useful to give a measure of volatility; for example, those related with fractals and Hurst exponent (see Grech and Pamula, 2008, among others).

References

Aite Group: Agency brokers & algorithmic trading: Here today, gone tomorrow? Impact Note 200611211, Boston (2006)

Aite Group: New world order: The high frequency trading community and its impact on market structure. Impact Report 200902251, Boston (2009)

Barberis, N., Thaler, R.: A survey of behavioral finance. In: Constantinides, G.M., Harris, M., Stulz, R.M. (eds.) Handbook of the Economics of Finance, 1st edn., vol. 1, ch. 18, 1053–1128. Elsevier, Cambridge (2003)

Brealey, R.A., Myers, S.C., Allen, F.: Principles of corporate finance, 9th edn. McGraw-Hill/Irwin, New York (2009)

Fama, E.: Efficient capital markets: a review of theory and empirical work. Journal of Finance 25, 383–417 (1970)

Friedman, M.: The case for flexible exchange rates. In: Essays in Positive Economics, pp. 157–203. University of Chicago Press, Chicago (1953)

Grech, D., Pamula, G.: The local Hurst exponent of the financial time series in the vicinity of crashes on the Polish stock exchange market. Physica A 387, 4299–4308 (2008)

Green, M.T., Fielitz, B.D.: Long range dependence in common stock returns. Journal of Finance Economics 4, 339–349 (1997)

Hillier, D., Ross, S., Westerfield, R., Jaffe, J., Jordan, B.: Corporate finance. European edition. McGraw-Hill, Berkshire (2010)

Mandelbrot, B., Hudson, R.: The (mis)behavior of markets. A fractal view of risk, ruin and reward. Profile Books, London (2004)

Peters, E.: Chaos and order in the capital markets. John Wiley & Sons, New York (1991)

Shleifer, A.: Inefficient markets: An introduction to behavioral finance. Oxford University Press, New York (2000)

Shleifer, A., Summers, L.: The noise trader approach to finance. Journal of Economic Perspectives 4, 19–33 (1990)

Determinants of Households' Risk

Francisco J. Callado Muñoz and Natalia Utrero González

Economics Department, Universitat de Girona
Campus Montilivi s/n
17071 Girona Spain
{franciscojose.callado,natalia.utrero}@udg.edu

Abstract. This paper deals with risk taking attitude and behaviour of households in Spain. It analyses whether business owning households are more risk tolerant than non owners, and which household features are conditioning risk taking perceptions and their relation to risk behaviour. For that, it uses data from the 2005 Spanish Survey of Household Finance (EFF). The paper improves the definition and measurement of risk behaviour and stress the relevance of owning a business or your home on your risk characteristics. Results will help to increase the understanding of risk tolerance and behaviour.

Keywords: households, risk attitude, risk behaviour, investment.

1 Introduction

Uncertainty is an important part of everyday life. Every little economic decision implies taking risks: financial, professional, political, regulatory, environmental or reputational. In fact the attitude towards risk is one of the main variables used in economic and financial analysis. For instance, from the point of view of economic theory, being risk averse or risk neutral will end up with different predictions on agents' actions. Therefore, understanding risk attitudes will clearly help in forecasting economic behaviour (Dohmen et. al. 2006).

Risk attitudes vary across individuals. Some personal characteristics might be more related to high tolerance to risk whereas others could imply less willingness to face risky decisions. Being able to identify the factors and determinants that lie behind these differences would improve the insight into this important economic concept. However, although risk tolerance conditions decision making, risk subjective perceptions may not be the only factor affecting economic behaviour. Being aware of the differences between risk perceptions and risk taking behaviour is highly important. First, these differences may indicate that some individuals do not understand risk and would be taking more or less risk that they actually desire (Schooley and Worden, 1996). Second, the existence of these differences would imply that risk attitude is not the only factor behind risk taking. There could be other variables, such as the context in which decisions are made, that can influence risk taking (Shoemaker (1993)). For instance, the European Markets in

A.M. Gil-Lafuente et al. (Eds.): Soft Comput. in Manag. and Bus. Econ., STUDFUZZ 287, pp. 201–217.
springerlink.com

Financial Instrument Directive (MiFiD) is a good example of this dichotomy. It requires financial advisors to identify customer risk preferences and to customize their advice accordingly. Typically, the identification takes place by way of self-disclosure individual's risk attitude and also by checking previous investment decisions. Understanding the degree of agreement between self-declared risk propensity and risk portfolio decisions would make easier the development of adequate financial planning services.

The analysis of risk tolerance has a long tradition in the literature. The majority of previous works has analysed the relationship between risk attitude and other personal characteristics. Friend and Blume (1975) propose a framework to measure risk tolerance that focus on the relationship between risk attitude and wealth. Riley and Chow (1992) find that risk aversion decreases with wealth, education and age. Hanna and Chen (1997) deal also with the links between risk tolerance, wealth and investment horizon through the use of historical investment data and simulations. Jianakoplos and Bernasek (1998) relate gender differences with household holdings of risky assets. Halek and Eisenhauer (2001) analyse risk aversion examining its relationship with different demographic groups based on characteristics such as age, wealth, education etc.

Together with the analysis of risk attitudes, previous studies have also addressed the analysis of risk taking behaviour focusing specially on portfolio composition[1]. This analysis has evidenced that in developed countries, housing is a major investment in households' portfolios (Cocco, 2005). In Spain 86% are owner occupied (Bover, 2010), 61% in France (Arrondel and Lefebvre, 2001), and 64% in US (Ioannides and Rosenthal, 1994). This result is interesting since due to financial constraints, investment in housing limit financial wealth to invest in other financial assets like bonds or stocks, specially for poorer households, Cocco (2005). Similarly, business ownership may also prevent agents from having a greater relative share of stockholdings (Heaton and Lucas (2000)). Therefore house or business tenure may affect final portfolio risk and composition. Nevertheless, Jin (2011) using an improved measure of risk, finds no conclusive results with respect to their effects on the risk of investment portfolios.

However, these results on portfolio risk and composition present two weaknesses. First, the measure of risk behaviour mainly used is based on percentage of risky asset owned (Xiao et al., 2001). However, when the portfolio includes different risky assets this measure of risk behaviour is not very appropriate. It assumes that all assets share the same risk features when each asset has different risk characteristics. In other words, with this measure, a household holding a 30% of its wealth in stocks would be considered to have the same risk as a household with 30% investment in treasury bonds. This is clearly not the case. Using the standard deviation of the household portfolio is an alternative to this proportion. It allows taking into account asset risk differences and distinguishing between household portfolios with the same percentage of risky assets (but different composition). Jin (2011) makes a first improvement in developing a measure based on standard deviation of three risky assets: stocks, private business and real state. Second, the analyses are partial since they do not incorporate together house and business tenure to analyse the effects on risk.

---

[1] See Campbell (2006) for a survey of main literature results.

Accordingly, the contribution of the paper is twofold. First, it improves the definition and existing measures of risk behaviour. Following Jin (2011), the risk taking behaviour will be calculated by means of the standard deviation of household portfolio, but the measure computed includes the main risky assets available for Spanish average investor: stocks, business and home ownership, hedge and pension funds, private equity and fixed rate extending previous classification.

Second, with the improved measure of risk taking behaviour, it tries to address the following questions: (i) what are the determinants of the differences of households risk taking and behaviour, (ii) to analyse the degree of integration between the risk attitude and the explicit risk taking behaviour of households (iii) finally, following the papers of Heaton and Lucas (2000) and Cocco (2005) analyse the effects of real state and private business holdings on risk attitudes and behaviour. For that, we use data from the 2005 Spanish Survey of Household Finance (EFF).

The rest of the paper is organised as follows. Section 2 reviews the studies on previous literature on the link between risk tolerance, financial behaviour and portfolio choice. In section 3, we present the data used. Section 4 presents the results. Finally section 5 concludes

2 Risk, Households and Portfolio Choice

The literature on risk taking behaviour analyses how household make investments. In particular, theoretical models traditionally distinguish between a risk-free asset and risky assets (Guiso et. al. (2002)). Traditionally, risky assets are considered to be stockholding, however, recently real state and business ownership have also been considered risky assets (Flavin and Yamashita (2002), Cocco (2005), Yao and Zhan (2005) and Jin (2011)). This inclusion is based on the evidence that housing and private business are substitute for stockholding, since investing in owner-occupied housing as well as private business holdings reduce the percentage of investment in stockholdings (Cocco (2005), Jin (2011) and Heaton and Lucas (2000)). Portfolio final composition will depend on the impact of real state and private business on portfolio choice, and therefore on risk taking.

The concept of risk aversion is usually analysed by the attitude toward risk, the risk taking behaviour or both. The attitude toward risk refers to people's basic preferences, traits or dispositions towards risk-taking and it is considered as independent of the situation or endowments whereas the risk taking behaviour refers more to how people behave in actual circumstances. In particular, risk attitude would be the driving force behind the behaviour of agents whereas risk-taking behaviour would be a description of the risk attitude and therefore the utility of agents. In an integrated framework Schoemaker (1993) suggest that risk attitude may affect risk-taking behaviour. However, Schooley and Worden (1996) show that both measures behave in a similar way indicating that households have a clear understanding of their relative risk taking. However they do not analyse the effects of real state or private business investments on risk.

Further, there is a wide agreement in the academic literature that the socioeconomic and demographic characteristics of households also have a word to say in determining subjective risk aversion. Wealth is one of the main determinants of risk aversion (Gollier, 2001). As discussed above, it is also one of the key characteristics of being a business owner and therefore it is also related to risk attitude of entrepreneurs (Say, 1975, Marshall, 1930, and Knight, 1971). Previous empirical papers find risk-taking behaviour positively related to household wealth (Schooley and Worden (1996) and Jianakoplos and Bernasek (1998)). It has also been documented that wealthier households' portfolios are heavily skewed towards risky assets, particularly investments in their own privately held business (Carroll, 2002). To better understand how households make risky decisions we will examine the relationship between risk attitude and behaviour and wealth. Accordingly, variables that may help increase family wealth, such as family income, could also affect the level of risk tolerance. Previous evidence showed that non-investment income (Sung and Hanna 1996) and total income (Grable and Lytton 1998) were positively related to the risk-taking attitude.

Education, knowledge and age are determinants of business start and ownership and can be also related to risk attitude of self employed (Say, 1975, Marshall, 1930, and Knight, 1971). Sung and Hanna (1996) confirmed that generally, people are more willing to take risks at a younger age, and Jianakoplos and Bernasek (1998) and Cohen and Einav (2007) found that age effects on risk-taking behaviour have a U-shape. Gender is also assumed to affect risk attitudes. Women are expected to be more conservative investors than men (Wang, 1997) and there is evidence supporting this view in financial decision-making (Barsky et. al., 1997, Donkers et. al. 1999 and Hartog et. al. 2002). Recent survey data suggest that wealth holdings of single women are less risky than those of single men of equal economic status (Jianakoplos and Bernasek, 1998, Sunden and Surette, 1998). Also, when asked about their attitudes toward financial risks, women seem to report a lower risk propensity than men (Barsky et al., 1997).

Finally, previous studies showed that the number of young dependents in a household has negatively affected the proportion of risky assets held by married couples (Jianakoplos, and Bernasek 1998).

Following the above discussion, we analyse risk attitude and risk behaviour and the consistency between both measures of risk. For that, we take into account whether households hold real state and private business or both in their portfolios.

In particular, we estimate the following model:

$$RM_{it}=\beta_0+\beta_1 BUS_{it}+\beta_2 H_{it}+\beta X_{it}+u_{it}$$

Where RM is the risk measure, namely risk attitude or risk behaviour, BUS and H are dummies for being a business owner and owning your home respectively and X is the set of socioeconomic characteristics. This model allows us to disentangle the role of having home and private business investments in household risk, after controlling for socioeconomic and demographic features. Furthermore, in some runs, risk attitude is included as a regressor to test the framework proposed by Schoemaker (1993).

3 Data and Variable Definition

We use the 2005 survey on household finances (Encuesta Financiera de las Familias (EFF)) from the Banco de España, similar to other countries' as Banca d'Italia survey (Survey on Household Income and Wealth (SHIW) and the US

Board of Governors Survey of Consumer Finances (SCF). EFF is multi-imputed dataset in order to enable analysis with complete-data methods[2]. The main feature of EFF is the oversampling of wealthy households (as SCF does). This is done due to the fact that the distribution of wealth is heavily skewed and some types of assets are held only by a small fraction of the population. In this sense, this oversampling achieves not only representativeness of the population but also of aggregate wealth. Besides, it allows for the study of financial behaviour at the top of the wealth distribution.

EFF gathers information on risk perceptions, real and financial assets on household grounds. Yet, information on personal features or about the type of employment situation is provided for all household members over 16. Because there is no information on the risk perception for each household member or the assets hold individually, the unit of analysis is the household.

3.1 Variable Definition

To account for risk attitude we use the direct perception of willingness to invest in risky assets. This is the response to the question "Which of the statements on this page comes closest to the amount of financial risk that you and your (spouse/partner) are willing to take when you save or make an investment? (1) Take *substantial* financial risks expecting to earn substantial returns; (2) Take *above average* financial risks expecting to earn above average returns; (3) Take *average* financial risks expecting to earn average returns; (4) *Not willing* to take any financial risks.". We have reversed the scale, so that higher values indicate higher tolerance to risk, making results easier to interpret.

The second risk measure is risk taking behaviour. In previous studies this variable is measured by the share of risky assets to total assets. However, as discussed above, this is not very appropriate since risky assets can have very different characteristics. Instead, in this paper we will calculate the standard deviation of the household portfolio as the measure of risk taking. In order to compute the portfolio risk we need to estimate the variance-covariance matrix of the risky assets. For that, we consider six different categories of risky assets: real state, public equity, private business, mutual funds, pension funds and bonds extending previous classification of risky assets (Guiso et.al. (2002), Campbell (2006) and Jin (2011)). In particular we include mutual funds, pension funds and bonds as additional categories of risky assets. Therefore, we improve the computation of risk behaviour in two ways, first by calculating the standard deviation of investment and second through the distinction of three additional risky assets. For that several indicators of the variance of each kind of asset are used. All indicators are computed in quarterly data. The House Price Index from the INE is used for real state investments. Variations in public equity are included by means of the Ibex35 from the Spanish stock exchange. As a proxy for private business we use the Ibex SmallCap. Data on Mutual and Pension funds variations are taken from the statistics published by the Spanish Association of Investment and Pension Funds (INVERCO). Finally, bonds are

[2] It provides 5 different imputations for each missing data.

measured by the AIAF (Spanish market of fixed rate) index of fixed rate. AIAF is Spain's benchmark market for Corporate Debt and Private Fixed Income[3]. Safety assets are: Checking, saving and money market accounts and vehicles.

Following the previous discussion, the independent variables are different household characteristics taken from the survey: home ownership status, household size, family income, education and marital status. A household is considered to own a business when any of its members has any kind of ownership in a firm. Related to other personal characteristics previously mentioned, information is not available on household grounds, but for the reference person, therefore we use the information on the reference person since she is considered the individual that chiefly deals with financial issues and holds the representation of the household[4] (similar strategy is followed by Schooley and Worden (1996), Guiso et.al. (2002) and Barasinska et al. (2008) among others).

Personal features include gender (dichotomus variable, 0 indicates male and 1 female), marital status (0 not married, 1 married or civil union), education (it takes values from 1, *illiterate* to 12 *postgraduate university education*) household size (members in the household), income (sum of labour and non-labour incomes for all household members), age (years), wealth (real assets plus financial assets less debt). As in other studies (Campbell (2006) and Jin (2011)), both age and wealth will be included also in quadratic terms to capture possible non-linear effects.

4 Methods and Results

As explained above EFF is a multi-imputed dataset. For each missing value, five imputed values are provided. These imputations are stored as five distinct datasets. To make inferences from the five multiply imputed (MI) datasets one has first to analyze each of the datasets and then combine the results. However, for explanatory analysis, it is enough to use one or two of the MI datasets. Therefore, we will use one MI dataset for descriptive statistics and combine results for regression analysis. In this latter case weights are taken into account because of the unequal probability of the household being selected into the sample given the oversampling of the wealthy in the EFF and geographical stratification. We follow Cameron and Trivedi (2005) that claim that weights should be used if regression is viewed as a tool to describe population responses.

4.1 Descriptive Analysis

Table 1, panel A summarizes the characteristics of households that own or do not own a business. The average age of the household business owners is forty seven years old, and they have an average level of education closer to 6 that corresponds

[3] Calculations have been made also with the yield index of public debt and results are similar. Linear Correlation between the two indices is greater than 90%.

[4] For more detail on the definition of the reference person, see CAPI questionnaire, EFF 2005.

to higher secondary school. A comparison between owners and non owners shows that households that own a business are slightly younger and better educated than non-owners. Business owners are with greater probability home owners, have a slightly larger family size and are more often married than non-owners. Households that own a business also have higher levels of income and wealth than non-owners. All differences are statistically significant except for gender distribution. Households with both males and females as heads have the same proportion of owning a business.

Table 1 Descriptive statistics of households who own and do not own business, weighted sample

	Panel A		Panel B	
Variable	Do not own business	Own business	Do not own home	Own home
Weighted percentage	0.8589	0.1410	18.69	81.30
Household Features				
Head's age				
Mean	**52.7286**	**47.2359**	**45.9179**	**53.3834**
standard deviation	(0.2397)	(0.3602)	(0.5699)	(0.22074)
Head's education				
Mean	**5.0640**	**5.8081**	5.2265	5.1558
standard deviation	(0.0461)	(0.1006)	(0.1018)	(0.0469)
Head's gender				
Mean	1.4502	1.4078	**1.5164**	**1.4277**
standard deviation	(0.0071)	(0.0150)	(0.0161)	(0.0069)
Head's marital status				
Mean	**0.6567**	**0.8332**	**0.5583**	**0.7099**
standard deviation	(0.0067)	(.0113)	(0.0160)	(0.0064)
Household size				
Mean	**2.6957**	**3.3932**	**2.6993**	**2.8159**
standard deviation	(0.0175)	(0.0341)	(0.0447)	(0.0169)
Home/Business owner				
Mean	**1.6111**	**1.7170**	**0.1067**	**0.1489**
standard deviation	(0.0113)	(0.**0212**)	(0.0100)	(0.0050)
Household income				
Mean	**9.8995**	**10.3611**	**9.6565**	**10.0341**
standard deviation	(0.,126)	(0.0274)	(0.0311)	(0.0123)
Wealth				
Mean	**11.4998**	**12.5554**	**8.5822**	**12.2077**
standard deviation	(0.0270)	(0.0408)	(0.0879)	(0.0130)

Bold numbers indicate significant differences at 0.01%.

Panel B presents the results distinguishing by home ownership. When comparing households that own their home with non-owners, results for marital status, household size, income and wealth are qualitative similar to those of panel A. Some differences are found in the case of age, education and gender. Home owners are older and more often males than non-owners. There are no significant differences with respect to education. Being a home owner means that you also have a greater probability of owning a business. Household income differences in panel A and B are similar. However, the difference in wealth is much more important in the case of home owners. Households that own their own house have more than 40% higher wealth than non-owners. This difference goes down to 10% when comparing business ownership. It seems that owning a house is more related to wealth than owning a business.

In order to compare risk-taking attitude and behaviour of business and home owners and non-owners we calculate the proportion of households that are willing to take different risk levels and the standard deviation of the household portfolio. Results are shown in table 2. Panel A presents findings for risk taking attitude. These numbers evidence that, in general, households that own a business are more willing to take risks than non-owners. This is in agreement with previous empirical evidence (see Sung and Hanna 1996; Grable and Lytton 1998, Grilo and Irigoyen 2006, Grilo and Thurik 2006 and 2008). However when we look at home ownership, there is no significant difference in risk attitudes between owners and non-owners.

Table 2 Risk tolerance and behaviour by ownership characteristics

Panel A: Risk tolerance level of business/home owners and non owners				
Willing to take (%)	Non owners	Business owners	Non home ownership	Home ownership
substained risk	0.57	2.69	1.31	0.77
above average risk	**1.42**	**4.81**	1.86	1.90
average risk	**13.14**	**23.72**	11.96	15.25
no risk	**84.85**	**68.76**	84.83	82.05

Bold numbers indicate significant differences at 0.01%

Panel B shows the results for risk behaviour. It can be observed that the percentage of households that own a business with lower risk portfolios is much lower than non owners. Accordingly, a higher percentage of business owners have the highest risky portfolios (share of risky assets between 75%-100%). All these differences are statistically significant. This evidence is in line with results of panel A. Results are completely different home owners and non-owners. Portfolios of households that own their home lie with greater probability in the three upper categories of risk taking (from 25% to 100%). The opposite result is found for the lowest risk portfolios (0%-25%), where non-owners represent around 80% of the sample. These divergent results in risk taking contrast with the non-significant difference found for risk attitudes.

Table 2 (*contiunued*)

Panel B: Risk behaviour of business/home owners and non owners				
Share of risky assets (%)	Non owners	Business owners	Non home ownership	Home ownership
0-25%	**29.37**	**7.52**	**78.79**	**14.82**
26-50%	**29.73**	**10.57**	**4.40**	**31.92**
51-75%	**32.64**	**6.90**	**4.07**	**34.37**
76-100%	**8.24**	**75.07**	**12.72**	**18.87**

Panel C: Risk behaviour by risk attitude				
Risk attitude		**Risk behavior**		
Non business owners	0-25%	26-50%	51-75%	76-100%
Below average risk	27.17	27.45	32.20	13.16
Average risk	27.79	28.81	18.01	25.38
Above average risk	19.99	22.74	19.23	38.02
Substained risk	15.65	31.64	24.51	28.18
Business owners				
Below average risk	6.75	15.61	4.88	72.73
Average risk	2.33	5.35	0.08	92.22
Above average risk	0	0	0	100
Substained risk	0	0	9.68	90.31
Non home owners				
Below average risk	80.26	4.16	4.58	10.63
Average risk	73.29	7.23	1.37	18.08
Above average risk	47.02	0	0.39	52.57
Substained risk	51.56	0.75	0	47.68
Home owners				
Below average risk	14.42	32.32	37.54	15.70
Average risk	18.31	30.73	19.50	31.44
Above average risk	10.23	22.71	19.12	47.91
Substained risk	0.02	35.40	29.70	34.86

Bold numbers indicate significant differences at 0.01%.

In order to analyse the relationship between risk attitude and risk behaviour we explore the links between the willingness to take risks of households that own or not a business or their home and their actual risk behaviour (Panel C). Results show a strong coherence between risk attitude and behaviour for business owners. Generally, the risk of portfolios held by households that own a business increases as the level of risk tolerance increases showing a proportional relationship. Non-owners present less consistency in their results. Although in some cases greater risk attitude corresponds to greater portfolio risk (76%-100% or 0%-25%), there is no general pattern among the non-owners. Then, results for this group are not so consistent and would seem to indicate that households do not perfectly understand the risks they are willing to take. Non home owners exhibit some coherence between risk attitude and behaviour. The lowest risk attitude categories show the greatest proportion of low risk portfolios whereas the opposite is found for the two highest attitude levels. However, households that report above average or sustained risk tolerance share approximately the same proportion in high and low risk portfolios. Results for the 4 categories draw a U shape with few medium risk portfolios.

In the case of home owners, some consistency between attitude and behaviour of home owners is present as well. Greater risk attitude means, in general, greater risk taking behaviour. Again, although reported risk attitudes were very similar between home owners and non-owners, the coherence with their risk taking is clearly different. Hence, it is interesting to analyse more deeply risk tolerance and behaviour to try to disentangle this conflicting evidence.

Finally table 3 presents the correlations among the main variables.

Table 3

	Bown	Age	Edu	Gender	Marital	Size	H own	Income	Wealth
Bown	1								
Age	-0.14*	1							
Edu	0.14*	-0.25*	1						
Gender	-0.08*	-0.05*	-0.12*	1					
Marital	0.14*	-0.09*	0.08*	-0.32*	1				
Size	0.24*	-0.36*	0.13*	-0.10*	0.49*	1			
H own	0.07*	0.19*	0,02	-0.09*	0.15*	0.06*	1		
Income	0.29*	-0.14*	0.49*	-0.19*	0.30*	0.36*	0.17*	1	
Wealth	0.23*	0.12*	0.36*	-0.19*	0.15*	0.07*	0.24*	0.52*	1

4.2 Regression Analysis

Multiple regression analyses are conducted to examine the factors associated with risk-taking attitude and behaviour. As independent variables, we introduce two dummy variables to identify if the household is a business owner or a home owner and a set of socioeconomic characteristics as control variables. As dependent variables, we have defined two alternative measures: risk attitude and risk behaviour. Since the former is an ordinal variable the model cannot be consistently estimated using ordinary least squares, therefore an ordered logit model is estimated. To account for risk behaviour we use the standard deviation of the portfolio hold and also the traditional percentage of risky assets for comparison purposes. Following Shoemaker framework, the risk tolerance measure is introduced when estimating risk behaviour. Panel A of table 4 presents the analysis of risk attitude.

Table 4 Risk Attitudes and Behavior

Panel A: Risk attitude, ordered logit model

	(1)	(2)	(3)	(4)
Business Owner	0.6556***	0.6280***	0.5990***	0.6083***
	(0.1383)	(0.1371)	(0.1401)	(0.1390)
Home Owner	0.0073	-0.0126	-0.0108	0.0005
	(0.0794)	(0.0792)	(0.0806)	(0.0813)
Age	-0.0177***	0.0473*	0.0438*	0.0224
	(0.0038)	(0.0267)	(0.0264)	(0.0283)
Age^2		-0.0006**	-0.0007**	-0.0007**
		(0.0003)	(0.0003)	(0.0003)
Education	0.0989***	0.0971***	0.0849***	0.0840***
	(0.0176)	(0.0174)	(0.0178)	(0.0176)
Gender	-0.3386***	-0.3314**	-0.2929**	-0.2792**
	(0.1156)	(0.1152)	(0.1168)	(0.1164)
Marital Status	0.1402	0.1376	0.1819	0.1572
	(0.1415)	(0.1423)	(0.1431)	(0.1403)
Household size	-0.1002*	-0.1317**	-0.1404**	-0.1420**
	(0.0532)	(0.0538)	(0.0545)	(0.0546)
Income	0.4433***	0.4213***	0.3201***	0.3310***
	(0.0931)	(0.0932)	(0.0895)	(0.0898)
Wealth	0.0884**	0.0860**	-0.3580***	-0.4815***
	(0.0334)	(0.0335)	(0.0578)	(0.0820)
Wealth^2	.		0.0315***	0.0292***
	.		(0.0041)	(0.0044)
Age*wealth				0.0033**
				(0.0015)
Constant	6.0028	7.1284	4.6084	3.6453***
	0.8540	(0.9750)	(1.0084)	(1.1066)

*, **, *** are statistically signficant at 10%, 5% and 1% respectively.

Panel A collects the results for risk attitude. The four columns present different specifications to capture any possible non-linear effects related to age and wealth variables as suggested above. In all runs, the business owner coefficient is highly significant and, as expected, has a positive effect on risk attitude, suggesting that households that own a business are more willing to take risks than non owners. At the same time, home ownership is not significant, supporting previous evidence from table 2, panel B. As previously discussed, several socioeconomic characteristics affect significantly risk-taking attitudes. Education and income present positive and significant coefficients indicating that more educated households with greater income are more prepared to take risks. On the contrary, being a woman and household size present negative and significant coefficients. Smaller households are more willing to take risks in line with Sung and Hanna (1996) and Jianakoplos and Bernasek. However, those households with female persons of reference are less inclined to take risks, indicating that women are more conservative when facing risks. This result is similar to Barskey et. al. (1997), Donkers et. al. (1999) and Hartog et. al. (2002). Age is statistically significant and presents a hump-shaped effect on risk attitudes. Younger households are less willing to take risks (column 1). However, when age squared is introduced, this result is reversed, showing an inverted U shape, meaning that the younger and older are less willing to take risks. This result is similar to Ameriks and Zeldes (2004), Poterba and Samwick (2001) and goes against evidence in Campbell (2006) for stockholding. The effect of wealth is also significant but, on the contrary, shows a U shape, meaning that poorer and richer households are more prepared to face risks. This result is in line with Campbell (2006) but contrary to Guiso et. al. (2003). The interaction term between age and wealth present a significant positive sign indicating that increases of age, wealth or both means a greater readiness to take risks.

Table 4, panel B collects the results for risk behaviour measured as the proportion of risky assets. Owning a business presents a positive and significant effect on risk behaviour. However, having a home tenure affects negatively risk behaviour. Contrary to results of the risk attitude regression, gender, and education are no longer significant. Age maintains the sign and the hump-shaped relationship also with respect to risk taking behaviour. Wealth is still significant and maintains the U shaped. Household size becomes relevant, indicating that bigger households tend to make riskier decisions. Finally, the risk attitude variable presents a positive and significant coefficient showing that households that declare to be more risk tolerance are taking significantly more risks.

Panel C collects the results for the new measure of risk behaviour. Results are slightly different. Looking at socioeconomic features, only age maintains sign and significance, and the inverted U shape as well. Therefore, socioeconomic characteristics are less important in determining risk behaviour when the risk measure is the standard deviation of returns on household portfolio. Owning a business and having a home tenure present a positive and significant effect on risk behaviour. In this case, not only is the sign of the coefficient, but also the magnitude, nearly double of panel B, what makes these variables more relevant in determining risk taking behaviour. Finally, risk attitude maintains the results found for the percentage of risky assets, but again the magnitude of this variable increases.

Table 4 (*contiunued*)

Panel B: Risk behaviour, Proportion of Risky Assets

	Proportion of Risky Assets			
	(1)	(2)	(3)	(4)
Risk attitude	0.0464***	0.0455***	0.0347***	0.0347***
	(0.0094)	(0.0094)	(0.0095)	(0.0095)
Business Owner	0.2743***	0.2712***	0.2636***	0.2641***
	(0.0150)	(0.0149)	(0.0149)	(0.0149)
Home Owner	-0.0937***	-0.0975***	-0.0979***	-0.0977***
	(0.0088)	(0.0088)	(0.0086)	(0.0087)
Age	0.0028***	.0119***	0.0110***	0.0105***
	(0.0003)	(0.0020)	(0.0019)	(0.0021)
Age^2		-0.0001***	-0.0001***	-0.0001***
		(0.00001)	(0.00001)	(0.00001)
Education	0.0008	0.0008	-0.0014	-0.0013
	(0.0017)	(0.0016)	(0.0016)	(0.0016)
Gender	-0.0113	-0.0092	-0.0038	-0.0035
	(0.0097)	(0.0097)	(0.0096)	(0.0096)
Marital Status	-.0226*	-0.0219*	-0.0166	-0.0168
	(0.0126)	(0.0127)	(0.0123)	(0.0124)
Household size	0.0189***	0.0141***	0.0131**	0.0131**
	(.0051)	(0.0052)	(0.0051)	(0.0051)
Income	0.0290***	0.0258***	0.0166**	0.0169**
	(0.0075)	(0.0073)	(0.0070)	(0.0069)
Wealth	0.0268***	.0263***	-0.0264***	-0.0295***
	(0.0022)	(0.0022)	(0.0055)	(0.0072)
Wealth^2			0.0040***	0.0040***
			(0.0004)	(0.0004)
Age*wealth				0.0001
				(0.0001)
Constant	-0.1861**	-0.3566***	-0.1314***	-0.1088
	(0.0775)	(0.0875)	(0.0861)	(0.0989)

*, **, *** are statistically signficant at 10%, 5% and 1% respectively.

Table 4 (*contiuned*)

Panel C: Risk behaviour, Standard Deviation of Household Portfolio

	Standard Deviation of Household Portfolio			
	(1)	(2)	(3)	(4)
Risk attitude	0.1272*	0.1253*	0.1284*	0.1278*
	(0.7619)	(0.0761)	(0.0774)	(0.0773)
Business Owner	2.6309***	2.6232***	2.6257***	2.6192***
	(0.1579)	(0.1581)	(0.1588)	(0.1589)
Home Owner	0.3783***	0.3693***	0.3707***	0.3676***
	(0.0542)	(0.0545)	(0.0545)	(0.0546)
Age	0.0021	0.0258**	0.0261**	0.0344***
	(0.0016)	(0.0101)	(0.0102)	(0.0102)
Age^2		-0.0002**	-0.0002**	-0.0002**
		(0.0008)	(0.0008)	(0.0001)
Education	0.0006	0.0006	0.0013	0.0010
	(0.0089)	(0.0088)	(0.0087)	(0.0086)
Gender	0.0078	0.0125	0.0112	0.0070
	(0.0519)	(0.0516)	(0.0517)	(0.0518)
Marital Status	0.0983	0 .0993	0.0976	0.1091*
	(0.0606)	(0.0607)	(0.0606)	(0.0609)
Household size	0.0227	0.0104	0.0107	0.0108
	(0.0281)	(0.0279)	(0.0279)	(0.0278)
Income	0.0191	0.0107	0.0136	0.0093
	(0.0387)	(0.0389)	(0.0387)	(0.0390)
Wealth	-0.0332***	-0.0347***	-0.0164	0.0389
	(0.012)	(0.0121)	(0.0383)	(0.0506)
Wealth^2			-0.0013	-0.0013
			(0.0027)	(0.0027)
Age*wealth				-0.0010*
				(0.0005)
Constant	1.2192**	0.7841	0.7023	-0.3010
	(0.4852)	(0.5163)	(0.5153)	(0.5359)

*, **, *** are statistically signficant at 10%, 5% and 1% respectively.

5 Conclusions

This study examines risk tolerance and behaviour of Spanish households using data from the 2005 Encuesta Financiera de las Familias (EFF) developed by the Bank of Spain. It does so improving the definition and measure of risk behaviour. Contrary to previous papers, the risk taking behaviour is calculated by means of the standard deviation of household portfolio. This measure allows to taking into account asset risk differences and distinguishing between household portfolios with the same percentage of risky assets (but different composition).

Considering that owning a business or having a home tenure could affect family risk taking decisions, special attention has been also devoted to the comparison between households who own businesses or their home and non owners. The findings can be summarized as follows: households that own a business are more willing to take risks and their portfolio has greater risk compared to households who do not own a business. The results for home ownership are a bit different. Although home owners have the same risk attitude than non-owners, their behaviour is riskier.

Looking at socioeconomic features, gender, education, income and household size affect risk-taking attitudes. Age is also a relevant factor and presents a hump-shaped effect on risk attitudes whereas wealth shows a U shaped, meaning that poorer and richer households are more prepared to face risks. Age and wealth interact together indicating that increases of age, wealth or both means a greater readiness to take risks.

Results are a bit different for risk taking behaviour. Gender and education are no longer significant. Age maintains the sign and the hump-shape relationship with respect to risk taking behaviour. However wealth is not significant. Finally, household size becomes relevant, indicating that larger households tend to make riskier decisions.

The findings of this study add to the analysis to further explain the risk decision-making of households and the characteristics that could shape their risk attitude and behaviour.

References

Ameriks, J., Zeldes, S.: How Do Household Portfolio Shares Vary with Age? Working paper. Columbia Business School (September 2004)

Arrondel, L., Lefebvre, B.: Consumption and Investment Motives in Housing Wealth Accumulation: a French Study. Journal of Urban Economics 50, 112–137 (2001)

Barasinska, N., Schäfer, D., Stephan, A.: Financial Risk Aversion and Household Asset Diversification, Discussion Papers of DIW Berlin 807, DIW Berlin, German Institute for Economic Research (2008)

Barsky, R.B., Juster, F.T., Kimball, M.S., Shapiro, M.D.: Preference Paramenters and Behavioural Heterogeneity: An Experimental Approach in the Health And Retirement Study. Quaterly Journal of Economics 112(2), 537–579 (1997)

Bover, O.: Housing Purchases and the Dynamics of Housing Wealth. Banco de Espanya WP1036 (2010)

Cameron, A.C., Trivedi, P.K.: Microeconometrics: Methods and Applications. Cambridge University Press (2005)

Campbell, J.: Household Finance. Journal of Finance 61, 1553–1604 (2006)

Carroll, C.: Portfolios of the Rich. In: Guiso, L., Haliassos, M., Jappelli, T. (eds.) Household Portfolio, pp. 1–24. The MIT Press, Cambridge (2002)

Cocco, J.: Portfolio Choice in the Presence of Housing. Review of Financial Studies 18, 569–597 (2005)

Cohen, A., Einav, L.: Estimating Risk Preferences from Deductible Choice. The American Economic Review 97(3), 745–788 (2007)

Dohmen, T.J., Falk, A., Huffman, D., Schupp, J., Sunde, U., Wagner, G.G.: Individual risk attitudes: New evidence from a Large, representative, Experimentally validated Survey. CEPR Discussion Paper 5517 (2006)

Donkers, B., van Soest, A.: Subjective Measures of Household Preferences and Financial Decisions. Journal of Economic Psychology 20(6), 613–642 (1999)

Flavin, M., Yamashita, T.: Owner-Occupied Housing and the Composition of the Household Portfolio. American Economic Review 92(1), 345–362 (2002)

Friend, I., Blume, M.E.: The Demand for Risky Assets. American Economic Review 75, 900–922 (1975)

Gollier, C.: The Economics of Risk and Time. MIT Press (2001)

Grable, J.E., Lytton, R.H.: Investor Risk Tolerance: Testing the Efficacy of Demographics and Differentiating and Classifying Factors. Financial Counselling and Planning 9(1), 61–74 (1998)

Grilo, I., Irigoyen, J.M.: Entrepreneurship in the EU: to Wish and not to Be. Small Business Economics 26(4), 305–318 (2006)

Grilo, I., Thurik, A.R.: Entrepreneurship in the Old and the new Europe. In: Santarelli, E. (ed.) Entre-preneurship, Growth and Innovation, pp. 75–103. Springer Science, Berlin (2006)

Grilo, I., Thurik, R.: Determinants of Entrepreneurial Engagement Levels in Europe and the US. Industrial and Corporate Change 17(6) (2008)

Guiso, L., Haliassos, M., Jappelli, T.: Introduction. In: Guiso, L., Haliassos, M., Jappelli, T. (eds.) Household Portfolio, pp. 1–24. The MIT Press, Cambridge (2002)

Guiso, L., Haliassos, M., Jappelli, T.: Household stockholding in Europe: where do we stand and where do we go? Economic Policy 18(36), 123–170 (2003)

Halek, M., Eisenhauer, J.G.: Demography of risk aversion. Journal of Risk and Insurance 68, 1–24 (2001)

Hanna, S., Chen, P.: Subjective and Objective Risk Tolerance: Implications for Optimal Portfolios. Financial Counselling and Planning 8(2), 17–26 (1997)

Hartog, J., Ferrer-i-Carbonell, A., Jonker, N.: Linking Measured Risk Aversion to Individual Characteristics. Kyklos 55, 3–26 (2002)

Heaton, J., Lucas, D.: Portfolio Choice and Asset Prices: the Importance of Entrepreneurial Risk. Journal of Finance 59, 137–163 (2000)

Ioannides, Y.M., Rosenthal, S.S.: Estimating the Consumption and Investment Demands for Housing and their Effect on Housing Tenure Status. Review of Economics and Statistics 20, 127–141 (1994)

Jianakoplos, A., Bernasek, A.: Are Women More Risk Averse? Economic Inquiry 36(4), 620–631 (1998)

Jin, F.: Revisiting the Composition Puzzles of the Household Portfolio: New Evidence. Review of Financial Economics 20(2), 63–73 (2011)

Knight, F.H.: Risk, Uncertainty and Profit. In: Stigler, G.J. (ed.), University of Chicago Press, Chicago (1921,1971)

Marshall, A.: Principles of Economics, 1st edn. Macmillan and Co, London (1890,1930)

Poterba, J.M., Samwick, A.A.: Taxation and household portfolio composition: US evidence from the 1980s and 1990s. Journal of Public Economics 87(1), 5–38 (2003)

Riley, W.B., Chow, K.V.: Asset allocation and individual risk aversion. Financial Ananlysts Journal 48, 32–37 (1992)

Say, J.-B.: A Treatise on Political Economy or the Production, Distribution and Consumption of Wealth, 1st edn. A.M. Kelley Publishers, New York (1803,1971)

Schoemaker, P.J.H.: Determinants of Risk-Taking: Behavioral and Economic Views. Journal of Risk and Uncertainty 6, 49–73 (1993)

Schooley, D.K., Worden, D.D.: Risk Aversion Measures: Comparing Attitudes and Asset Alloca-tion. Financial Services Review 5(2), 87–99 (1996)

Sundén, A.E.: Gender Differences in the Allocation of Assets in Retirement Savings Plans. American Economic Review Papers and Proceedings 88, 207–211 (1998)

Sung, J., Hanna, S.: Factors Related to Risk Tolerance. Financial Counselling and Planning 7, 11–20 (1996)

Wang, H.: Empirical Analysis of Household Portfolio Allocation Based on a Rational Expectation Model. Ph.D Dissertation. The Ohio State University (1997)

Yao, R., Zhang, H.: Optimal Consumption and Portfolio Choices with Risky Housing and Borrowing Constraints. Review of Financial Studies 18(1), 197–239 (2005)

Xiao, J.J., Alhabeeb, M.J., Hong, G.-S., Haynes, G.W.: Attitude Toward Risk and Risk-Taking Behaviour of Business-Owning Families. The Journal of Consumer Affairs 35(2), 307–325 (2001)

Discussing Relationship between Asian and US Financial Markets

Mohammed K. Shaki and María Luisa Medrano

Department of Business, National University
Department of Business Administration, Rey Juan Carlos University
Paseo Artilleros S/N 28032 Madrid, Spain
mshaki@nu.edu, marialuisa.medrano@urjc.es

Abstract. This paper analyses the relationship between Asia's stock markets with US main stock indices. The study focuses on the stock interaction and informative transmission among of nine stock markets in Asia (Japan, Hong Kong, Indonesia, South Korea, Malaysia, Philippines, Singapore, Thailand and Taiwan) and three stock markets in US (Dow Jones, NASDAQ and S&P500). The collected weekly data is from Inform Winner Plus 2000 and the study period is from the first week of January 1990 to last week of June 2007. This empirical study adopts different econometric methods such as ADF unit root, cointegration test, vector error correction model, an impulse response function and Granger Causality to find out what kind of relations exist between Asian and US financial markets.

Keywords: Financial Markets, Asia, US, Dow Jones, NASDAQ, S&P500.

1 Introduction

Stock markets directly impact to each country's economic development. Any stock market's volatility could concentrate investors' attention on it. Investors always want to know what kind of messages hint in the stock variation and hope obtains abnormal return from equity market. This is the reason why lots of researches have tried to predict stock markets. Hsiao (1981) used bivariate AR to study relationship between monetary support and stock price among six countries' stock markets (Australian, Japan, Hong Kong, Singapore, Philippines, and Thailand). Eun and Shim (1989) used VAR and Johansen Cointegration test to test global nine big stock markets relationship which include New York, Tokyo, London, Toronto, Frankfurt, Zurich, Sydney, Hong Kong, and Paris. Fisher and Palasvirta (1990) used the cross-spectral analysis to test dynamic stock markets relationship among the majority of 23 countries.

Each country's economic development has moved toward liberalization, internationalization, and localization since 1990. Asia has experienced important local development and its stock markets have such a vigorous growth that the total trading volumes are almost closely with US or UK stock market. Asian stock markets

A.M. Gil-Lafuente et al. (Eds.): Soft Comput. in Manag. and Bus. Econ., STUDFUZZ 287, pp. 219–234.
springerlink.com

are belonging to small to middle size opening economic except for Japan. Each country has higher international trade dependence on each other. Those countries are also easily influence by big trading partner's economic variation. This study wants to focus on the relationship between stock interaction and informative transmission among of nine stock markets in Asia and three stock markets in US.

2 Literature Review

Prior studies are more emphasis on US, Japan, Europe countries stock market relationship. However those markets have strength of stock system and maturity of transaction process, most of Asia countries' market just start growing up and do not have integrity of equity system. Those Asia equity market are easily impact by developed countries' stock, especially small size of stock market. The things could happen in developed nations that might not appear in emerging countries. For answering research questions the study adopts unit roots, cointegration test, vector autoregressive model, Granger causality.

Eun and Shim (1989) used VAR and Johansen Cointegration test to test global nine big stock markets relationship which include New York, Tokyo, London, Toronto, Frankfurt, Zurich, Sydney, Hong Kong, ad Paris. The study period was from December 1979 to December 1985. The findings were those nine stock markets have existed highly interaction. Researchers also found trace out the dynamic responses of one market shock to another. Especially shocks to U.S. are rapidity transmitted to other nations, whereas other nations can not have the significant explains U.S. movement. This seems U.S. stock market is the leader of the rest of the countries.

Fisher and Palasvirta (1990) used the Spectrum Analysis to test dynamic stock markets relationship among the majority of 23 countries. The study period is from 1986 to 1988. The findings show those nations' stock market have highly interdependence and US stock index price has lead almost every countries indices. US with Malaysia and Singapore have highly interdependence each other, however with Japan appears lower interdependence that means both of nations' market more independent each other.

Ko and Lee (1991) indicated comparing US, Japan with Asia four dragon which are Hong Kong, South Korea, Singapore, and Taiwan daily stock return. The findings show South Korea and Taiwan stock markets have lower cross-correlation with other nations' because both of countries have blue law for foreign investors. Japan, Hong Kong, Singapore, and U.S. have significant correlation each other that means those countries' capital market are connecting closely and look like cointegrative market.

Cheung and Mak (1992) indicated Asia stock markets which include U.S., Japan, Hong Kong, Singapore, Taiwan, South Korea, Thailand, Philippines, Malaysia, and Australia don't have efficient informative transmission and U.S. stock market is leader of other nations except for Taiwan, South Korea, and Thailand. Japan does not have a leader position in the Asia market. Researchers adopted error correction model and granger causality to test those stock markets' short and long-term dynamic relationship.

Arshanapalli and Doukas (1993) adopted cointegration test to discuss the linkage and the dynamic international stock markets pre- and post- financial crisis 1987 period. The findings were in sharp contrast with previous research which had discovered strong interdependence among national stock markets prior to October 1987. The findings also indicated the degree of international co-movements among stock price indices has increased substantially except for Japan for post-1987 period. It also found Japan stock market did not have any linkage with US, France, UK, and German markets during 1987 crisis period.

Ghosh (1999) examined the debacle of the Asian-Pacific stock markets by utilizing the theory of cointegration to investigate which developing markets are moved by the markets of Japan and the United States. The findings indicated Hong Kong, India, Malaysia, and South Korea have long-term relations and Japan, India, Philippines, and Singapore have other long-term relations. Whereas Taiwan and Thailand do not have been influence by U.S. or Japan stock market.

Gerrits and Yuce (1999) use vector autoregressive model (VAR) to examine the relationship interdependence of daily stock price variation among of UK, Howland, US, and German. The findings indicated US has an impact phenomena to three Europe countries, and those European countries are interdependent each other in short and long-term period.

Cheng, Francis, and Alan (2004) investigated the causal links between the world's largest stock markets, namely the U.S. market, the U.K. market, and the Japanese market, over various time horizons. The findings indicated there is a significant bi-directional causal effect among the three markets and there is not one global causality relation prevailing over all time scales.

Nieh, Lin, and Chan (2005) discussed on the study of the regional integration in Pacific basin and attempt to explore the potential leader in Asia-Pacific Financial Center. It was by means of analyzing the long-term and short-term interrelationships among the stock markets of China, Hong Kong, and Taiwan after Asian Financial Crisis. The findings did not support a long-run equilibrium relationship among three places around Taiwan Strait by the Johansen cointegration test. However, their short-run dynamic relationships are found by VAR model analysis. The article concluded all the above statements saying that Hong Kong is shown to have the potential leading position in the future Asian-Pacific Financial center.

3 Methodology

In this paper we want to answer to the following questions:

1. Do those stock markets have long-term consistence trend?
2. Do those stock markets have cointergrative phenomena existence?

3. Do those stock markets have autoregressive phenomena existence each other?
4. Do those stock markets have causality relations each other?

For trying to answer those questions, we will use unit root, cointegration test, vector autoregressive model, and Granger causality. The weekly data are collected from Informed Winners Plus 2000, and the study period is from first week of January in 1990 to fourth week of June in 2007. The study also adopts EViews to deal with the whole hypothesis test.

3.1 Unit Roots

Any time series data need to test unit root to make sure data stationary or not. It has many ways to test unit root for example DF (Dickey-Fuller test, 1976), ADF(Augumented Dickey-Fuller test, 1979), PP(Phillips-Perron test, 1988). In this study researchers adopt ADF to test unit root. The reason study adopts ADF because it assumes time series is high level $AR(p)$. ADF uses lagged length (p) to eliminate correlation of error-term series to let $AR(p)$'s error-term to be white noise, and then process testing.

ADF test assumes series has high level of $AR(p)$ and check it has drift or not. It has two ways to produce true process:

1. $AR(p)$ is non-drift and unit root series

$$y_t = \beta_1 \Delta y_{t-1} + \beta_2 \Delta y_{t-2} + \ldots + \beta_{p-1} \Delta y_{t-p+1} + y_{t-1} + \varepsilon_t \tag{1}$$

2. $AR(p)$ is drift and unit root series

$$y_t = \beta_1 \Delta y_{t-1} + \beta_2 \Delta y_{t-2} + \ldots + \beta_{p-1} \Delta y_{t-p+1} + \alpha + y_{t-1} + \varepsilon_t \tag{2}$$

First thing of ADF uses OLS method to estimate regression which has standard, intercept, and intercept & trend three methods. In this study estimated regression adopts intercept without trend.

$$y_t = \alpha + \beta_1 \Delta y_{t-1} + \beta_2 \Delta y_{t-2} + \ldots + \beta_{p-1} \Delta y_{t-p+1} + \gamma y_{t-1} + \varepsilon_t \tag{3}$$

Its hypothesis is:

$$H_0 : \gamma = 1$$
and
$$H_1 : |\gamma| < 1$$

its statistics has two ways:

$$Z_{IF} = \frac{T(\hat{\gamma}-1)}{1-\hat{\beta}_1-\hat{\beta}_2-\ldots-\hat{\beta}_{p-1}}$$
$$t_{\hat{\gamma}} = \frac{(\hat{\gamma}-1)}{SE(\hat{\gamma})} \tag{4}$$

T is sample size, and $\hat{\beta}$, $\hat{\beta}_2 \ldots \hat{\beta}_{p-1}$, $\hat{\gamma}$ are estimated by the formula (3)'s β_1 $\beta_2 \ldots \beta_{p-1}$, γ by using OLS. SE is standard error for γ.

If the result accepts H_0 that means the series y_t is unit root (series is non-stationary), otherwise reject H_0 means y_t does not have unit root and series is stationary I(0)

In this study choosing lagged length is adopted by Akiake Information Criteria (AIC). Its method uses lagged length to treat sample's function and its purpose to minimize function and let residual sum of square minimization.

$$I_p = \log \hat{\sigma}_p + p\frac{2}{T} \tag{5}$$

T is sample size, $\hat{\sigma}_p^2$ is estimated regression's lagged term p of sum of square error.

3.2 Cointergration Test

If study relates time series, the topic are often touching about multi-variation relations and correlation. The study often adopts cointegration test, error correction model, or Granger causality to find variables existent lead- lag relations or not.

Engle and Granger (1987) pointed out that a linear combination of two or more non-stationary series may be stationary. If such a stationary linear combination exists, the non-stationary time series are said to be cointegrated. The stationary linear combination is called the cointegrating equation and may be interpreted as a long-run equilibrium relationship among the variables. The purpose of the cointegration test is to determine whether group of non-stationary series are cointegrated or not.

It has two ways to test cointegrated relationship: one is Engle and Granger two stage cointegrated test, and the other is Johansen (1988) brought up maximum likelihood approach. First approach is the Engle-Granger cointergration test which consists of a two-stage procedure. In the first stage, the residual error is tested for stationary.

$$y_t = \alpha + \beta x_t + \varepsilon_t \quad (6)$$

Variables y and x might individually be non-stationary but if the estimate of their residual error is stationary, y and x are said to be cointegrated. It implies that y and x form a long run relationship and the regression is not spurious. This approach has shown that any cointegrated series has an error correction representation. Therefore, if the residual error of the estimation in the first step is stationary, the error correction model can be estimated. In the second stage, the error correction model is estimated, which represents the short run dynamics of the model.

The second approach is Johansen method to test cointegration relationship. Johansen proposes two tests to determine the number of cointegration vectors. The first is the likelihood ratio test based on the maximal eigenvalue and the second is the likelihood ratio test based on the trace test. The power of the trace test is lower than the power of the maximal eigenvalue test. If the null hypothesis of no cointegration vector can be rejected, it indicates that there is a long run relationship among the variables in the model. As a result, the error correction mechanism can be presented.

Trace test:

$$H_0 : r \leq q(\text{at most q integrated vector})$$
$$H_1 : r > q(\text{at least q+1 integrated vector})$$

$$LR = -2\ln(\theta) = -T\sum_{i=r+1}^{p}\ln\left(1-\hat{\lambda}_i\right) \quad (7)$$

T is sample size, $\hat{\lambda}_i$ is estimated of characteristic root.

Maximum eigenvalue test:

$$H_0 : r = q(\text{q integrated vector})$$
$$H_1 : r > q+1(\text{q+1 integrated vector})$$
$$LR = -2\ln\langle\theta; r|r+1\rangle = -T\ln\left(1-\hat{\gamma}_t\right) \quad (8)$$

Above both of tests assume variables do not exist cointegration that means r =0. If after testing reject null, then adding number of vector to test till completely accept null hypothesis. When the result has one or more significant characteristic root existence that means variables have long-term equilibrium relationship.

3.3 Error Correction Model

When economic variables have cointegration relationship that means variables have long-term equilibrium relationship. However in the short term those variables might be had deviation phenomena. Banerjee et al., (1993) developed error correction model (ECM) to solve variables short term deviation phenomena. The model shows below:

$$\Delta y_t = \beta \Delta x_t + \alpha (y_{t-1} - \gamma x_{t-1}) + \varepsilon_t \tag{9}$$

Whereas $y_{t-1} - \gamma x_{t-1}$ is error correction item, γ is variable x and y long-term equilibrium relation. β is Δx and Δy short-term relation. α is adjusting speed of error item.

When $\alpha > 1$ means ECM will fast correct to equilibrium phenomena that indicates strong signal from adjusting to eliminate deviation of error term. When $\alpha \leq 0$ means weak signal from adjusting to eliminate deviation of error term.

3.4 Vector Autoregressive Model

Vector Autoregressive Model (VAR) is Sim (1980) reported the model of analysis time series. The approach is not like traditional econometric and treating every variables are endogenous in the system as a function of the lagged values of all of the endogenous variables in the system. The model is:

$$y_t = \alpha + \sum_{i=1}^{m} A_i y_{t-i} + \varepsilon_t \tag{10}$$

$$E(\varepsilon_t) = 0$$

$$E(\varepsilon_t \varepsilon_t') = \sum \neq 0$$

$$E(\varepsilon_{t,} y_{t-1}) = 0$$

y_t : ($n \times 1$) vector of endogenous regression and has jointly covariance stationary criteria's linerly stochastic process. y_{t-m} is ($n \times 1$) m level of lagged variables.

n: endogenous variables
m: time lagged length

A_i : ($n \times n$) vector of coefficients that is transmitting function of economic phenomena
α : ($n \times 1$) vector of constant

ε_t : it is structural disturbance and ($n \times 1$) forecasting deviation that means random of shocks or impulses. It is also non-series correlation.

$E(\varepsilon_t) = 0$: each of regressions' expect mean of error is equal to zero.

$E(\varepsilon_t \varepsilon_t^{'}) = \sum$: it is diagonals and ($n \times 1$) vector of covariance. If its value is not equal to zero that means interactive functions' contemporaneously error correlated each other.

$E(\varepsilon_{t,} y_{t-1}) = 0$: it means each function has independent time series criteria and error and lagged-term is independent each other.

3.5 Impulse Response

A shock to the one variable not only directly affects itself but is also transmitted to all of the other endogenous variables through the dynamic (lag) structure of the VAR. An impulse response function traces the effect of a one-time shock to one of the innovations on current and future values of the endogenous variables.

On the impulse response model processes calculating based on vector autoregressive model (VAR) that original model uses Wald decomposition theorem to orthogonal transform data to represent by vector moving average. That means each variable can be represented by random shock current and lag period of linear regression.

$$Y_t = \alpha + \sum_{i=1}^{m} \gamma_t y_{t-i} + \varepsilon_t \tag{11}$$

$$Y_t - \sum_{i=1}^{m} \gamma_t y_{t-i} = \alpha + \varepsilon_t \tag{12}$$

$$(1 - \gamma_1 L - \gamma_2 L^2 - \ldots - \gamma_n L^n) Y_t = \alpha + \varepsilon_t$$

$$Y_t = (\alpha + \varepsilon_t)(1 - \gamma_1 L - \gamma_2 L^2 - \ldots - \gamma_n L^n)^{-1}$$

$$Y_t = \alpha^{'} + \sum_{i=0}^{\infty} \gamma_t \varepsilon_{t-i} \tag{13}$$

Then function adds lower triangular matrix A, $AA^{'} = A$ and new function is:

$$Y_t = \alpha^{'} + \sum_{i=0}^{\infty} \gamma_t A A^{'} \varepsilon_{t-i} \tag{14}$$

Then function can be simplified to be:

$$Y_t = \alpha^{'} + \sum_{i=0}^{\infty} \gamma_t D_i W_{t-i} \,. \tag{15}$$

From above function that VAR model can transform to be represented by moving average method, for example (8). Whereas $D_i = \gamma_t A$, $W_{t-i} = A$ are not current correlation and they are random white noise serially uncorrelated. From above functions that can investigate each variable could be represented by current and few lag periods' innovation. If study wants to investigate one variable is shocked by other independent variables' influence and uses continuous response function to observe that variable's changeable of size and varieties. That means an impulse response function traces the effect of a one-time shock to one of the innovations on current and future values of the endogenous variables.

3.6 Granger Causality

Granger causality is part of VAR model and its method based on Granger (1969) to define its causality that if extra adding one message variable x to explain more behaviors of variable y. And it can reduce variable y's conditional variance. Based on above phenomena that calls x Granger causes y. Otherwise it changes to be y Granger causes x. If above both of phenomena exist that represents x and y have feedback relations. If both of phenomena exist that represents x and y have independent relations.

4 Findings

This chapter examines hypotheses based on collecting market indices from Inform Winners Plus 2000. First step we test each variable's unit root to make sure data stationary or not. Second step we uses Johansen cointegration test to examine those variables have long term equilibrium relationship. Third step we adopt error correction model to discuss variables short term interactive relationship. Forth step research we use VAR to predict each variable movement influences the other variables. Fifth step we adopt Granger causality to distinguish variables causality relationship.

4.1 Descriptive Statistics of Stock Indices Return

Before starting explains the findings that needs to descript basic statistics of stock indices return. It basically understands those stock indices' return characteristic. Kurtosis is a measure of whether the data are peaked or flat relative to a normal distribution. And Skewness is a measure of symmetry, or more precisely, the lack of symmetry.

The results indicate all stock indices' kurtosis are more than 3 that means more of the variance is due to infrequent extreme deviations, as opposed to frequent modestly-sized deviations. Most of stock indices' skewness appears negative values that indicate data are skewed left. However Indonesia, Malaysia, and Philippines' skewness are positive values for indicating data that are skewed right.

All of data sets do not fit to normal distribution and then research needs to consider testing series autocorrelation. From std. dev. appears all of stock indices are almost the same that means each of stock volatility alike.

4.2 ADF Unit Root

Before processing time series model the study needs to test each variable's unit root to check data stationary or not, otherwise research uses non-stationary to test time series model that could cause spurious regression and bias test and make study to be valueless. This study adopts ADF unit root to test data stationary or not. Table 1 indicates all of the stock indices accept null to appear unit root for every one that means I(0) is not stationary in all of countries based on 1% significant test. Then all of stock indices run first difference and the results appear all of data reject unit root null hypothesis to accept alternative hypothesis no matter which kinds of ways and testing based on 1% significant test. It means data are stationary after run first difference I(1) based on 1% significant test.

Table 1 Stock Indices of Unit Root (ADF)

	Level I(0)	First Difference I(1)		
		Intercept	Trend and Intercept	None
Hong Kong	-0.369039	-21.52444***	-21.55476***	-21.46360***
Japan	-0.751601	-21.93646***	-21.97096***	-21.94203***
Singapore	1.079216	-21.15583***	-21.30303***	-20.99883***
South Korea	0.466471	-22.96705***	-23.07564***	-22.83827***
Taiwan	-1.294857	-21.64898***	-21.67128***	-21.66071***
Thailand	-0.663790	-20.93400***	-20.93330***	-20.90421***
Dow	-0.304155	-23.47997***	-23.56426***	-23.44005***
NASDAQ	-1.615498	-22.26634***	-22.24490***	-22.29121***
SP500	-0.939720	-24.75913***	-24.80652***	-24.77649***
Indonesia	3.074634	-21.65989***	-22.20190***	-12.89476***
Malaysia	0.777315	-19.09733***	-19.16239***	-18.97359***
Philippines	1.868151	-19.72463***	-20.24983***	-19.66863***

4.3 Cointegration Test

According to above analysis, It indicates all of the stock indices are I(1) stationary, then study needs to use Johansen cointegration test to check those stock indices have long term equilibrium relations existence. Before study adopts cointegration test that needs to check each stock index's lag length. This study adopt AIC (Akaike information Criterion) to get each stock index's lag length. Table 2 shows the each stock index's the fitness of lagged length. Lag length is 1 in Hong Kong, Japan, Singapore, Dow, NASDAQ, and Philippines. Taiwan's lag length is 2 and South Korea and S&P500's are 3. Lag length in Malaysia is 4 and Thailand and Indonesia are 5.

Table 2 Panel A: Fitness of lagged Length for Return of Stock Index (Based on AIC)

		Japan	
Lag Length	AIC	Lag Length	AIC
1	-4.156398***	1	-4.332686***
2	-4.154640	2	-4.329009
3	-4.151058	3	-4.324682
4	-4.152098	4	-4.322660
5	-4.150391	5	-4.315911
Singapore		South Korea	
Lag Length	AIC	Lag Length	AIC
1	-4.423751***	1	-3.520527
2	-4.417050	2	-3.531653
3	-4.414781	3	-3.537688***
4	-4.418487	4	-3.535776
5	-4.411748	5	-3.532136
Taiwan		Thailand	
Lag Length	AIC	Lag Length	AIC
1	-3.785973	1	-3.893885
2	-3.787509***	2	-3.888110
3	-3.783891	3	-3.901075
4	-3.782349	4	-3.900730
5	-3.777207	5	-3.904090***

Table 2 ***(Continued)*** Panel B: The fitness of lagged Length for Return of Stock Index (Based on AIC)

DOW		NASDAQ	
Lag Length	AIC	Lag Length	AIC
1	-4.750824***	1	-3.585157***
2	-4.747068	2	-3.579467
3	-4.743817	3	-3.580170
4	-4.742349	4	-3.578144
5	-4.736333	5	-3.574325
S&P500		Indonesia	
Lag Length	AIC	Lag Length	AIC
1	-4.685171	1	-3.984275
2	-4.680078	2	-3.986666
3	-4.687751***	3	-3.983084
4	-4.685643	4	-3.988736
5	-4.682606	5	-3.990983***
Malaysia		Philippines	
Lag Length	AIC	Lag Length	AIC
1	-4.500034	1	-4.196489***
2	-4.495770	2	-4.190956
3	-4.498232	3	-4.188851
4	-4.504435***	4	-4.183237
5	-4.500168	5	-4.178292

From table 3 cointegration test appears all of stock indices are 1% significant no matter what is trace and Max-Eigen statistic indicate those indices have existed long term equilibrium relations. The reason all of stock indices have existed long term consistence relations that next step needs to test ECM to make sure those indices' short term relationship.

Table 3 Panel A: Unrestricted Cointegration Rank Test

Hypothesized Number of Cointegrating Equations	Eigenvalue	Trace Statistic	5% Critical Value	1% Critical Value
None	0.284431	1018.945	NA	NA
At most 1 **	0.255085	873.0262	277.71	293.44
At most 2 **	0.238558	744.6305	233.13	247.18
At most 3 **	0.211825	625.8025	192.89	204.95
At most 4 **	0.194427	522.0190	156.00	168.36
At most 5 **	0.170503	427.7554	124.24	133.57
At most 6 **	0.158845	346.2512	94.15	103.18
At most 7 **	0.145098	270.8324	68.52	76.07
At most 8 **	0.122351	202.4811	47.21	54.46
At most 9 **	0.121112	145.5793	29.68	35.65
At most 10 **	0.107826	89.29255	15.41	20.04
At most 11 **	0.086713	39.54765	3.76	6.65

Table 4 Panel B: Unrestricted Cointegration Rank Test

Hypothesized Number of Cointegrating Equations	Eigenvalue	Max-Eigen Statistic	5% Critical Value	1% Critical Value
None	0.284431	145.9190	NA	NA
At most 1 **	0.255085	128.3957	68.83	75.95
At most 2 **	0.238558	118.8280	62.81	69.09
At most 3 **	0.211825	103.7835	57.12	62.80
At most 4 **	0.194427	94.26358	51.42	57.69
At most 5 **	0.170503	81.50418	45.28	51.57
At most 6 **	0.158845	75.41883	39.37	45.10
At most 7 **	0.145098	68.35126	33.46	38.77
At most 8 **	0.122351	56.90184	27.07	32.24
At most 9 **	0.121112	56.28673	20.97	25.52
At most 10 **	0.107826	49.74490	14.07	18.63
At most 11 **	0.086713	39.54765	3.76	6.65

4.4 Error Correction Model

According to the Granger Representation theorem, when variables are cointegrated, there must also be an error correction model (ECM) that describes the short-term dynamics or adjustments of the cointegrated variables towards their equilibrium values. ECM consists of one-period lagged cointegrating equation and the lagged first differences of the endogenous variables. Using the Vector Autoregression (VAR) method, we can estimate the ECM. It is necessary to test lag length before running ECM. Table 4 indicates the fitness of lag length is 3 for ECM based on AIC.

Table 4 The Lag Length for Error Correction Model

Lag Length	AIC
1	-50.94671
2	-51.68349
3	-51.95024***
4	-51.77949
5	-51.46102

The results of the Vector Error Correction for all of stock indices, indicates most of stock indices' cointegrating equations are significant existence except for Indonesia, South Korea, and Malaysia that mean those indices' short-term dynamics or adjustments of the cointegrated variables towards their equilibrium values.

For example Japan stock index with all of stock indices indicates Japan index movement is influenced by the other stock indices that means if Japan index diverges from other indices linear relations in short term, then it will be adjusted by next length. It is easily to say that movement of Japan index could be predicted vector error correction model. As dependent variable is Japan index that would be influenced by one lag length of Hong Kong and S&P500 indices, one and two lag lengths of Indonesia, Thailand and Taiwan indices, and one, two, three lag lengths of Philippines and Singapore indices.

Although cointegrating equations for Indonesia, South Korea, and Malaysia are not significant existence, however they are also influenced from some of stock indices. For example, Index of South Korea is influence by one and two lag length of Indonesia and S&P500 indices, two lag lengths of DOW index, one and three lag length of Malaysia, three lag lengths of Philippines, and one, two, three lag length of Thailand.

The results of Vector error correction model accord with Johansen cointegration rank test that all of stock indices have short and long term equilibrium relations.

4.5 Impulse Response

A shock to one variable not only directly affects itself but is also transmitted to all of the other endogenous variables through the dynamic (lag) structure of the VAR.

An impulse response function traces the effect of a one-time shock to one of the innovations on current and future values of the endogenous variables.

This study adopts impulse response to look at each stock index impulse response varieties and last effect. The results indicate most of Asia's returns of stock index are influence from Japan and Hong Kong's impulse of stock index returns. Especially for the short term period Asia's returns of stock index have more volatility than long term period.

On other side Dow and S&P500 do not have obviously impulse from Asia's stock indices except for NASDAQ is impulse from Japan, Hong Kong. That is because US stock indices are leader of world and those behavior are more independent than other nations' index.

The results of Impulse Response for return of all markets analysed indicate US stock indices do not have influence from Asia national return of stock index and only NASDAQ has a little bit impulse from Japan and Hong Kong because two countries' economic development are more depend on high-tech industry as similar as NASDAQ. As looking at most of Asia's returns of stock index are highly influence from Japan and Hong Kong's impulse response. The most of shock periods are one to two weeks that means one stock does not have longer time influence from other national index.

4.6 Granger Causality

Granger causality is part of VAR model and its method based on Granger (1969) to define its causality that if extra adding message in one variable to explain more behavior of others variable. That means study could use Granger Causality to figure out which variable is Granger cause or which one is does not Granger cause or both of variables are no relations and independent each other or both of variables are closely relations and feedback each other.

The results of the analysis of Granger Causality indicate return of Hong Kong stock index Granger causes Indonesia, South Korea Malaysia, Philippines, Singapore, Thailand, Taiwan, and NASDAQ's stock index. Return of Indonesia stock index Granger causes Hong Kong and Malaysia's stock index. Return of South Korea stock index Granger causes Indonesia, Malaysia, and Philippines's index return. Return of Malaysia stock index Granger cause Singapore and Thailand's stock index. Return of Philippines stock index Granger causes Hong Kong, Indonesia, and South Korea stock index. Return of Singapore stock index Granger causes Hong Kong, Indonesia, Malaysia, Philippines, and Taiwan stock index. Return of Thailand stock index Granger causes Hong Kong, Indonesia, Malaysia, Philippines, and Taiwan stock index. Return of Taiwan stock index Granger causes Indonesia, Philippines, and Malaysia stock index.

US stock markets which includes DOW, NASDAQ, and S&P500 Granger cause most of Asia's stock return except for Malaysia.

The results of the analysis also shows that US stock indices are the leader in the world and Hong Kong index is the leader in Asia.

5 Conclusions

This study tries to understand the relationship between Asia's stock markets which include Japan, Hong Kong, Indonesia, South Korea, Malaysia, Philippines, Singapore, Thailand, and Taiwan with US main stock indices which are DOW, NASDAQ, and S&P500. The collected weekly data is from Inform Winner Plus 2000 and study period is from first week of January in 1990 to last week of June in 2007.

This empirical study adopts ADF unit root to check whether variables are stationary or not, cointegration test to determine whether group of non-stationary series are cointegrated or not, vector error correction model to solve variables are short term deviation phenomena. Vector autoregressive model is not like traditional econometric and treating every variables are endogenous in the system as a function of the lagged values of all of the endogenous variables in the system. An impulse response function traces the effect of a one-time shock to one of the innovations on current and future values of the endogenous variables. And Granger Causality to figure out which variable is Granger cause or which one is not Granger cause or both of variables are no relations and independent each other or both of variables are closely relations and feedback each other.

ADF Unit Root. The study used ADF unit root to analyze 9 Asia's stock and 3 US's indices. From the raw data found that non-stationary existed in all of stock indices. After raw data using first difference the results found all of variables achieve stationary.

Cointegration Test. The trends of global stock markets have kept stable relations. This study indicates 9 Asia's stock markets and 3 main US stock indices have existed linear and long term equilibrium relations among them.

Vector Error Correction Model. The study indicates most of stock indices' cointegrating equations are significant existence except for Indonesia, South Korea, and Malaysia that mean those indices' short-term dynamics or adjustments of the cointegrated variables towards their equilibrium values.

Impulse Response. From the result indicates US stock indices are lead for the world that they do not shock from other markets. It is proof that US economic development is a leader in the world and its influence will direct or indirect impact to global markets. As the results found region stock indices Thedevelopment have higher relations than other area market. Markets are belonging to the same region that has similar economic development. Most of Asia's stock markets are more innovation from Japan and Hong Kong than US indices.

Granger Causality. Granger causality defines its causality that if extra adding message in one variable to explain more behavior of others variable. In Asia Hong Kong stock market Granger causes to most of this region stock indices. It is the reason Hong Kong has returned to China since 1997 that it has closely relation with China. It also can say to represent China economic development because good companies' stock in China would like to public in Hong Kong market. Meanwhile many foreign companies use Hong Kong to be set their branch in Asia for entering China market. These reasons create important position of Hong Kong in Asia. The finding also indicates US stock indices have major role in the world especially for DOW that its index dynamic movement has essential indictor in the global stock markets.

References

Arshanpalli, B., Doukas, J.: International Stock Market Linkages: Evidence from the Pre- and Post-October 1987 Period. Journal of Banking and Finance 17, 193–203 (1993)

Banerjee, A., Dolado, J.J., Galbraith, J.W., Hendry, D.F.: Co-intergration, error correction, and the econometric analysis of Non-stationary Data, pp. 1–13. Oxford University Press (1993)

Cheung, Y.L., Mak, S.C.: The international transmission of stock ket fluctuation between the developed markets and the Asian Pacific markets. Applied Financial Economics 2(1), 43–47 (1992)

Dickey, D.A., Fuller, W.A.: Distribution of the estimates for autore-gressive time series with a unit root. Journal of the American Statistical Association 74, 427–431 (1976)

Engle, R.F., Granger, C.W.J.: Cointergration and Error correction. representation, estimation and testing. Journal of Econometrics 55, 251–276 (1987)

Euu, C.S., Shim, S.: International Transmission of Stock Market Movements. Joumal of Financial & Quantitative Analysis 24(2), 241–256 (1989)

Fisher, K.P., Palasvirta, A.P.: High Road to a Global Marketplace:the International Transmission of Stock Market Fluctuation. The Financial Review 25, 371–394 (1990)

Ghosh, A.: Who moves the Asia-Pacific stock markets U.S. or Japan. Empirical evidence base on the theory of cointegration. The Financial Review 34, 159–170 (1999)

Granger, C.W.J.: Investigating causal relations by econometric models and cross-spectral methods. Econometrica 37, 424–438 (1969)

Granger, C.W.J.: Developments in the study of cointegrated economic va-riables. Oxford Bulletin of Economics and Statistics 48(3), 213–222 (1986)

Johansen, S.: Statistical analysis of cointegration vectors. Journal of Economic Dynamics and Control 12, 231–254 (1988)

Ko, K.S., Lee, S.B.: A Comparative Analysis of The Daily Behavior of Stock Returns: Japan, The US, and The Asian NICs. Journal of Business Finance 18, 219–234 (1991)

Nieh, C.C., Lee, C.F.: Dynamic Relationship Between Stock Prices and Exchange Rates for G-7 Countries. Quarterly Review of Economics and Finance 41(4), 477–490 (2001)

Phillips, P.C.B., Perron: Testing for a Unit Root in Time Series Regression. Biometrica 75, 335–346 (1988)

Sims, C.A.: Macroeconomics and Reality. Econometrica 48, 1–48 (1980)

A General Solution to the Mortgage Loans in Mexico

María Berta Quintana León and José Serrano Heredia

Faculty of Accounting and Administrative Sciences, University of Michoacán,
Ave. Francisco J. Mujica S/N Edificio AII Col. Felicitas del Rio C.P. 58030, Morelia,
Michoacán, México
qulb560320@hotmail.com, ppepe10@yahoo.com

Abstract. This article presents an analysis that constitutes an approach to the reality on a general solution for the case of mortgage credit in Mexico. Propose ways of real solution, between the financial institutions-debtors, knowing that there are ways to build a real solution, it is believed that is imperative that we be decided and have a real desire for the settlement to a future that is not seen better. Miss more time is to play that there are no longer options or alternatives in the future, or more complicated and costly. The current solutions have become an important part of the problem, because a real solution only have been delayed, postponed. It is usually thought, that the decisions of the mortgage credit belong to the economic and financial sphere; the decision is influenced by the availability of funds, in addition to other restrictions, and institutional policies.

Keywords: Mortgage credit, Model, Interest, Growth, Financing, Housing.

1 Introduction

Housing is a durable good; it is perhaps the most important physical assets and their price is often several times higher than the income of the applicants. For this reason, the acquisition of housing, in most cases, it is only viable through a long-term credit, mortgage credit, the guarantee is the own house acquired, which permits it to defer the pressure that the price on the income, and thus constitute the modality credit that normally we use in order to buy a home.

The mortgage market changed substantially in the past decade. During the early 2000s the number and range home loans options available to consumers increased considerably. The rapid expansion and subsequent collapse of the mortgage and housing markets continues to spur vigorous debate over consumer's ability to navigate the mortgage market and whether some lending institutions and loan products are in borrower's best interest (Collins, 2005).

Housing cost burden, or the proportion of income dedicated to housing, is linked with long-term economic success, high cost burdens are particularly problematic for lower income households, resulting in more crowded, lower quality residences and longer, more expensive commuting distances to jobs (Redstone Akresh and Diaz McConnell, 2009).

A.M. Gil-Lafuente et al. (Eds.): Soft Comput. in Manag. and Bus. Econ., STUDFUZZ 287, pp. 235–249.
springerlink.com

In Latin American Countries, government housing policies have emphasized promoting homeownership in formal housing markets as the best way of satisfying the housing needs of the population (Jean-Jacques, 2009).

Many families in Mexico and other countries cannot afford it in cash, representing an opportunity to acquire it without making a single time a significant amount of money, meaning that a commitment with who is giving the credit. The mortgages carry in general a lower interest rate that the personal loans, and the time limits for return are much longer, usually 15, 20 or 30 years. The presumption that mortgage markets for low-income borrowers and neighborhoods are undeserved by lenders has led to a variety of increased government interventions on the supply side of the housing market.

Firms engaged in more low-income mortgage lending have higher costs than those engaged in less low-income lending, which is consistent with higher credit risk for low-income loans. Nevertheless, these firms are no more profitable than those that do less low-income lending (Phillips-Patrick, Malmquist and Rossi, 1997).

Mortgage loans, are individual loans to finance the acquisition of a housing, are long-term mortgage between 15 and 30 years, with interest rate which is usually fixed, variable or combination of both (Martinez, 2005). Normally the payments require an income that represents no more than a certain percentage of 25% or 30 %, hiring credit means being committed to pay that amount for many years. At this time in Mexico, approximately $10 pesos for every $1000 pesos of credit monthly (1 American dollar for every 80 dollars of credit). Access to decent housing that may help to improve the levels of welfare and quality of life of the population has become a problem because of population growth complainant about this satisfier and the few sources of funding, in addition to the high costs of intermediation.

In general, riskier borrowers pay higher interest rates to compensate for their higher probability of default, whereas safer borrowers pay lower interest rates, financial markets use credit ratings to asses borrower s probability of default and therefore, determine the corresponding interest to be charged, the lack of credit ratings inhibits the ability of financial institutions to internalize possible information asymmetries, and limits the prospects for economy activity to develop (Barboza and Trejos, 2009).

Several studies have shown that financial systems are more efficient and competitive in countries with low sector entry barriers; making lower active equilibrium lending rates possible (Bhattacharya and Thakor, 1993).

One of the most widely debated sectors in terms of economic competition is the financial system – particularly banks. Although worldwide trends in this sector have been moving towards concentration, the available studies conclude that this is not impairing competition. Such trends can be seen in Mexico, where the banking sector has tended towards apparent concentration resulting from mergers between financial intermediaries, with emphasis on national institutions being absorbed by foreigners (Avalos and Hernandez Trillo, 2008).

- Preserving stability and mitigate the risks of financial institutions in housing, requires the attainment of financial resources in the long term. The Mexican economy differs in many ways from other economies around the world. However, that doesn't make it very different in terms of economy analysis. An assessment of macroeconomic conditions requires an evaluation of fiscal and monetary conditions that has to be done whatever country is being analyzed. In the case of Mexico, is a country that has only recently consolidated its macroeconomic stability much of this success is due to the stability and widespread availability of banking credit (Gomez-Alcala, 2008).

It is therefore necessary to have the existence of:

- Appropriate instruments for a collection of resources with time limits, compatible with those of placements (long term).
- Sources of resources whose nature is such, that tends to be targeted toward long-term investment.

The development of primary mortgage markets, reveals, to a greater or lesser extent, an obvious shortfall in attracting long-term resources for financing housing and, at the same time, a growing recognition of the need to generate new alternatives for the mobilization of financial resources toward the housing sector that link the mortgage lending market for housing to the capital market. Since then, the need for the housing finance systems to develop instruments that mobilize savings for long-term institutional from capital markets does not rule out the traditional scheme of financial market intermediation, but rather to complement this. In this context, is the development of such instruments as the securitization of mortgage assets; although it is not the only way to capture the capital market. The securitization of mortgages consists in the packaging of mortgage assets individual with certain payment flows, which support the issuance of title-standardized values and negotiable that to be purchased by investors in the secondary market, provide liquidity to the originators. The market for financial service has been changing, banks have been permitted to offer additional underwriting services, the market for residential mortgage credit has been transformed, through securitization and the declining role of savings and loans associations (Alton, 1992).

There is extensive empirical evidence that credit risk is an important factor that financial institutions must deal with. The solvency crisis of financial systems is, in fact, the consequence of the accumulation of problem loans over time (Salas and Saurina, 2002).

This work considers that, Mexico as a first step of a definitive solution of the problem of mortgagees-financial institutions (mainly commercial banks), must establish an alternative economic model viable and differently from the current to produce a sustained economic growth with equity, with macroeconomic instruments for overcoming the problems of the real economy in the short term with an economic strategy of long-term, to make the least costly solution giving a better liquidity and solvency to society and in particular to the debtors. While we do not acknowledge the size of the problem, we will have to make do with small solutions, such as until now, delay to implement a public policy, support plan, the economic situation had become more complicated and the more time afternoon in resolved, the financial crisis will be relocated to what is known as "real economy", which is why the markets react to the low.

There are important macroeconomic variables on the housing prices and the stock of houses sold on the national and regional levels. The macroeconomic variables produce cycles in housing prices and houses sold. The housing market is very sensitive to shocks in the employment growth and mortgage rate at both the national and regional levels. In particular, regional housing prices reflect regional employment growth, as well as national mortgage rates. The economic variables have a different impact on the dynamic behavior of housing prices and the number of houses sold in different regions at different time periods and that these economic aggregates alone cannot explain the fluctuations in real estate values and construction levels that occurred in some regions (Baffoe-Bonnie, 1998).

A late decision hurt more what originally had happened, and makes it more likely that there are problems in the Mexican economy, many people recommend act quickly and implement the necessary measures without losing any time, and "saving" the banks not the bankers. Over the past 15 years the banking system in Mexico has undergone major transformations and its development reveals a traumatic track record (Rivera and Schatan, 2008), that is why the state must return to assume its role, more efficient, with better responses especially during times of crisis. The comprehensive proposal of principles of solution of the mega problem between financial Institutions (banks) and mortgagees (nonperforming loan portfolio) set out:

- Objectively evaluates the financial possibilities of the real banking system.
- Really considered the financial possibilities of the treasury and the cost/benefit prosecutors of the solution.
- Adjusts to the specific realities of debtors.

2 General Basic Principles of Solution

A solution should try to answer two questions: what economic and financial measures should be taken? Who (and how) must pay for the operation of the financings?

One of the major causes of this problem, is the very serious and disastrous situation in the national economy, which on many occasions this in an emergency situation and not a short-term, which suffers from structural deficiencies that require a rapid solution and fund, radical, which will enable all move forward, since a worker, employee, currently earn on average less than 1/4 part than it received 25 years ago and despite what it may say have not been generated jobs at the pace required by the nation, and each day we are more Mexicans in working age. The current economic growth has been lower than 3% on average, therefore GDP per capita has been reduced, and there are only the parastatal of PEMEX, CFE (Mexican Petroleum and Federal Electricity Commission), which produce a great part of the GDP, the other has already been sold.

Notwithstanding the financial crisis, housing finance remains a booming industry globally with strong growth potential in developing economies. Especially in the subset of countries dubbed emerging markets, the sector's risk environment and available risk management options have changed dramatically in the past two decades reflected by low inflation, the development of funding markets, deregulation, communications development, and the globalization of financial services. However, elevated global house price risk and mortgage loans defaults that already materialize in some of the new markets are the price to be paid for fast grow, while access to housing finance for low-income households still remains at a nascent stage (Hans-Joachim, 2011)

Nowadays, it is (external debt) approximately 175.000 million dollars between private initiative and government (the total domestic and external public debt reached at around 150.000 million dollars and its cost represents around 43.000 million dollars, and it is dedicated 2.5 times more money to the payment of the debt that public education, for which payments are made every year 15.000 million dollars of interest that now represents 6% of GDP in the current account. This means that if the balance of trade was balanced, there would be a continuing shortage of 6% of GDP in the current account to harbor with loans and equity investments with interest rates of 9 to 12% per year, which are very high in international standards, this is due to the high risk of that Mexico might be declared in moratorium or suspension of payments. Therefore the interests of 15.000 million dollars and a write-down of the same amount would represent approximately 10% of the national GDP, needing dollars or other hard currency to pay, import (Mexico had a current account deficit equivalent to 0.5 % of the Gross Domestic Product (GDP) in the first quarter of 2011, in the midst of a deterioration in the external accounts derived from the global crisis, as reported by the Bank of Mexico, what is equivalent to 1.075 million dollars, an amount more moderate than the 2.576 million the same period of the year 2008, equivalent to 1.0 % of GDP, which are derived mainly from:

- Exports.
- Services (tourism).
- Credits (interest payments).
- Foreign productive investment.
- Speculative direct investment.

A model of mortgage loan, in several countries, whose traditional machinery and even dominant of the formal financial sector in housing finance, is depository institution retail. In this model, an institution purchases savings of companies and individuals, some homeowners and provides loans to homebuyers. Taking savings of non home buyers, deposit-taking institutions can access a wide variety of funds to a cheap cost. There are various types of depository institutions, including commercial banks that offer a full range of banking services, savings banks that cover the housing sector, and specialized financial institutions in the housing (societies of builders, savings and loan associations) that focus on primarily their loans in the housing (Lea, 2008).

The modernization of the banking system, its growing internationalization bank disintermediation processes and the importance of competition in ensuring the industry's efficient performance raise a series of institutional problems of prime academic and public-policy interest (Rivera and Schatan, 2008)

A key feature of this model's deposit is:

- Anchor, is mainly of the deposits retail savings people, but also these institutions can issue bonds.
- The interest rate is usually variable in the short term to the deposits, compared with the long-term loans for housing

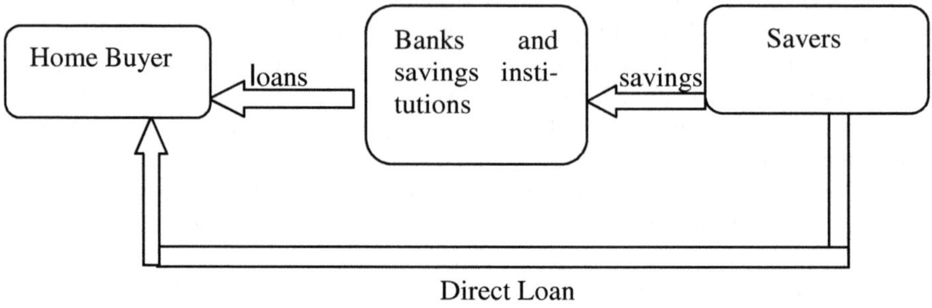

More clearly, the model's most traditional mobilization of financial resources for housing is the one that is based on the intermediation of the Financial System. In the participating banks or financial institutions in multiple object - that is to say, which are financed by various activities, including housing - and still prevails in many Latin American countries, financial institutions specializing in mortgage loans for housing, as a primary source of funding for this purpose, take household savings, primarily through deposits in relatively short periods. The credits generated by these institutions remain as assets on their balance sheets throughout the period of validity and are also responsible for the administration and collection of those credits. In other words, perform all the duties of the mortgage process: collection of resources, origination and administration of loans (Gonzales Arrieta, 2005).

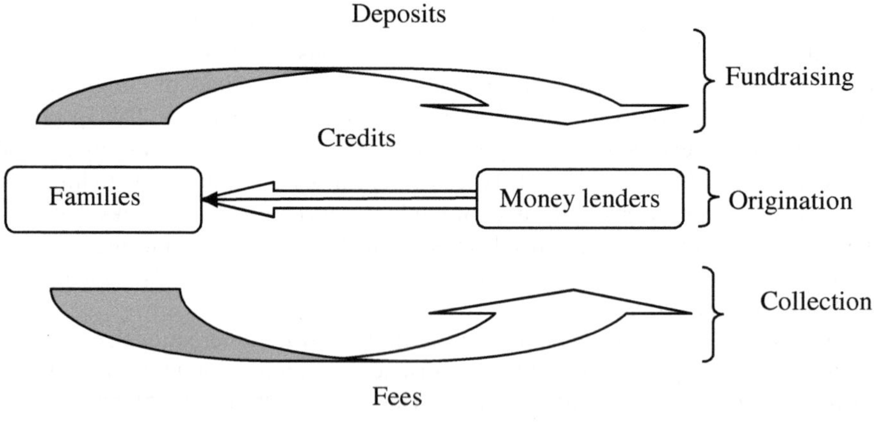

An emerging feature of mortgage markets is the functional separation in which specialists develop several functions to mortgage loans, this model of mortgage loan; it involves a financial institution that develops the increased role of service, call origination, mooring and risk management (Lea, 2008)

Commercial banks in developing economies are traditionally reluctant to enter lending to low-income customers, reflecting high transaction cost, perceived low creditworthiness of low-income borrowers and inadequate product and underwriting imposed by regulation barriers, and lack of competition associated with subsidies for a few incumbents, and cartel-style market structures (Hans-Joachim, 2011).

3 A Comprehensive Settlement

A realistic solution to the problem for the banks and debtors must comply with the following general scheme:

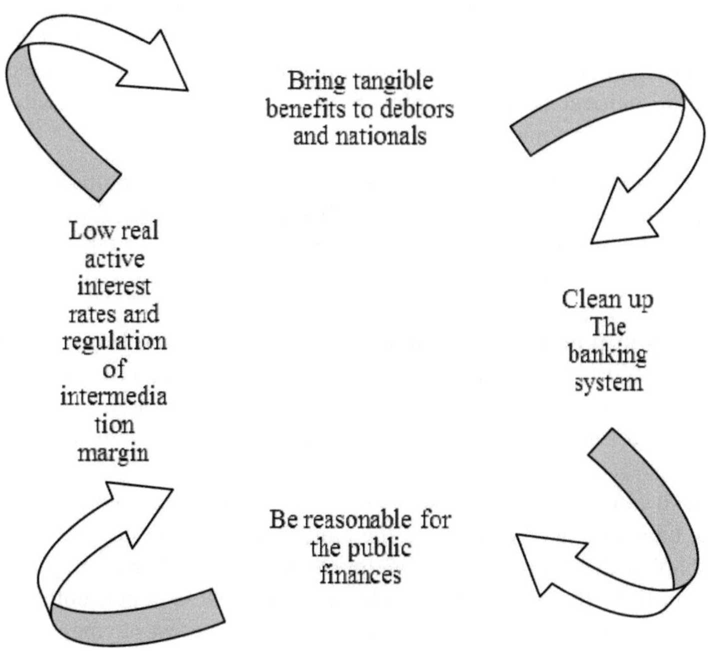

This proposal, would yield better results than the current, this is demonstrated in several ways such as; empirical universal evidence that show:

- There is not a single country that has had a sustained economic growth under a neoliberal strategy.

- Countries that have implemented this strategy have suffered enormous damage and have had to rectify to revive growth.
- Countries that have implemented economic strategies of the type of the proposal of this work have been successful.

Empirical Mexican evidence, that proves:

- The discordance between the great declared objectives of the neoliberal strategy and its actual results.
- The harmful effects of the neoliberal strategy in the real economy
- Excessive social costs of the strategies, neo-liberal, contrary to the ultimate end of all economic strategy truly national, which is the rise of the welfare.

4 Explanation

The selection of policies needs to consider different attributes with the occurrences of different scenarios in the future, and search for the policy that better adapts to the different situations that could occur (Gil-Lafuente and Merigó, 2010). We describe the step process to follow, in order of solve this issue. First, we present the general decision approach and then a comment to each of them with additional data in some cases.

1. *General renegotiation of Mexico's foreign debt*, with rescheduling of payments of principal and interest, in accordance with the actual financial capacity of the country and with a grace period to allow return to economic growth.

This is necessary condition for being able to generate jobs and, consequently, the people have the money to be able to cope with their debts, and not to fall into non-performing loan portfolio or with real possibilities of exit from it, as the non-performing loan portfolio is due largely to the people with bank debits that he lost his job or that their incomes remain the same, but with a purchasing power less and are unable to find a better-paying job, simply because it does not exist in our country.

Therefore it is real capacity to serve its external debt and that is attractive for the creditor banks without calling for the cancellation of the debt or reduction to more than half of the interest, but if by approximately 30 %, reducing the interest rate of the credit to the developing world until the 5 % -6% per annum as maximum. This renegotiation is indispensable because of interest payments on foreign debt during the next 5 years will represent more than 5% of annual GDP and the depreciation committed represent another 6% of annual GDP, so that if Mexico is committed to serving their terms current external debt, will need to provide more than 10% of GDP.

Mexico should pursue grow at a minimum rate of 5% to generate at least 1 million jobs a year new defendants by new generations (not counting 0.2 million migrants per year to the United States) requires a coefficient of gross fixed investment of 24% of GDP. This means that to serve its external debt in the terms currently hired and at the same time achieves a growth of 5% per annum, Mexico will require a coefficient of gross domestic savings of 34% of GDP, and coefficient has never been reached in Mexico, the maximum has been made of the 22.4 % of GDP under conditions of growth. The choice is; or we paid in the current terms but do not grow at the levels required or not paid in the current terms but if we can get grow. The government has been refinanced their depreciation, but this have generated and will continue to difficulties as:

Mexico may not even pay the interest on their external debt, 5% of annual GDP in round numbers, and at the same time grow at 5% annual, the most that Mexico can pay to abroad is a 2% of GDP, as a net flow of external debt (feasibly financed with foreign direct investment or physical, not speculative).

The strategy of to borrow to pay maturity leads to a frequent need to borrow urgent short-term for very high interest, or to perform any placements of role extremely onerous, as experience has shown Mexican and international.

2. *Reduce the debt service and to defer as much as possible the payment of principal.*

3. *Exchange with banks, debt for growth*, paying interest as they are always calculated and when the annual growth exceeds national 4 %, less than this may not transfer resources to the outside. Provide an annual payment of (The excess or shortfall on the interest would be accumulated or deducted from the capital):

60% Of interest, if the country grows less than 2%
80% " " between 2% - 4%
120% " " between 4% - 6%
150% " " higher than the 6%

4. *Mexico not to declare moratorium* only establishes a procedure where payment is subject to the distribution of resources between Mexicans and foreigners, while faster grow Mexico more money they will receive their creditors. If Mexico would grow 7% per annum it would pay the 150 %, that is to say, all the interests (15.000 million dollars) plus 50% of this amount to the main (7.500 million).

5. *An Exchange-rate policy competitive* that avoid in the future an overvaluation of the currency coin (peso), from a level of balance that must be taken as exchange floor, taking for this 2 options:

As soon as the Bank of Mexico has the reserves necessary to regulate the rate of exchange, it must leave a free floating regime or adopt a new floating band, with periodic adjustment of the floor and ceiling exchange agreement with the differential in the inflation rates of Mexico and the United States, which would be in a gradual manner and not drastic.

While there is no foreign exchange reserve sufficient to regulate the exchange rate on a floating band, the Bank of Mexico must be taken to prevent price of the dollar down from the floor exchange rate balance adjusting this periodically on the basis of the differential inflationary (a type of change permanently competitive.

This will help to avoid competition from imported products will ruin to Mexico's productive capacity, and at the same time it will avoid that Mexico should continue its spiral of indebtedness to pay excessive imports of goods.

6. *A Pragmatic Trade Policy* to use the maximum margins of maneuver to regulate our foreign trade, applying (as they do in the United States and Canada and countries with successful development) tariffs, technical standards, safeguards and provisions against unfair trade practices, to which you have right according to the NAFTA (agreement signed by the governments of Canada, Mexico, and the United States, creating a trilateral trade block in North America, came into force on January 1, 1994. In terms of combined GDP of its members, as of 2010 the trade bloc is the largest in the world) and the GATT-WTO (The General Agreement on Tariffs and Trade, was negotiated during the UN Conference on Trade and Employment and was the outcome of the failure of negotiating governments to create the International Trade Organization (ITO). Was signed in 1946 and lasted until 1993, when it was replaced by the World Trade Organization-WTO in 1995), undertaking some negotiations to introduce safeguards in branches of production and the investment of high vulnerability and that are relevant to Mexico because of its economic importance or as generators massive job.

In other words, do not open the borders indiscriminately; set rates, measures, subsidies to monitor the flow of goods and NOT unprotect NATIONAL INDUSTRY. Critically review the trade policy with the countries with which are recorded the larger deficit (China, $27,000 millions; Korea, 12,000 million; Japan, 14,000 million; European Union, 19,000 million) for the purpose of negotiating compensatory financing or invoke the safeguard clauses of the free trade treaties.

7. *Implement real industrial development policies* and for promoting agriculture supported in a true planning for appropriate macroeconomic policies (competitive exchange rate, interest rate that encourages investment, trade policy pragmatic). With certain instruments, such as:

Instruments for promoting general economy

- Infrastructure construction.
- Development of scientific research - technical.
- Education and job training.

Specific instruments of promoting sectoral

- Tax incentives to new industries, technological innovation, and technology transfer.
- Credit Support to small and medium-sized industry, studies to support feasibility, through the national bank of development, market studies.
- Vigorous Promotion of external products with substitution of imports with a priority attention of the internal market which includes the production of mass consumption goods non-tradeables, guaranteed prices for agricultural commodities.

All of this would increase the domestic supply of products and of exportable goods by reducing pressure on the external sector, at the same time that would generate jobs and income with multiplier effects on investment, production and employment.

8. Maintain a flow of foreign exchange balanced, which will depend on:

A. - Productive activities	B. - Financial adjustments
Trade balance	Treasure bonds (dollars not generated)
Balance of services	Foreign direct investment

There is now approximately a real deficiency in the trade balance of 18.000 million dollars and the balance of services of 10.000 million dollars with a direct investment of 6.000 million representing 2% of GDP, the financial adjustment to have a flow of foreign exchange balanced would be 20.000 million (treasure bonds) not to resort to the use of international reserves.

9. Search for having a checking account even though negative but manageable, as well as a prudent management but flexible public finances which will allow for the use of instruments of income-public spending to regulate the economic cycle and promote development, but without falling into populist excesses. They should be done grow the domestic market to foreign direct investment come by reducing uncertainty, since the necessary today is of the order of 30% of GDP, with a need for domestic savings of more than 25% of GDP in order to have a prosperous economy.

10. Breaking down the rate of interest through the elimination or severe reduction of the current account deficit which will decrease until a level healthily fundable with foreign direct investment or physical, the requirements of external savings and hence the need to pay surcharges of interest in liquid resources. The reduction of financial intermediation margins from its current level to its historical average of 7% real, through a financial agreement concluded by the government, bankers and productive sectors that at the same time inject prudent liquidity to the economy, restructure bankruptcy according to the real capacity to pay of the debtor and to give oxygen to banks that they were in a likely virtual technical bankruptcy by the accumulation of portfolio unplayable. Should ease monetary policy to use prudently our room for maneuver in the expansion of domestic credit (difference between the monetary base and the value in national currency of the international reserves of the bank of Mexico).

11. Do not reduce current spending, but keep it at least to its level in the year 1994 and with an increase in public investment in 1% of GDP compared with 1994, using sources non-inflationary financing, through the progressivity of the tax burden on the layers of very high-income (increasing taxes on the very rich, with personal income increased to 200.000 dollars, which may mean additional tax revenue of a 2 % -3% of GDP), These must be favorable to the states so that they can expand their spending on infrastructure and economic development, to do this:

Is necessary a fiscal balance in order to operate with a budget deficit moderate, much more during an economic emergency for growth (countries of successful economic development such as Germany, currently with a fiscal deficit of 4.5 % of GDP, Japan, Italy, operate with a flexible approach in the management of public finances, which is indispensable to regulate economic cycle). A moderate expansion of public investment that will allow triggering economic growth, each percentage point of GDP in public construction would generate around 300.000 direct jobs, with enormous multiplier effects on the economy, with a prudent expansion of the current expenditure would meet pressing needs.

12. *Avoid the fall of the internal market* for mass products through the conservation of the purchasing power of wages to the level of 1994 (proceeding, after surpassed the economic emergency, to its gradual recovery toward wage levels prior to the neoliberal model).

That is why it is necessary indexing wages to prices, with quarterly adjustment, retroactive, avoiding both the impoverishment of the working population, such as the vicious circle recessive caused by the fall of the internal market, with a negotiating strategy. To persuade creditors to assume an attitude of cooperation in order to overcome the root of the Mexican financial crisis and accept treatment of external debt consistent with economic growth and consequently acquire real capacity to pay.

Foreign investors and creditors are aware of the external risk and exchange, brokerages do not in fact give Mexico the investment grade high but always the country risk rating speculation, that's why charge high interest rates that include a currency risk premium, it is necessary to assume part of its risk. Mexico should not return to the voluntary market capital only marginally if there is no economic growth that is why it is necessary and feasible renegotiation and rescheduling of payments of principal and interest, according to the financial capacity of the country and with a grace period to revive growth.

Renegotiating with the International Monetary Fund (IMF) and the united states, to see that recessionary policies contained in the letter of intent and in the framework agreement, have already been experienced in Mexico with enormous economic and social costs that it is no longer possible in the country use the same and that in that way were not generated own resources to serve the external debt. This will not be possible to implement it if the Mexican government continues to believe that the neoliberal strategy is the only option.

13. Must be motivated domestic savings (financial savings, economic savings), by providing positive real interest rates. As you know if the real interest rate is negative it is better to consume now that save, as a real comparison the real interest rate of 28-day Cetes (Certificates of the treasury)was 16.65 per % (30/September/ 1997), 8.03 % (2/December/ 2008), currently is 4.26 (4/March/ 2012).

On the other hand, the interest rate on savings accounts for people lower and middle class, is very low and is an incentive to consume the negative real interest rates, so the money is spent as soon as it is received, this must be removed, a good instrument are the Udis (investment units) since they generate a small minimum actual yield higher than a saving any or promissory note, but these instruments are too rigid, this is the result of poor planning of the unequal economic structures that encourage the immediate consumption and eliminates any possibility of savings.

14. *Avoid the concentration of power*, at the end of 2011, even the country's capital (Federal District) concentrated 2/3 of the total savings, producing only ¼ part of the national GDP, which only helps to concentrate the economic and political power that is not good for the country.

15.*The savings should be used for productive investment* and the financial system must be subordinated to the interests of the real economy, while maintaining the commercial banking system in private hands but locking it to the real surveillance and regulation of the state, which must submit to public control interest rates and establish drawers credit discarding the neoliberal myth that the market sets interest rates active and passive, when in fact it is of oligopolistic prices influenced only by the financial groups

16. *Deploy an anti-inflationary strategy* that, with priority on the reactivation and the sustained growth of the real economy, containing inflation gradually reducing it to a strip of 0 % -5 %, (6.17 % 20/may/2009, 6.08 30/January/2009, 5.34% 14/January/2012) by deploying the previous economic policy proposals that will increase the supply of products and services along with the demand is therefore to discard the obsession to rapidly reduce the rate of inflation, as well as the pernicious monetarist orthodoxy that knows only stabilize prices through fiscal policies monetary and credit and wage severely recessive who contract consumption depress sales, hold back investment, reduce output and make it fall employment, which produces only a disinflation ephemeral unsustainable in the long term.

17. *The government should never return to acquire a bank that cannot resolve their problems* and want to keep them via forms of financing. In doing so, it must resell them that it is better and cheaper to keep them.

18. *Requires banks to improve their management and business management*, as well as modify its rationality short-term profitability, applying a true financial planning in line with the reality and current situation prevailing in the country.

5 Conclusions

In the world, housing finance is growing at an unprecedented rate. In the last decade, the mortgage debts have increased. The majority of the developed economies have experienced a sharp rise in the levels of debt used to finance.

This is also the case for a significant number of emerging markets such as Mexico, despite the fact that housing financing remains underdeveloped in many parts of the many parts of the developed world. Macroeconomic instability with inflation and high interest rates, low growth, legal systems weak that don't adequately protect the interests of the lenders, in addition to an underdeveloped infrastructure markets for housing financing, also as underdeveloped and poor banking regulations and capital markets.

An economy and stable growth, low interest rate, seem to be the key in the expansion of the mortgage market. In the past, interest rates in many countries have fallen significantly with a parallel increase in mortgage debt, in spite of this, the rate of mortgages remains high in emerging markets. In the past decade, a consensus has emerged "conventional commercial banks are unable or very slow to provide access to sources of funding in emerging markets". To meet with adequacy mortgage loan demand in the next 20 years and to develop alternative mechanisms of financing, it is essential to develop a market for de-mortgages in which to participate various institutional investors, domestic and foreign. Some structural reforms, privatization of pension funds, the introduction of life insurance, are providing a resource base for promising long-term investment and the emergence of institutional investors. Also, the international capital market can provide an additional demand for securitized bonds and induce the adoption of more demanding standards.

References

Alton, R.: Gilbert.: The changing market in financial services. In: Proceedings of the Fifteenth Annual Economic Policy Conference of the federal reserve bank of St. Louis. Springer, St. Louis (1992)

Arellano, R., Castañeda, G., Hernández, F.: The fragility of the financial system and its implications on the current account. Mexican economy, new era. Second Half II(2) (1993) (in spanish)

Avalos, M., Hernandez Trillo, F.: Banking Competition .ECLAC, Mexico, United Nations (2006) (in spanish)

Barboza, G., Sandra, T.: Micro Credit in Chiapas, México: Poverty Reduction Through Group Lending. Journal of Business Ethics 88 (supp. 2), 283–299 (2009)

Beliakov, G., Calvo, T., Pradera, A.: Aggregation functions: A guide for practitioners. Springer, Berlin (2007)

Calva, J.L.: Crisis of the debtors, UNAM. Institute of Economic research, Mexico (1997) (in spanish)

Michael, C.J.: Mortgage Mistakes? Demographic Factors Associated with Problematic Loan Application Behaviors; from the issue entitled "Thematic Issue on Family Finance". Journal of Family and Economic Issues 32(4), 586–599 (2005)

Figueira, J., Greco, S., Ehrgott, M.: Multiple criteria decision analysis: State of the art surveys. Springer, Boston (2005)

Fodor, J., Marichal, J.L., Roubens, M.: Characterization of the ordered weighted averaging operators. IEEE Transactions on Fuzzy Systems 3, 236–240 (1995)

Gil-Aluja, J.: The interactive management of human resources in uncertainty. Kluwer Academic Publishers, Dordrecht (1998)

Gil-Lafuente, A.M., Merigó, J.M.: Decision making techniques in political management. In: Lodwick, W.A., Kacprzyk, J. (eds.) Fuzzy Optimization. Springer, Berlin (2010)

Hans-Joachim, D.: Economic research in a financial group in Mexico. Business Economics 43(1), 55–59 (2008)

Gonzales Arrieta, G.M.: Mortgage credit and access to housing for lower-income households in Latin America, vol. 85. Cepal magazine, Mexico (2005) (in Spanish)

Hans-Joachim, D.: Housing Finance in Emerging Markets. In: Regulation and Access to Finance, pp. 83–117. Springer (2011)

Baffoe-Bonnie, J., John: The dynamic impact of macroeconomic aggregates on housing prices and stock of houses: A National and Regional Analysis. The Journal of Real Estate Finance and Economics 17(2), 179–197 (1998)

Lall, S.V., Freire, M., Yuen, B., Rajack, R., Helluin, J.-J.: Urban Land Markets: Improving Land Management for Successful Urbanization. Springer, Heidelberg (2009)

Lea, M.: Structure and Evolution of Housing Finance Systems. ch. 2 (2008), http://ihfp.wharton.upenn.edu/2009Readings/Lea%20-%20Structure%20and%20Evolution%20of%20Housing%20Finance%20Systems.pdf

Martinez, J.L.: Interest Rates and recent developments of bank credit in Mexico (1995-2002). Editorial porrua, Mexico (2005) (in spanish)

Phillips-Patrick, F., Malmquist, D., Rossi, C.: The economics of low-income mortgage lending. Journal of Financial Services Research (1997)

Redstone Akresh, I., Diaz McConnell, E.: Housing Cost Burden and new lawful immigrants in the United States. Springer, USA (2009)

Rivera, E., Schatan, C.: Competition Policies in Emerging Economies Lessons and Challenges from Central America and Mexico. Springer, New York (2008)

Salas, V., Saurina, J.: Credit Risk in Two Institutional Regimes: Spanish Commercial and Savings Banks. Journal of Financial Services Research 22(3), 203–224 (2002)

The Spanish Savings Banks: Analysis of Its Efficiency in Its Reorganization Strategy

Milagros Gutiérrez Fernández and Ricardo Palomo Zurdo

Department of Financial Economy and Accounting
University of Extremadura, University CEU San Pablo, Spain
mgutierrezf@unex.es, palzur@ceu.es

Abstract. In this paper, we analyze the evolution of saving banks from the point of view of their overall importance in order to interrelate them and be able to assess their progress in the decade preceding 2010, the year, together with 2011, in which the largest reorganization of this sector has taken place, going from 46 to only 15 institutions (including CECA), as a result of the survival of some of them and the merger processes or the creation of Institutional Protection Systems (IPS). Using a methodology based on data envelopment analysis, we evaluate their comparative efficiency and their position within the merger processes which start from 2010, showing significant differences between organizations.

Keywords: Savings banks, IPS, efficiency, data envelopment analysis; bank reorganization, mergers.

1 Introduction

This paper aims to analyze the degree of performance in achieving efficiency on behalf of the Spanish savings banks. These financial institutions of the social economy, with legal basis, represent half of the Spanish banking sector, and given its business model and the concentration of its activity in the home sectors, a considerable number of them were affected by the financial crisis that started in 2007 - 2008, but also by other factors like non-regulation, new technologies and internationalization (Howcroft, Ul-Haq and Hrmmerton, 2010; Värlander and Julián, 2010).

With a market share of 57% of Spanish mortgage offices and a network of over 25,000 branches, according to the Bank of Spain, savings banks have finalized the largest and most intense process of reorganization or restructuring that has taken place in Spanish banks, from which important consequences will be derived, among them the necessary rationalization of the branch network and a rethinking of the business model by some institutions with a strong territorial presence, an essentially domestic policy area and a social retail banking character.

There are 15 organizations, groups or savings banks concentration processes as of January 2012 that includes all the previous 46 organizations (including the

A.M. Gil-Lafuente et al. (Eds.): Soft Comput. in Manag. and Bus. Econ., STUDFUZZ 287, pp. 251–265.
springerlink.com

CECA)[1]. They have emerged mainly by the grouping of savings banks originating within the same region, although, in some cases they have crossed their regional boundaries, being a prime example, particularly, the biggest merger by assets: in the case of the IPS (Institutional Protection System) "Bankia" composed by seven institutions among which Cajamadrid and Bancaja stand out due to their large volume of assets.

The working hypothesis is to verify, through data envelopment analysis or DEA, if the grouping criteria that have led to mergers or IPS that have taken place (especially during the spring of 2010), subject to a clear time and institutional pressure by the Bank of Spain, as well as the need to comply with the deadline given for the implementation of FROB[2], have taken into account efficiency and similarity criteria with regards to the level of performance achieved during the previous decade to these processes by the organizations that have established links with each other in 2010 and 2011.

Most of the researches studying the efficiency of banks are based on comparisons prior and after the merger, analyzing the effects. However, the aim of this study is not to analyze the extent to which there is or has not been an improvement in the efficiency of the organizations that have merged or formed an IPS, but to see if such mergers were taken into account considering economic efficiency and similarity criteria in the performance of the organizations that have merged. For that reason, we use the main nonparametric estimation technique for border efficiency, the data envelopment analysis or DEA, as applied to savings banks in Spain during 1999-2009.

Therefore, the paper has been divided into four separate sections, in addition to this first introductory section. The second section provides an analysis of the evolution ensued by the savings banks sector in Spain, especially as it affects the reorganization through mergers that these entities have carried out over the past ten years. The third section justifies the methodology used in the study to refer to several previous papers, both domestic and international, which have analyzed the efficiency of the entities which constitute the banking sector. The fourth section proceeds with the empirical study and results are discussed. The fifth and final section provides the conclusions of the study.

2 The Spanish Savings Banks: Evolution and Reorganization Processes

According to sources from the Bank of Spain, the 45 savings banks (in addition to CECA) control a market share equivalent to half the Spanish banking sector (50.16% of deposits, 47.43% of loans and 57% of mortgage loans), having experienced a spectacular growth in their sales network and an extraordinary

[1] The Spanish Confederation of Savings Banks (CECA) is the representative body or employers' association, which also acts as a savings bank. Given its uniqueness, it has been excluded from the analysis in this work.

[2] The Bank Restructuring and Reordering Fund created by Royal Decree 9/2009, initially endowed with 9,000 million euro, expandable to 99,000 million.

process of expansion outside their areas of origin, particularly intense since the late nineteen nineties.

However, the general trend has been severely threatened by the effects of the financial crisis (Figueira and Nellis, 2009), which has been the decisive catalyst for the reorganization and corporate reshape[3]. Thus, in 2010 and 2011 there have been intensive merger processes which, in some cases, have been formalized through the establishment of joint stock companies with shares from the savings banks that make up the IPSs, which opens the door to future capitalization or "privatization" allowing capital from institutional and private investors. In this sense, the fundamental legal structure remains reduced to the management of the social work that characterizes these companies, which is the reason for having a dual structure that contemplates the financial activity framed in a shareholders company and, in parallel, survives the essence of the savings bank as a fundamental legal form for the management of social work.

The results of the restructuring process of Spanish are shown in figure 1. In total, the number of savings banks has diminished of 45 existing entities in December, 2009 to 14 entities or groups in January, 2012. The formula most used in the current bank reordering has been that of the institutional systems of protection, in which the members support their own identity and their operative capacity in the commercial plane, as well as their economic independence.

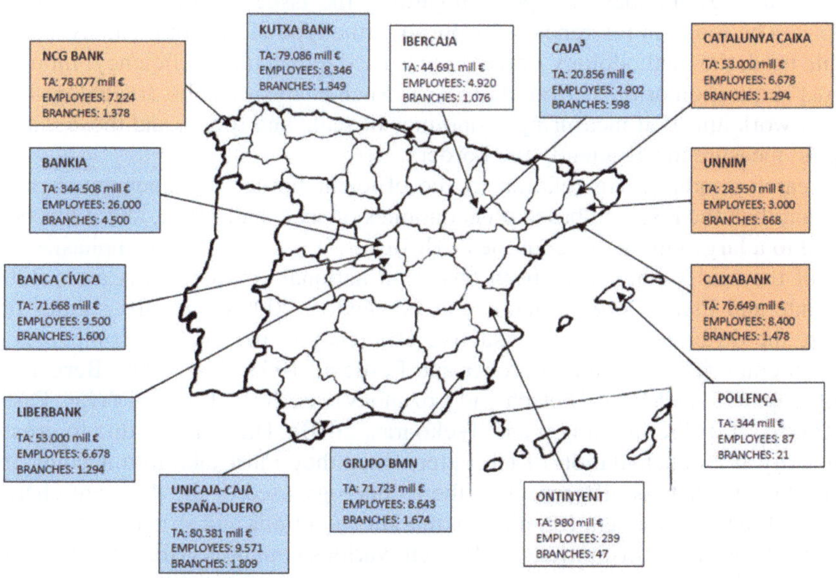

Source: Own elaboration (Blue: IPS; Orange: Merger; White: Without process).

Fig. 1 Main magnitudes and headquarters of the processes of integration of the savings banks

[3] In any case, there is a wide variation in the balance sheet, profitability indicators and bad debts of these organizations. Thus, while some continue to stand out for their healthy situation, other cross serious difficulties and two of them have required the intervention of the supervisor authority.

3 Methodology of the Analysis

Once the process of restructuring that has taken place in the savings banks has been studied, it is considered important to determine the efficiency of such entities. Thus, according to Berger, Hunter and Timme (1993) it is precisely in situations of change, as is the case today, when regulation changes that were previously prohibited or limited allow new possibilities for expanding the entities. It is the moment when this question tends to arise with greater intensity, despite of being affected by the above circumstances when it is more difficult to analyze it (Calvo, 2002).

Generally, research studying the efficiency of banks is based on comparisons made prior to and after the merger, analyzing their effects (Ismail, Davidson and Frank, 2009). However, the aim of this study is not to analyze the extent to which there is or has not been an improvement in the efficiency of the organizations that have merged or formed an IPS, but to see if such mergers were taken into account considering economic efficiency and similarity criteria in the performance of the organizations that have merged (Chao, Yu and Chen, 2010).

The analysis of the economic efficiency of financial institutions is one of the issues has been the subject of extensive study work on the banking sector, both from a national and an international perspective, giving rise to an abundant literature covering a large number of aspects regarding the issue. Thus, various research works can be found in reference to its purpose, the type of institution analyzed, the sample used, the methodology or techniques used to calculate efficiency, inputs or outputs in question or the period of analysis. Such analysis can be divided into two groups: work aimed at measuring economies of scale and scope, and those that attempt to measure the efficiently the borders.

Research conducted in relation to the efficient border has been much more abundant than the research based on economies of scale and scope, and it has been applied to a large number of countries[4]. However, it is important to emphasize that most of the said studies have been based on national comparison (Calvo, 2002; Belmonte and Plaza, 2008; Maudos and Fernández, 2008), since the studies that analyze the differences in the productive efficiency in the banking companies of different countries are limited (Fecher and Pestieau, 1993; Yeh, 1996; Berger and Humphrey, 1997; Chaffai, Dietsch and Lozano-Vivas, 2001; Lozano-Vivas, Pastor and Hasan, 2001; Kontolaimou and Tsekouras, 2010). This may be due to greater complexity in the calculation of the latter, since they must take into account important factors such as differences in the technology used or the different characteristics of the environment (regulation, market conditions and structure, economic conditions, etc.) that institutions undergo in various countries (Bikker, 2001). Not taking them into account can carry to erroneous conclusions on the efficiency, the power to confuse productive inefficiencies with advantages or adversities in the middle in which the banking companies operate that prevent them from developing the activity optimally (Lozano-Vivas, 2001).

---

[4] The works of Berger, Hunter and Timme (1993), Berger and Humphrey (1997), Lozano-Vivas (2001) or Calvo (2002), among others, contain a good summary of the literature on studies of efficiency in financial institutions in several countries.

This paper will study the economic efficiency of the sector of savings banks using for it the main technique of estimation not parametric of efficient border: data envelopment analysis or DEA. This methodology was developed initially by Charnes, Cooper and Rhodes (1978), and as a nonparametric technique it does not require the prior association of a determined functional form to its function of production, for which is an adequate technique for those companies that do not have a marked commercial end, in which they intervene a multitude of outputs and of inputs, and among the ones that certain technological homogeneity exist (López et al., 2002). Likewise, the search of an efficient border that permit to discriminate among the diverse positioning of each company keeping in mind input/output variables turn out to be very convenient given the objective of the current investigation. Keeping that in mind, the application of DEA is considered adequate for the study of the economic efficiency of the Spanish savings banks.

4 The Measure of the Performance and Efficiency of the Savings Banks by Means of the Data Envelopment Analysis DEA

When applying the DEA methodology to the analysis of the efficiency it should be followed by a series of steps. Based on activity data (inputs and outputs) available for an assembly of productive activities, it will be decided which is the assembly of feasible activities. Next, the index of efficiency whose value is desired to calculate should be defined, for which it will be necessary to build a mathematical program that is able to estimate the said index from the available data and the proposed technology.

From said algorithm of calculation and using mathematical programming techniques, envelopment will be built in the linear lines that will determine the border formed by the companies that hypothetically will be considered as efficient. The inefficiencies of the rest of companies will be evaluated measuring the distance that separates them from said border.

In short, we can say that DEA technique allows us to determine a multidimensional ratio that provides a classification, a hierarchy or ranking of companies in function of their different levels of performance (Escobar and Guzmán, 2010).

4.1 The Election of the Banking Input and Output

It is one of the most compromised tasks of the efficiency analysis. The savings banks are banking companies, which is why the identification and the measurement of the inputs and outputs that determine their efficiency is not simple. This is due to their multi-product character, therefore their benefits originate both from its function as intermediary agent between investors and plaintiffs of funds, but also to offer their clients a range of not necessarily complementary services to the own function of mediation (Lozano-Vivas, 1992). Likewise, it should be considered that the product of the banking companies is intangible, heterogeneous and of joint and interdependent production, this complicating the previous tasks.

The said particularities of credit companies involve being able to be conceptualized from two perspectives: that of production and that of mediation (Humphrey, 1985). The study object case will be based on the focus of mediation, proposed by Sealey and Lindley in 1977, therefore, continuing authors as Calvo (2002), this seems more appropriate when is desired to measure the efficiency of the banking company.

At the moment of selecting the inputs and outputs that are going to be incorporated in the model, the anxiety of the investigator should be directed to the inclusion of a reduced number of the same (Cooper et al., 2000), therefore otherwise it will be more difficult to find the most efficient group (Belmonte, 2007).

This way, the following inputs originating from the collecting of clients' funds have been considered, of the utilization of human and physical capital, and of other necessary expenses for the exploitation. Concretely in the following variables will be used:

- *Deposits from clients:* in spite of the lack of existing consensus on the consideration of the role of the deposits as bank input or output (Humphrey, 1992), in this work it is started from the fact that the deposits are necessary resources to carry out the functions of own mediation of the savings banks, for which will be treated as inputs, just like in the works of Marco and Moya (1999 and 2000), Bernad, Fuentelsaz and Gómez (2009), Casu and Girardone (2010), Kontolaimou and Tsekousas (2010) or Escobar and Guzmán (2010). Taking into account the specific elements of deposits from the clients instead of the total of the deposits received by the banking company responds to the desire to center the analysis on the orientation to the client, typical to the savings banks.

- *Staff expenses*: The savings banks are characterized by an intensive labour force, traditional retail banking business model, which is why the cost (quantified by means of staff expenses) should be considered as input of the model, the staff being a productive resources in obtaining the banking output. This variable has been utilized to a large extent from the studies that analyze the efficiency of the banking companies, like those carried out by Maudos and Pastor (2000), Pastor and Serrano (2000), Belmonte (2007), Escobar and Guzmán (2010) or Belmonte and Plaza (2008), among others.

- *Amortization costs:* the use of the immobilized material is included in the model through the "Amortization" of the profit and loss account, which intends to express the consumption of physical capital that can be associated to the obtaining of banking output. This same focus is utilized in the investigations carried out by Lozano (1997), Maudos and Pastor (2000), Calvo (2002), Belmonte (2007) or Belmonte and Plaza (2008).

- *Other administrative expenses*: this variable has not been traditionally employed as input in the study of the banking efficiency, nevertheless, is considered prominent therefore its adequate management provides more efficiency; and because its value is relevant. Precisely, the new technologies have impacted the reduction of these prices.

Regarding the definition and measurement of the banking output there is a great controversy because there is no optimum measure; however, the considered variables are very in accordance with those utilized in other works, so the following outputs will be considered in the model:

- *Gross margin:* the utilization of the gross margin[5] as banking output results somewhat as being an anomalous if are analyzed the studies that have been exposed previously. However, main output has been taken into account as for being the value of the production of reference for the Bank of Spain in its efficiency calculation. Likewise, it is about a matter that includes both the result of the activity of mediation (the so called traditional business) as the result of the transactions with other products and services for which are paid management fees, for which, taking into account, especially the relevance of its inclusion in the model proposed.

- *Credits for clients*: this variable was used in most of the studies that tried to measure the technical efficiency of the banking companies (Lozano, 1997; Maudos and Pastor, 2000; Maudos et al., 2002; García-Cestona and Surroca, 2006; Belmonte, 2007; Escobar and Guzmán, 2010; Belmonte and Plaza, 2008; Kotolaimou and Tsekousas, 2010). This derives from the function of mediation carried out by the banking business between the offering agents and plaintiffs of funds, grasping deposits and granting credits.

4.2 Transformations in the Considered Variables

In order to calculate the different levels of efficiency of the organizations under study, it has required public data from the corresponding savings banks for the financial periods included between 1999 and 2009. After removing the original data from each period, they came together to consider the study to the 45 existing savings on December 31, 2009 (excluding SASB), taking into account that some of them are results of mergers that occurred in the analyzed time period[6]. In such cases, the merging banks have been regressively taken into account for the purpose of the corresponding exercise, although this is likely to create occasional fictitious entities.

Furthermore, the data has been deflated to work with adequately homogeneous series using two deflators: the labour costs of the financial activities, employed for homogenizing the staff costs, and the added value of the financial intermediation, used to deflate the rest of considered variables.

[5] It should be kept in mind that the presented empirical study is based on an analysis of the years between 1999-2009, which is why it should be carried out with caution and severity by the entrance in force of the Circular 6/2008, of November 26, of the Bank from Spain to companies of credit, of modification of the Circular 4/2004, of December 22, on norms of public and confidential financial information, and models of financial situations. Thus, for the years 2008 and 2009 the gross margin (once "other revenues of exploitation" and "other exploitation expenses" are deducted) will be considered as output.

[6] In particular, the following processes have been considered: the absorption of C.A. and P.M. Pamplona Municipal by the Navarra in year 2000; the merging of C.A. Provincial of Pontevedra with Vigo and Ourense in the same year, with the consequent formation of C.A. of Vigo, Ourense and Pontevedra (Caixanova); the intergration of C.A. and Carlet Loans in Bancaja in the year 2001; and the merger of P.M. and C.A. of Huelva and Sevilla with San Fernando Bank of Sevilla and Jerez in the year 2007, creating P.M. and San Fernando C.A. of Huelva, Jerez and Sevilla (CAJASOL).

4.3 Model Specification

Considering the transformations mentioned above, the efficiency levels of the organizations under study (saving banks) will be calculated following the DEA methodology in its CCR model, whose formulation would be the following:

$$Max\ \phi_o$$

Subject to: (1)

$$\sum_{j=1}^{n} \lambda_j y_{rj} \geq \phi_o y_{ro}$$
$$\sum_{j=1}^{n} \lambda_j x_{ij} \leq x_{io} \qquad j = 1,2, \dots, o, \dots, n$$
$$\lambda_j \geq 0$$
$$r = 1,2, \dots, s; \qquad i = 1,2, \dots, m$$

By observing the model above we can check that it is a question of maximizing the efficiency of the unit being assessed by comparing DMUo with all the real or fictitious DMUs which produce the same or more than the entity being assessed by consuming less or the same respectively as DMUo.

In any event, the weightings obtained as a program solution (λ_j), must be greater or equal to zero, and they express the specific weight which each DMU has within the DMU_0 peer group. Thus, if $\phi_0 = 1$, the DMU being assessed (DMU_o) will be efficient, when the program finds no other units (real or fictitious) which produce more or the same and consume the same or less that entity "o".

In any event, one of the main constraints of the CCR model formulation is that it assumes constant performance to scale implying some correspondence between the input level used and the output produced (Belmonte, 2007). Given this limitation, Banker et al. (1984) proposed an alternative model, the BCC model, based on the hypothesis of variable returns to scale. This model has the advantage of enveloping more accurately the observations, with the costs that the above involved in terms of overestimation of the efficiency of the smaller and larger observations (Dyson et al. 2001). Notwithstanding, it also has the disadvantage that its application does not achieve the same efficiency levels when the problem is formulated using different approaches, the main characteristic of the CCR model (Thanassoulis 2001).

Considering the above reasons, the performance (efficiency) levels of the Spanish credit unions being examined will be calculated using formulated the DEA methodology in its CCR model. In order to do this, the authors have developed specific software in an Excel spreadsheet using the macro solver. This macro uses the Generalized Reduced Gradient algorithm (GRG), version GRG2, for solving mathematical programming problems.

4.4 Analysis of the Results

The obtained results are presented through a series of graphs that intend to interrelate the level of performance (calculated using the method mentioned above) and the size of the 45 existing entities before restructuring (14 after the restructuring process occurred in the 2011 exercise). Likewise, values also the dispersion in the measure of performance from the individual standpoint, aggregate, and intra-group.

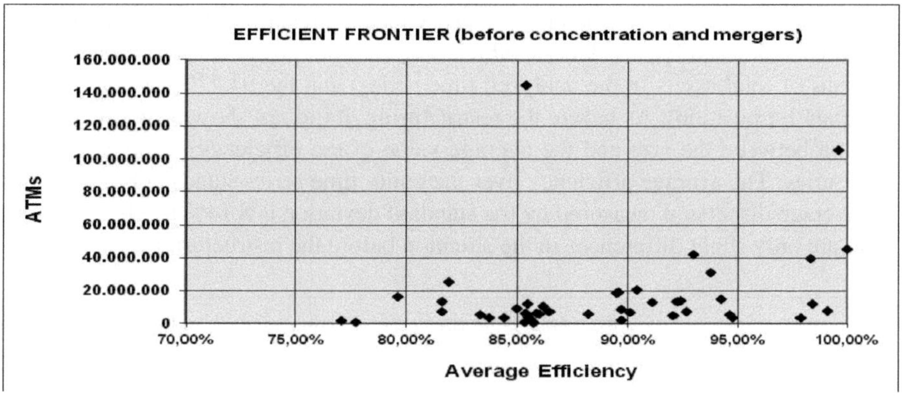

Source: Own elaboration. ATM: Average Total Assets (thousands of Euros).

Fig. 2 Relationship between average efficiency and average size of the period of 1999-2009 (before restructuring)

Figure 2 shows very different efficiency values for all the 45 entities counted prior to the restructuring process. The obtained results are ranging from 77% (Manlleu, with the 41st position for average total assets of the time series 1999-2009) and 100% (Bankia, which occupies the 3rd position in the size of the sector). Despite this initial observation, it has not seen a significant correlation between the size and the average efficiency over the considered time series. The average efficiency of the whole in this period is situated at 88.79% and the average measured dispersion by the standard deviation is 7.29% for the sector.

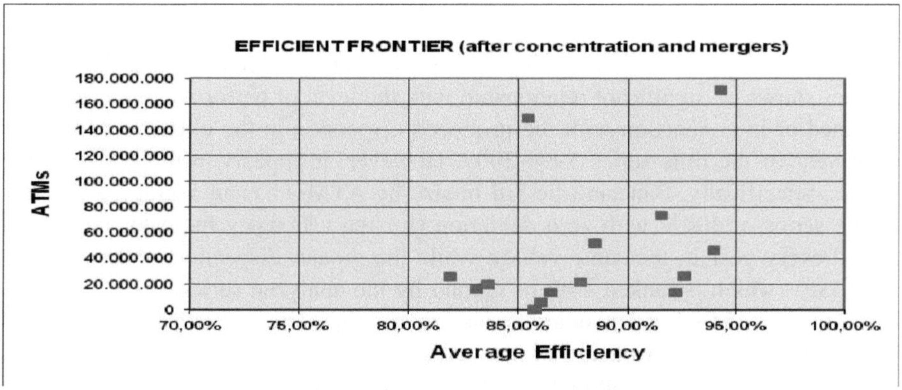

Source: Own elaboration. ATM: Average Total Assets (thousands of Euros).

Fig. 3 Relationship between average efficiency and average size of the period of 1999-2009 in added value (applied to the resulting entity after the restructuring.)

On the basis of the concentration process developed in 2010-2011, Figure 3 shows efficiency values less diverse than before the restructuring; although for its calculation it has taken the same 1999-2009 data series for treating them of aggregate form for the entities and resulting groups of this restructuring. The results are included between 83.63% (the entity "New Unicaja", resulting from the merger between Unicaja-Caja España-Duero, which is ranked 8th position out of a total of 18, in terms of total assets in the analyzed time series) and the 94.27% (Bankia, which occupies 1st position). As before the restructuring, it did not show a significant correlation between the size and the average value of the efficiency in the considered time series. The average efficiency over the entire time series stands at 87.57%; and the average dispersion measured by the standard deviation is 8.14% for the sector, so there are only slight differences in the situation before the restructuring.

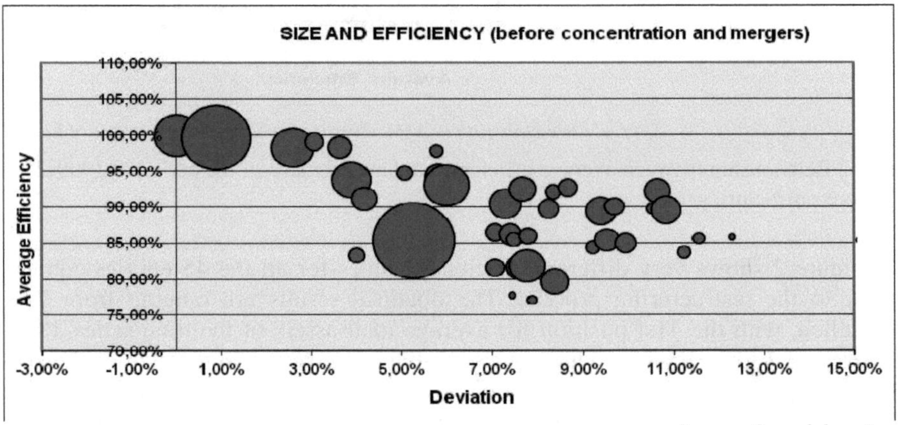

Source: Own elaboration

Fig. 4 Relationship between the average efficiency, deviation and size in the period 1999-2009 (before restructuring)

From another point of view, Figure 4 allows you to appreciate that the size of entities shows no significant relationship with the level of performance (efficiency) obtained or in connection with the produced dispersion in the obtained results for each year of the time series 1999-2009. However, it appears in the two extreme values. Specifically, Bancaja (the 3rd in size by ATMs) has an average efficiency in the series of 100% with zero deviation (for the efficiency indicator has maintained every year in the same value); while the greater dispersion corresponds to Caja Jaén (which is ranked 44th or second by the analyzed series of average total assets), with 15.09% of standard deviation for the series of corresponding efficiency data for each period and with an average efficiency of 85.32%.

Figure 5 allows us to appreciate that the size of the entities resulting from the process of restructuring of the sector occurred in 2010-2011, taking 1999-2009 time series data as the basis for calculating aggregate, it has not significant relation

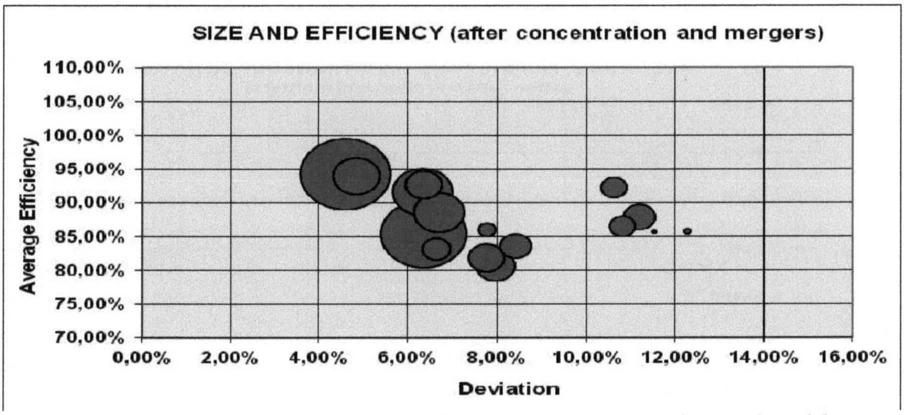

Source: Own elaboration

Fig. 5 Relationship between the average efficiency, dispersal and size in the period 1999-2009 (applied to the resulting entity after the restructuring.)

to the level of performance (efficiency) obtained, although somewhat less dispersion is detected with respect to the average efficiency in the case of larger group sizes. With the restructuring it has reduced the dispersion in the performance indicator as expected of group processes or merger. Specifically, Bankia (1st position by ATMs) reached the higher average efficiency in the series with a value of 94.27% and, also, has the lowest standard deviation (4.60%), while the greater dispersion corresponds to Caja Ontinyent (17th position by average total assets among the 18 organizations or groups), with 12.27% standard deviation for the series of corresponding efficiency data for each period and an average efficiency of 85.76%. Therefore, again a correlation can be seen between size and performance for extreme values of the sample.

Figure 6 addresses the deviation measurement or dispersion in the level of performance through the analyzed time series, but now applied to all groups and entities resulting from the restructuring process in 2010-2011. It is appreciated that in some processes, intra-group dispersion (ie, which measures the deviation between the average efficiency of each entity and the average for the group or entity constituted) is quite small, indicating greater homogeneity in the degree of performance during the 1999-2009 series of exercises before the merger process. Obviously the dispersion is zero in the case of entities that have remained outside the process and minimal in the case of a merger between La Caixa and Caixa Girona. Specifically, the slightest deviation intragroup -after referred to as- is observed for NovaCaixa-Galicia (merger of Caixa Galicia with Caixanova with a deviation of 0.34%); while the highest intra-group dispersion (which may indicate greater heterogeneity in the performance achieved in the years before the merger) corresponds to Marenostrum (the IPS consists of Caja Murcia, Penedés, Sa Nostra and Granada, with a value of 7.38%).

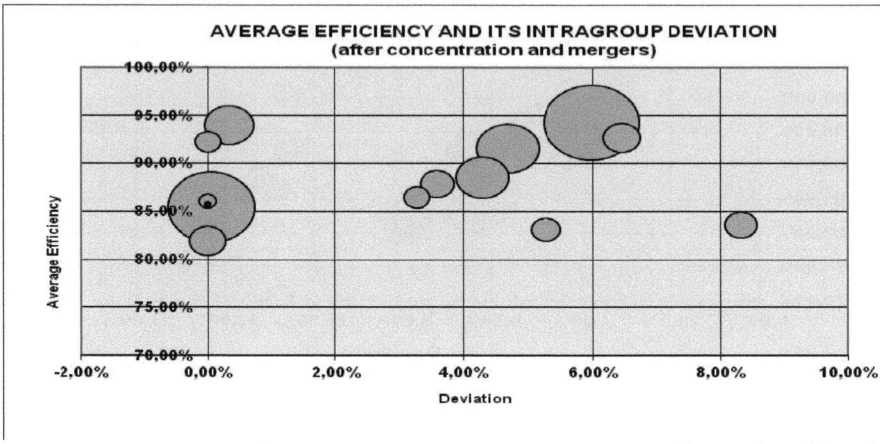

Source: Own elaboration

Fig. 6 Relationship between the average efficiency and intra-group dispersion (after restructuring)

Figure 7 takes only the data of performance or efficiency and total corresponding assets for the year 2009 (the latest data available and, at the same time, the restructuring process prior to 2010), adding to it to obtain the result in the new map of Spanish savings banks, made up of 14 entities or groups. There was no significant relationship between the size and efficiency gained, with values in the range included between 75.65% for Banca Cívica (the merger of Caja Navarra, Burgos, General de Canarias and Cajasol), ranked 8th in size) and 94.78% for Nova-CaixaGalicia (the merger of Caixa Galicia and Caixa Nova).

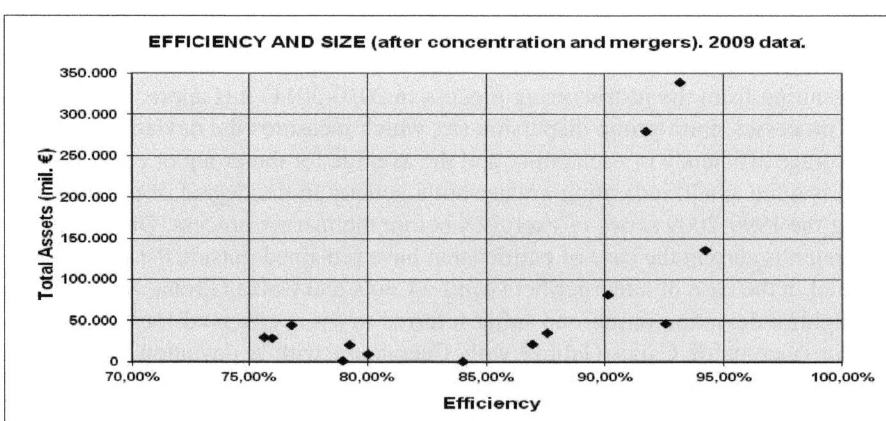

Source: Own elaboration

Fig. 7 Relationship between efficiency and size after the restructuring of the sector

5 Conclusions

In the previous sections we have tried to assess the performance efficiency of Spanish savings banks based on the justification of its important influence in the Spanish financial sector and financial institutions for consideration as social economy because of its legal nature and purpose social.

Strongly affected by the effects of the financial crisis that began in 2007 and characterized by a business model type mostly retail, residential (minimum international presence) and dependent on its sales network is especially important to try to track trends and situation to understand the starting point found in the recent merger process occurring in 2010-2011 that has led to a rapid restructuring from 45 entities to 14 entities and merger processes (cold mergers or IPS).

Based on the reference to studies on the measurement of efficiency, one of the objectives of this work has been applied to the technique of data envelopment analysis to the subsector of Spanish savings banks during the period 1999-2009, differing from most studies of this kind, which have focused on the analysis of the scale and scope of economies of this entities class. In this sense, the main contribution of this work is the use of the method used to measure the efficiency and the time period in which the analysis is performed, which refers to a full decade, including the years of the financial crisis.

It has succeeded in demonstrating that in the restructuring process of the savings banks occurred during 2010 and 2011 there has been no clear distinction between entities that had a similar experience in terms of performance or efficiency gains over the financial years 1999 and 2009.

Nevertheless, it seems to be a reduction in dispersion in the performance indicator by combining the 45 original entities and to give rise to the current 14. Additionally, you can detect some groups (IPS) that count, at the time of its formation and based on their past records between 1999 and 2009, with a lower degree of dispersion in the standard measure of efficiency, this may mean a better fit or link between those parties is in their best future performance.

In any case, it is clear that the process of grouping has been made subject to a clear time pressure by the supervisory authorities besides being influenced by certain political conditions resulting from territorial linkages and autonomous[7].

In this sense, the analysis of the results of the coming years will provide information that will substantiate or refute these hypotheses.

Moreover, it should be noted that mergers and the formation of various IPS occurred between late 2009 and 2011 will have to focus, necessarily, in rationalizing the branch network and in the necessary management of economies of scale, which will result in an improvement or at least in the preservation of their level of performance.

---

[7] In Spain there are 17 regions and two autonomous cities and the current legislation carries a strong political influence in the corporate bodies of the savings banks associated with each territory.

References

Banker, R.D., Charnes, A., Cooper, W.W.: Some models for estimating technical and scale inefficiencies in data envelopment analysis. Management science 30(9), 1078–1092 (1984)

Belmonte, L.J.: El sector de las cooperativas de crédito en España. Un estudio por comunidades autónomas. Publicaciones CES Andalucía, Sevilla (2007)

Belmonte, L.J., Plaza, J.A.: Análisis de la eficiencia en las cooperativas de crédito en España. Una propuesta metodológica basada en el análisis envolvente de datos (DEA). CIRIEC-España: Revista de economía pública, social y cooperativa 63, 113–133 (2008)

Berger, A.N., Humphrey, D.B.: Efficiency of financial institutions: International survey and directions for future research. European Journal of Operational Research 98(2), 175–212 (1997)

Berger, A.N., Hunter, W.C., Timme, S.G.: The efficiency of financial institutions: a review and preview of research past, present and future. Journal of Banking and Finance 17 (2-3), 221–249 (1993)

Bernad, C., Fuentelsaz, L., Gómez, J.: El efecto del horizonte temporal en el resultado de las fusiones y adquisiciones: el caso de las cajas de ahorros españolas. Revista Europea de Dirección y Economía de la Empresa 18(4), 49–68 (2009)

Bikker, J.A.: Efficiency in the European banking industry: an exploratory analysis to rank countries. Cahiers Economiques de Bruxelles 172, 3–28 (2001)

Calvo, M.: Dimensión y eficiencia: en el caso de la banca en España. Tesis doctoral, Universidad de La Laguna, Tenerife (2002)

Casu, B., Girardone, C.: Integration and efficiency convergence in EU banking markets. Omega 38(5), 260–267 (2010)

Chaffai, M.E., Dietsch, M., Lozano-Vivas, A.: Technological and environmental differences in the European banking industries. Journal of Financial Services Research 19(2), 147–162 (2001)

Charnes, A., Cooper, W.W., Rhodes, E.: Measuring the efficiency of decision making units. European Journal of Operational Research 2(6), 429–444 (1978)

Confederación Española de Cajas de Ahorros.: Anuario estadístico de las cajas de ahorros. Servicio de Estudios de la CECA, Madrid (several years)

Cooper, W.W., Seiford, L.M., Tone, K.: Data envelopment analysis: a comprehensive text with models, applications, references and DEA-solver software. Kluwer Academic Pulishers, Norwell, Massachusetts (2000)

Dietsch, M., Lozano-Vivas, A.: How the environment determines banking efficiency: A comparison between French and Spanish industries. Journal of Banking and Finance 24(6), 985–1004 (2000)

Dyson, R.G., Allen, R., Camanho, A.S., Podinovski, V.V., Sarrico, C.S., Shale, E.A.: Pitfalls and protocols in DEA. European Journal of Operational Research 132(2), 245–259 (2001)

Escobar, B., Guzmán, I.: Eficiencia y cambio productivo en las cajas de ahorro españolas. Revista CIRIEC-España 68, 183–202 (2010)

Fecher, F., Pestieau, P.: Efficiency and competition in OECD financial services, in The measurement of productive efficiency: Techniques and applications, pp. 374–385. Oxford University Press, New York (1993)

Figueira, C., Nellis, J.: Bank merger and acquisition activity in the EU: much ado about nothing? Service Industries Journal 29(7), 875–886 (2009)

García-Cestona, M.A., Surroca, J.: Evaluación de la eficiencia con múltiples fines. Una aplicación a las cajas de ahorro. Revista de Economía Aplicada 14(40), 67–89 (2006)

Grifell-Tatjé, E., Lovell, C.A.K.: The sources of productivity change in Spanish banking. European Journal of Operational Research 98(2), 364–380 (1997)

Howcroft, J.B., Ul-Haq, R., Hammerton, R.: Bank regulation and the process of internationalization: a study of Japanese Bank entry into London. Service Industries Journal 30(8), 1359–1375 (2010)

Humphrey, D.B.: Flow versus stock indicators of banking output: effects on productivity and scale economy measurement. Journal of Financial Services Research 6(2), 115–135 (1992)

Humphrey, D.B.: Costs and scale economies in bank intermediation. Handbook for banking strategy, pp. 745–783 (1985)

Humphrey, D.B., Pulley, L.B.: Bank's Responses to Deregulation: Profits, Technology, and Efficiency. Journal of Money, Credit and Banking 29(1), 73–93 (1997)

Ismail, A., Davidson, I., Frank, R.: Operating performance of European Bank mergers. Service Industries Journal 29(3), 345–366 (2009)

Kontolaimou, A., Tsekouras, K.: Are cooperatives the weakest link in European banking? A non-parametric metafrontier approach. Journal of Banking and Finance 34, 1946–1957 (2010)

López, J.S., Appennini, A., Rossi, S.P.R., Di Salvo, R., Mazzilis, M.C., Guidi, A., De Recherches Financières, S.U.E.: Italian mutual banks. Soc. Univ. Europ. de Recherches Financières (2002)

Lozano-Vivas, A.: Un estudio de la eficiencia y economías de diversificación del sistema bancario español. Revista Española de Financiación y Contabilidad 73, 855–880 (1992)

Lozano-Vivas, A.: La eficiencia del sistema bancario español en el marco de la Unión Europea. Ekonomiaz 48, 318–343 (2001)

Lozano-Vivas, A., Pastor, J., Hasan, I.: European bank performance beyond country borders: What really matters? Review of Finance 5(1-2), 141–152 (2001)

Marco, M.A., Moya, I.: Contraste de un indicador de eficiencia agregado y la estimación paramétrica. Aplicación al sector de crédito cooperativo español. CIRIEC-España, revista de economía pública, social y cooperativa 33, 155–173 (1999)

Marco, M.A., Moya, I.: Factores que inciden en la eficiencia de las entidades de crédito cooperativo. Revista Española de Financiación y Contabilidad 24(105), 781–808 (2000)

Maudos, J., Pastor, J.M.: La eficiencia del sistema bancario español en el contexto de la Unión Europea. Papeles de economía Española 84/85, 155–168 (2000)

Maudos, J., Fernández, J.: El sector bancario español en el contexto internacional: evolución reciente y retos futuros. Fundación BBVA, Bilbao (2008)

Maudos, J., Pastor, J.M., Perez, F.: Competition and efficiency in the Spanish banking sector: the importance of specialization. Applied Financial Economics 12(7), 505–516 (2002)

Pastor, J.M., Serrano, L.: Efficiency, endogenous and exogenous credit risk in the banking systems of the Euro area. Applied Financial Economics 15(9), 631–649 (2005)

Sealey, C.W., Lindley, J.T.: Inputs, outputs, and a theory of production and cost at depository financial institutions. Journal of Finance 32(4), 1251–1266 (1977)

Thanassoulis, E.: Introduction to the theory and application of data envelopment analysis: a foundation text with integrated software. Springer, Heidelberg (2001)

Yeh, Q.J.: The application of data envelopment analysis in conjunction with financial ratios for bank performance evaluation. Journal of the Operational Research Society, 980–988 (1996)

Sports and Tourism

Non-equity Agreements in the Hospitality Industry: Analysis from the Perspective of the Forgotten Effects

Onofre Martorell-Cunill[1], Anna M. Gil-Lafuente[2], Antoni Socias Salvà[1], and Carles Mulet Forteza[1]

[1] Department of Business Administration, University of Balearic Islands, Crta. Valldemossa, Km 7.5. Campus UIB. Building G.M de Jovellanos. 07122 Palma de Mallorca, Spain
[2] Department of Business Administration, University of Barcelona, Av. Diagonal 690, 08034 Barcelona, Spain
{onofre.martorell,asocias,carles.mulet}@uib.es, amgil@ub.edu

Abstract. This paper uses fuzzy incidence matrixes to determine the causes that have led non-equity strategies to become the preferred growth strategies for international expansion of those hotel chains with the most number of rooms in the word. Both direct cause and second generation causes that arise in the majority of the socio-economic cases have been identified. In fact, determining the second generation effects (or forgotten effects) is one of the main contributions of this study as it shows that those causes that are usually not foreseen, at least in the first instance, affect notably in the expansion strategies of hotel chains in the medium and long term.

Keywords: Growth strategies, Delphi analysis, Fuzzy matrix, Direct effects, Forgotten effects.

1 Introduction

Hotel chains follow a wide variety of different growth strategies. They can expand through franchise agreements, management contracts, hotel ownership, leaseholds, mergers and takeovers, joint ventures or a combination of any of the former (Okumus, 2004).

In recent years, a growing number of studies highlight how non-equity strategies, like franchise agreements or management contracts, have revolutionized the international hotel industry. For instance, Martorell and Mulet (2010) show that, in 2005, franchise agreements represented 60% of all rooms operated by the world's 35 hotel chains with the biggest room portfolios, while management contracts accounted for 20% of the corresponding total. Other papers, suchs as Erramilli et al. (2002) believe that a further major decision to be taken by companies expanding into a foreign market is to choose from among the various non-equity entry options.

A.M. Gil-Lafuente et al. (Eds.): Soft Comput. in Manag. and Bus. Econ., STUDFUZZ 287, pp. 269–285.
springerlink.com

Thanks to these new developments in growth strategies, large US and UK hotel groups have consolidated and increased their position of leadership among the top ranks of the world's biggest hotel chains. All this has been possible because these chains are the prime exponents of growth systems that do not require capital investment. If these systems ensure greater, swifter growth, then given the high flexibility that they offer compared with other growth strategies (Martorell et al., 2008), we can see that US and UK chains are better prepared to compete by doing so in a more efficient way.

In contrast, studies do not abound that explore the reasons and factors behind the spectacular growth of these non-equity growth strategies during the last decade. It must be remembered that, according to Alexander and Lockwood (1996), in the hotel industry in the 1990s, no single growth strategy prevailed over any of the others, and so hotel chains were prepared to follow different systems of growth, while today the world's leading hotel chains mainly expand through non-equity systems.

There are several arguments to support the view that non-equity agreements are currently the favourite growth strategy used by hotel chains:

- Growth strategies that are not based on capital transactions (non-equity modes) are most popular among consumer-services firms (such as hotels and restaurants) as compared to professional-services businesses (for example consulting organisations) (Erramilli, 1990).
- Given the characteristics of the hotel industry, such as a willingness to transfer know-how or the easy codification of hotel management and control systems for use in hotel franchises, growth strategies can be adopted that do not require capital-based transactions (Contractor and Kundu, 1998a).
- Given the low earning capacity of hotel chains that mainly base their growth strategies on hotel ownership or leaseholds plus shareholders' growing interest in investing in increasingly profitable hotel chains, hotel ownership and the actual running of the hotel are coming to be seen as two separate roles. As a result, the tendency to expand by extending a chain's portfolio of owned property has begun to change in favour of strategies non-equity, particularly since one of the main goals of hotel chains has become to accelerate the expansion process (Altinay, 2005).
- The hotel sector, as a service industry where it is possible to separate capital investment from management skills, is a clear example of how internationalisation can be modelled without considering shareholder investment (Contractor, 1990; Contractor and Kundu, 1998a, 1998b). Indeed, in the hotel industry, non-equity modes account for 65.4% of multinational properties worldwide.

The main purpose of this study is therefore to identify the causes and factors that have led to the growth of non-equity contractual systems in the international hotel industry.

To achieve this goal, this study is divided into the following four sections. The first section summarizes the main growth strategies that are used in the hospitality industry. The second section describes the methodology used in the development of the study. The third and fourth section lay out the empirical analysis and the interpretation of results and the conclusions of the study.

2 Growth Strategies in the Hotel Industry: The Theoretical Framework

2.1 Hotel Franchise Agreements

There have been numerous debates on the issue of business format franchising definitions but a broad definition, framed to meet the points commonly raised in debates, has defined franchising as: «the granting of a license for a predetermined financial return by a franchising company (franchisor) to its franchisees, entitling them to make use of a complete business package, including training, support and the corporate name, thus enabling them to operate their own businesses to exactly the same standards and format as the other units in the franchised chain» (Stanworth and Curran, 1999).

Originating in the USA, franchising emerged as a powerful new way of facilitating the growth of service organizations. The origins of modern-day hotel franchising can be traced back to the 1950s when Holiday Inn established itself as the economy segment's primary business format franchisor. Hotel companies that applied stricter operating standards than had previously been common among independent hotels have subsequently expanded and grown in the economy segment by means of franchising. Lashley and Morrison (2000) pointed out that business format franchising has become an established global enterprise trend within the service sector, in general, and, more specifically, within the hospitality services sector. Meanwhile Ingram (2001) stated that developing franchises through new franchise operations would not just be common to the service sector (hotels, restaurants, pubs etc.), but that it would also extend to other new businesses.

In the hospitality industry, a franchise strategy could be as simple as granting a license to a small company with one single accommodation unit, or it could be as complex as master franchising, where a company is given rights to develop a certain brand name in a particular region of the world.

A franchise contract usually covers a period of twenty to thirty years in exchange for the payment of a fixed yearly sum, with the extension of the agreement on its expiry if both parties are satisfied with their mutual collaboration. However, the franchisor carries out regular inspections to ensure full compliance with rules regarding corporate unity and the production process. Irregularities of either or both kinds constitute sufficient grounds for the contract's termination, due to the possible damage that the franchisor's reputation may suffer as a result of the hotel's non-compliance.

As well as giving initial assistance to their franchisees, hotel franchisors also offer them a finance programme for the business and advice on the design and construction of the hotels. A review of relevant literature indicates that such support services are a prime factor in franchising decisions.

Three key documents must be drafted before a franchise is granted. Firstly, an operating manual must be created, with instructions regarding the management process once the franchise has started operating. A franchise contract must then be drawn up, stipulating the legal obligations of both parties. Lastly, a franchise prospectus must be drafted aimed at recruiting the right kind of franchisee. Once the three documents have been drawn up, the process of recruiting and training the new franchisees begins, together with the selection of the premises and subsequent launch of the franchise outlets (Stanworth et al, 2004).

With a franchise contract, franchisors request the payment of royalties from their franchisees. In many cases, these fees are the second highest expense faced by most franchisees after their top expense, personnel costs. More specifically, the franchisor charges an initial fee, which is normally a fixed sum per room. In addition, it also charges a yearly royalty and an annual fee for advertising and marketing, which involves a percentage sum based on room revenue. Finally, it also charges a reservation fee that usually involves a fixed sum for each reservation made. As discussed by Bhattacharyya and Lafontaine (1995), the contract between franchisor and franchisee is typically linear, including an initial fee to join the organization (between $20 000 and $50 000) and a percentage of revenue meant to reimburse for advertising, marketing, and the use of centralized reservation systems (between 6 and 9.25%). In addition, it may also be costly for motels to provide the services that are required to meet the organization's quality standards.

2.2 Hotel Management Contracts

Business management contracts can be traced back to practices in the British colonies. The concept was subsequently developed into a management contract system in the United States by the motel industry of the 1960s. In the hotel trade itself, management contracts were introduced as a result of losses incurred by Hilton Hotels Corporation, which ran a hotel in Havana in the 1960s under a leasehold agreement when the Cuban Revolution occurred. From this event, they learnt that, in developing countries, hotel investment entails big risks. Taking advantage of the firm's international positioning, it converted most of its leasehold agreements into management contracts.

A hotel management contract is basically an agreement between a hotel management company and an owner company, under the terms of which the former runs the hotel. The owner does not take operational decisions but is responsible for supplying the necessary capital and for the payment of expenses and debts. The management company receives a fee for its services and the owner normally keeps the remaining profits after expenses have been deducted.

With a management contract, the hotel chain has full operational control over earnings and expenditure, but the financial burdens (and even the responsibility for meeting any operational costs) are the owner's. In the early days of management contracts, this was true to such an extent that it was the owner who had to provide all the inventories and working capital, as well as continuing to inject capital in the event of losses when operations went wrong. This was fully in the management company's favour and quite clearly it was an unfair type of agreement, because the management company's business profits were implicitly guaranteed thanks to the payment of a fixed basic fee that was paid regardless of the real profits. This meant that the management company did not need to worry much about keeping costs down. However, because it is becoming increasingly common for big hotel chains to offer their services under management contracts, the rivalry that this engenders means that owners and developers can be choosier, demanding a fairer division of operational and financial responsibilities, even demanding an amount of capital from the company running the hotel. The main thing that differentiates management contracts in the 1960s and management contracts now is hotel owners' increasing capacity to negotiate their involvement in management, budgeting and marketing plans.

Each management contract must be designed to suit each particular situation and special care must be taken with the agreement of fees. The contract's stipulations should include the following details: what capital, if any, must be provided by the management company, budget and expenditure limits, accounting and financial conditions, the contract's length, a renewal clause (if appropriate), the conditions for the cancellation of the contract, services to be provided by the management company, and the minimum yearly amounts to be spent on advertising, maintenance, and replacing furnishings, equipment and other similar and/or related items. It should be made very clear in the contract that this is an agreement between an owner and a hired management company and that it is not a leasehold contract. The owner supplies the hotel (i.e. the building, furnishings, decoration, equipment and working capital) and is financially and legally liable for the business. The management company agrees to run the hotel, paying all expenses on behalf of the owner. It retains a commission from the revenue it makes, handing over the surplus to the owner.

This capital can be paid in various different ways. The management company can be asked to pay the pre-opening costs and/or to supply the initial working capital required, and this might even be extended to include the cost of furnishings and/or complementary items and/or equipment. Some financial contributions may take the form of a strategic alliance between the owner and the management company, where both supply a certain amount of cash and sign a joint mortgage on the property.

The payment system for management contracts tends to consist of a basic type of fee and profit-related ones. This has led to the development of three general payment systems: a basic fee (a percentage of the hotel's total turnover), a basic fee and an incentive fee (a percentage of the GOP), and a basic or incentive fee, depending on the higher of the two.

2.3 *Hotel Ownership and Leaseholds*

Ownership means the total or partial acquisition of a hotel. Its main advantage is the fact that the owner or owners retain the hotel's entire profits. However, due to the high level of financial assets that are needed for this growth strategy, it is almost unviable in the case of hotel chains the size of United States' ones, for instance, where Wyndham Worldwide alone has more hotel beds than an entire country with a big tourism industry like Spain.

A leasehold contract entails renting a hotel for a period of time that is normally never less than three years, subject to automatic renewals. In the case of the hotel industry, the object of the contract is the leasehold of the business itself rather than the actual hotel, including all its belongings, fittings and fixtures and equipment. Normally the leaseholder is a hotel group, and so the leasehold tends to involve the use of the said chain's corporate image and production process. Annual amortization fees are usually paid by the lessor, and the latter is also responsible for maintaining and keeping the building in optimum condition.

As for the financial remuneration paid by the leaseholder to the lessor, several forms of payment are possible but the most common examples are the payment of a fixed yearly sum, which is reviewed annually in accordance with certain conditions, normally the price index; a percentage payment of the hotel revenue (about 5%); a percentage payment of the cash flow generated (about 15%); or either of the last two systems combined with a fixed sum of money.

At the lessor's instigation, it is customary for an agreement to be made whereby the leaseholder pays a fixed yearly sum to cover any replacement costs for the hotel's different services. Consequently, an initial inventory is usually made of all the belongings, fixtures and fittings that are leased, with a view to the settlement of any possible future disputes.

The hotel owner is entitled to inspect the property in order to check the condition in which the facilities have been kept. Once the contract has expired, the hotel chain must return the buildings, fixtures and fittings to the owner in the same state they were in at the beginning of the contract. The hotel chain usually pays for any repairs needed to keep the buildings, fixtures and fittings in good repair during the period in which the contract remains in force, although generally speaking it is not responsible for any repairs of an extraordinary nature.

3 The Methodological Framework for the Study

To pinpoint the reasons that have led the world's major hotel chains to opt for non-equity agreements as their main growth strategy, a Delphi analysis was conducted. This analysis was performed in six stages. During the first stage, a panel of experts was asked to name the main factors that have allowed this top group to expand so heavily in recent years.

In this respect, we have rounded off the list of factors from the Delphy analysis with some additional ones, identified from a review of literature. In the following stage, the experts were given their colleagues' opinions in order to try and reach a consensus on the results. During the initial stage, the questionnaire was sent out to 14 experts. Later the number was reduced to 10, because some of their results were not coherent with the outcome of the first analysis. The results of this second stage and the literature review are shown in Table 1. Once the reasons or causes had been determined, their influence (effects) on the choice of growth strategy used by these chains was assessed. To determine these effects, the experts were also asked, during a third stage, to use proxy variables to create an effects matrix. In this case, the experts chose 'increased sales', 'taking advantage of manpower and work effectiveness' and 'greater, improved access to capital markets' as the proxy variables used to refer to equity-based growth strategies, and 'increased profitability', 'speeding up the hotel chain's growth' and 'countries with a culture gap' as the proxy variables for non-equity growth strategies.

Table 1 The main causes that affect the type of growth strategy chosen by hotel chains

CAUSES
Control of new markets
Economies of scale
Restructuring investments
Boosting international competitiveness
Economic liberalization
Making the most of cultural and linguistic barriers.
Staff training
Increasing the hotel chain's size
Technological progress
Better brand recognition

Because each of these causes influences the different effects to be achieved to a differing extent, an experton-type was then performed with a view to constructing a direct incidence matrix (Gil-Lafuente and Anselín, 2009; Gil-Lafuente, and Barcellos de Paula, 2010; Gil-Lafuente and Luis-Bassa, 2011). The concept of incidence is associated with the effects of the elements of one set on the elements of another or on themselves. Incidence is a subjective notion and it is generally hard to measure, so we will use fuzzy matrices to assign values (awarding a numeric value from a suitable scale), ranging from no incidence (zero) to the highest possible incidence (one) (Gil, 1995). Because this system of assigning values is more detailed, compared with classic binary methods, it offers a truer vision of the reality we wish to analyse.

For this reason, the experts were asked, during the fourth stage, to express their opinions by choosing real interval values [0,1], based on the following scale: 0 no influence; 0.1 virtually no influence: 0.2 hardly any influence; 0.3 a very weak influence; 0.4 a weak influence; 0.5 a medium influence; 0.6 a considerable influence; 0.7 quite a strong influence; 0.8 a strong influence; 0.9 a very strong influence; 1 a full impact. Subsequently, second-order influences were defined in order to capture equally important effects that had not been taken into account (forgotten effects).

The results of the direct or first-generation incidence matrix (matrix A) by the experts are shown in Table 2. To establish the accumulated first-generation and second-generation (or indirect) effects, we use Kaufman and Gil's algorithm (1988). To do this, the same experts that developed the matrix shown in Table 2 were asked, during a fifth stage, to express their opinions on the possible direct influences of each of the initiatives (causes) shown in the rows of the this matrix on both each other and all the other factors. Thus, each cause that influences a hotel chain's growth strategy is a cause and an effect of itself and all the other causes. The results of this new matrix (matrix B) are shown in Table 3 (Gil-Lafuente and Gil, 2009).

Table 2 Results of the direct incidence matrix (Matrix A)

	Increased sales	Taking advantage of manpower and work effectiveness	Greater, improved access to capital markets	Increased return	Speeding up the hotel chain's growth	Countries with a culture gap
Control of new markets	0,3	0,7	0,9	0,8	0,7	0,1
Economies of scale	0,3	1	0,2	0,8	0	0
Restructuring investments	0,7	0,8	1	0,5	0,4	0,1
Boosting international competitiveness	0,9	0,9	0,5	0,6	0,5	0,7
Economic liberalization	0,2	0	1	0,4	0,7	0,1
Making the most of cultural and linguistic barriers.	0,5	0,3	0,1	0,8	0,8	1
Staff training	1	1	0,2	0,6	0,5	0,4
Increasing the hotel chain's size	1	0,4	0,8	0,2	1	0,7
Technological progress	0,4	0,8	0,3	0,2	0,2	0,6
Better brand recognition	0,8	0,3	0,6	0,8	0,9	0,5

Table 3 Results of matrix B (cause-cause relations)

	Control of new markets	Economies of scale	Restructuring investments	Boosting international competitiveness	Economic liberalization	Making the most of cultural and linguistic barriers.	Staff training	Increasing the hotel chain's size	Technological progress	Better brand recognition
Control of new markets	1	0,1	0,3	0,7	0,9	0,9	0,4	0,7	0	0,9
Economies of scale	0	1	0,5	0,8	0,6	0,1	0,8	0,9	1	0,9
Restructuring investments	0,2	0,6	1	0,4	0,1	0	0,8	0,4	0,9	0,3
Boosting international competitiveness	1	0,6	0,3	1	1	0	0,6	0,1	0,2	0,8
Economic liberalization	1	0,1	0,3	0,8	1	1	0,2	0,6	0,1	0,2
Making the most of cultural and linguistic barriers.	0,8	0,2	0,6	1	0,8	1	0,4	0,8	0	0,2
Staff training	0,1	1	0,8	0,9	0	0	1	0,2	0	0,9
Increasing the hotel chain's size	0,9	1	0,2	0,6	0	0	0,3	1	0	1
Technological progress	0,1	1	1	0,7	0	0,7	0,8	0	1	0,3
Better brand recognition	1	0,9	0,9	0,6	0	0	0,1	1	0	1

Finally, during the sixth stage, the experts were also asked to express their opinions on possible relations between the effects of the first matrix on these same effects and on all the others. The results of this new matrix (matrix C) are shown in Table 4.

Table 4 Results of matrix C (cause-effect relations)

	Increased sales	Taking advantage of manpower and work effectiveness	Greater, improved access to capital markets	Increased return	Speeding up the hotel chain's growth	Countries with a culture gap
Increased sales	1	0,3	0	0,9	0,1	0
Taking advantage of manpower and work effectiveness	0,4	1	0	1	0,1	0
Greater, improved access to capital markets	0	0	1	0,6	0,6	0
Increased return	0	0,2	0,2	1	0,3	0
Speeding up the hotel chain's growth	0,8	0,1	0,1	0,7	1	0
Countries with a culture gap	0	0,1	0,3	0,7	0,9	1

To find the accumulated first and second-generation effects, it is now sufficient to develop a new fuzzy incidence matrix, which we will call D, derived by composing matrices B, A and C, using the max-min composition. The results of this composed matrix (Matrix D) are shown in Table 5 (Gil-Lafuente and Gil, 2009).

Table 5 Results of the composition of matrices B, A and C (Matrix D)

	Increased sales	Taking advantage of manpower and work effectiveness	Greater, improved access to capital markets	Increased return	Speeding up the hotel chain's growth	Countries with a culture gap
Control of new markets	0,8	0,7	0,9	0,8	0,9	0,9
Economies of scale	0,9	1	0,8	1	0,9	0,7
Restructuring investments	0,8	0,8	1	0,8	0,6	0,6
Boosting international competitiveness	0,9	0,9	1	0,9	0,8	0,7
Economic liberalization	0,8	0,8	1	0,8	0,9	1
Making the most of cultural and linguistic barriers.	0,9	0,9	0,8	0,9	0,9	1
Staff training	1	1	0,8	1	0,9	0,7
Increasing the hotel chain's size	1	1	0,9	1	1	0,7
Technological progress	0,8	1	1	1	0,7	0,7
Better brand recognition	1	0,9	0,9	0,9	1	0,7

4 Factors That Have Fostered an Increase in Non-equity Growth Strategies in the Hotel Industry

In continuation, we will analyse the main reasons why non-equity growth strategies have become the most widely used in the international hotel trade. To do so, from the data obtained by our panel of experts, we shall analyse the extent to which some (direct or first-generation and second-generation) causes have influenced the change in the development strategy used by the world's top hotel chains.

Matrix D contains the accumulated first and second-generation influences. To isolate the effects of second-generation influences, also known as detecting indirect causes, a procedure must be found via which the direct influences inferred from the original matrix (matrix A) can be separated from the accumulated influences that appear in matrix D. Several procedures may be used to achieve this goal. The simplest way, which we believe to be the most suitable for this study, is to find the algebraic difference between matrices D and A. In this way, an indirect fuzzy incidence matrix (matrix E) can be obtained, as shown in Table 6.

This new matrix only brings to light second-generation effects. To interpret the results properly, the highest values of this matrix must be observed, because values close to 1 indicate the presence of a forgotten (or second-generation) effect, while values close to 0 indicate the opposite. Values equal to or higher than 0.8 (showing, at minimum, a strong second-generation influence) indicate cause-effect relations that had not initially been taken into account by the experts consulted or that had only been taken into slight consideration. In accordance with the results, a backtrack procedure is then established. This consists of finding all the values in fuzzy matrix E which meet the chosen criterion (in this case, values of over 0.8), removing the column and row where this value is located from matrix A and amending it with the value obtained in matrix E (the value of the forgotten variable or one that was not taken into account). All this leads to the development of a new first-order incidence matrix, matrix F (see Table 7).

Table 6 Results of indirect or second-generation incidence matrix (Matrix E)

	Increased sales	Taking advantage of manpower and work effectiveness	Greater, improved access to capital markets	Increased return	Speeding up the hotel chain's growth	Countries with a culture gap
Control of new markets	0,5	0	0	0	0,2	0,8
Economies of scale	0,6	0	0,6	0,2	0,9	0,7
Restructuring investments	0,1	0	0	0,3	0,2	0,5
Boosting international competitiveness	0	0	0,5	0,3	0,3	0
Economic liberalization	0,6	0,8	0	0,4	0,2	0,9
Making the most of cultural and linguistic barriers.	0,4	0,6	0,7	0,1	0,1	0
Staff training	0	0	0,6	0,4	0,4	0,3
Increasing the hotel chain's size	0	0,6	0,1	0,8	0	0
Technological progress	0,4	0,2	0,7	0,8	0,5	0,1
Better brand recognition	0,2	0,6	0,3	0,1	0,1	0,2

This new matrix was given to the experts for them to ratify or rectify their opinions. In this study, in a high percentage of cases the experts answered by rectifying their opinions, bringing them in line with the presented evidence, since only some of them cast any doubt on whether 'control of new markets' could be influential in increasing or reducing the cultural gap between countries, therefore boosting the use of non-equity strategies.

When the initial matrix developed by the experts, matrix A, is analysed, it can be observed that the main reasons that have led to the growth in non-equity contractual agreements by the world's top hotel companies are 'making the most of cultural and linguistic barriers' and the 'better brand recognition'. In contrast, the main influences on the use of equity-based growth strategies are 'restructuring investments, 'staff training', 'economic liberalization, 'boosting international competitivenessl' and the 'technological progress', while the last two factors can also affect, to a lesser extent, in the use of non-equity strategies. Meanwhile, 'control of new markets', 'economies of scale' and 'increasing the hotel chain's size' are all variables that might influence the use of either equity or non-equity modes. As a result, the panel of experts decided that most of the causes that were analysed allowed hotel chains to expand through equity-based growth strategies.

In contrast, if we analyse the last incidence matrix (matrix F), we can see that the experts consulted had 'forgotten' several relations between each of these causes and the mode of expansion chosen by the hotel chains. Most of them can lead to a proliferation of non-equity contractual agreements. Thus after analysing this last matrix, it can be observed that the experts only managed to capture all the effects generated by the 'restructuring investments', 'staff training' and 'boosting international competitiveness' variables, since no effect was overlooked in the

Table 7 Results of the new direct incidence matrix (Matrix F).

	Increased sales	Taking advantage of manpower and work effectiveness	Greater, improved access to capital markets	Increased return	Speeding up the hotel chain's growth	Countries with a culture gap
Control of new markets	0,3	0,7	0,9	0,8	0,7	0,8
Economies of scale	0,3	1	0,2	0,8	0,9	0
Restructuring investments	0,7	0,8	1	0,5	0,4	0,1
Boosting international competitiveness	0,9	0,9	0,5	0,6	0,5	0,7
Economic liberalization	0,2	0,8	1	0,4	0,7	0,9
Making the most of cultural and linguistic barriers.	0,5	0,3	0,1	0,8	0,8	1
Staff training	1	1	0,2	0,6	0,5	0,4
Increasing the hotel chain's size	1	0,4	0,8	0,8	1	0,7
Technological progress	0,4	0,8	0,3	0, 8	0,2	0,6
Better brand recognition	0,8	0,3	0,6	0,8	0,9	0,5

analysis of these three variables. This demonstrates that the consulted experts had initially captured all the relevant influences for these effects in their first assessment, since in the analysis no second-generation effects were generated that display a strong influence or higher (values equal to or higher than 0.8). Consequently, the initial matrix drawn up by the experts (matrix A) has not altered with regard to these two effects.

Thus it can be said that these variables – 'restructuring investments', 'staff training' and 'boosting international competitiveness' – are factors that solely affect hotel chain growth through equity strategies. In contrast, the 'technological progress' and 'economic liberalization' variables, which were only important in equity-based expansion, were also found to have a strong second-generation effect on an increased earning capacity (in the case of 'technological progress') and in modifying the cultural gap (in the case of 'economic liberalization'); that is, on factors that affect growth through non-equity strategies. Consequently, these variables are also important in determining hotel chain growth through non-equity systems, and so they are no longer exclusive to equity-based growth, particularly in the case of the 'technological progress' variable.

In fact, from our analysis of the first-generation effects of the 'technological progress' variable, it could already be seen that this factor did not play a very decisive role in the choice of growth strategy followed by hotel chains, since it only has a strong first-generation effect in the case of taking advantage of manpower and work effectiveness, not having any important first or second-generation effect on an increase in turnover or on improved access to capital markets. In contrast, the 'economic liberalization' variable continues to be mainly influential in making chains opt for an equity-based growth strategy, since the analysis shows that the experts had also overlooked the effects that this variable has on taking better advantage of manpower and work effectiveness, with this effect mainly having an impact on international hotel growth through an equity-based formula. Consequently, while it is true that this variable also exerts a certain influence on the decision to opt for non-equity growth, it is mainly influential in encouraging an equity-based growth strategy.

Matrix F also shows that the 'making the most of cultural and linguistic barriers' and 'brand recognition' variables only affect non-equity growth strategies, because they have no significant second-generation influence on any of the factors that affect equity-based growth.

Matrix F also reveals that most forgotten or second-generation effects are influential in non-equity growth strategies. In fact, the only second-generation effect that influences equity-based growth is the one referred to above in relation to modifying the cultural gap ('economic liberalization'), with the five remaining ones affecting non-equity growth strategies. From all this, the predominant influence of variables not fully defined in the analysis by our panel of experts could be clarified. Thus 'control of new markets' allows hotel chains to expand through non-equity growth formulae, not only because it can push up a chain's earning capacity, but also because it can allow the chain to take advantage of a bigger cultural gap between countries.

The same occurs with economies of scale. As well as the increased earning capacity to which this strategy can give rise, it also allows hotel chains to speed up growth. If, in the hotel business, economies of scale are associated with logistics or common architectural designs, and these can be shared by a franchise network and local partners at a relatively low cost in terms of the transfer of know-how, then these economies can be obtained without direct investment in the ownership of real estate or even without the presence of executives from the parent company (Contractor and Kundu, 1998a). An important aspect is that originates in the variable 'increasing the hotel chain's size'. According to the initial matrix developed by the experts (matrix A), this variable was only important when this top group expanded through equity modes. However, matrix F shows that this variable also plays a considerable role in boosting a hotel chain's earning capacity, and this, as commented on previously, is influential in the adoption of non-equity growth strategies. Something similar is true in the case of 'technological progress'. As well as allowing better advantage to be taken of manpower and work effectiveness (as shown in the direct incidence matrix developed by the experts, matrix A), it also enables a hotel chain to boost its profits (Matrix F). In the hotel industry, a brand name is considered to be strategic advantage normally owned and controlled by the parent company (Dunning and McQueen, 1981). This advantage also boost the possibility of successful alliances (Contractor and Kundu, 1998a). Companies with well-known trade names tend to increase the number of franchise operations and management contracts, thus also managing to modify their market share.

As for the variable relating to an increase in the size of a hotel chain, it can be seen to have already been very influential as an incentive for non-equity growth approaches. It also allows hotel chains can accelerate their growth. Thus the results of our analysis of this variable confirm those of other studies, for example, Gatignon and Anderson (1988) and Agarwal and Ramaswami (1992), indicating that policies that involve close control over operations in newly penetrated markets are less frequent in the case of large-scale foreign investment. This argument is based on the idea that, given the scale of international operations in the hotel industry, hotel chains are forced to accept partners so as to share the high cost of investment. This tends to compel hotel chains to accept many partners, thus reducing their control over operations, leading to a greater number of non-equity operations.

5 Conclusions

This study has helped to explain the main direct and indirect causes of the marked proliferation in non-equity growth strategies by the world's top chains.

Through the incidence matrices, different cause-effect relations have been revealed that cannot easily be directly discovered through intuition or experience, which is why they have come to be known as second-generation effects. In highly simplified situations, it is possible to determine not just direct relations but second-generation ones through intuition, but during the course of normal activities by hotel chains, numerous factors are interrelated in various different ways, sometimes acting as causes and other times as effects. As a result, a complicated network of relations is created that makes it practically impossible to detect each and every one of the indirect links among all of them. It is therefore natural for us to overlook some relations using intuitive means, although they can be detected using the techniques presented in this study because they help us to reach conclusions that transcend the initial opinions of our panel of experts. True, there will always be some grey areas; unexplored avenues and variables that are not taken into account, because this is characteristic of human activities, given how dynamic they are and the fact that they are subject to sizeable, often unforeseen changes. Above all, this study has brought to light variables that encourage a greater trend toward non-equity agreements by the world's top hotel companies. These include control of new markets, economies of scale, economic liberalization, making the most of cultural and linguistic barriers, increasing the size of the hotel chain, technological progress and greater recognition of the hotel chain's brand name. Each of these variables has different indirect effects, not taken into account by the experts, which act as a greater or lesser incentive for hotel chains to opt for non-equity expansion stratagems.

Having analyzed the objectives of this study, it can be concluded that non-equity agreements have become the most popular growth mechanisms for the world's hotel chains, since it leads the field by a long way compared with other development policies.

Growth strategies that do not involve capital shareholdings are the only means of reaching the required size for effective economies of scale and scope (Contractor and Kundu, 1998a). In perceived risky environments, contractual arrangements are extremely attractive, as the high capital costs associated with hotel property development are thus borne by the real estate companies, although there may be a requirement for a minority equity stake so as to gain the contract in the first instance.

From the forgotten effects matrix, it can be seen that, today, most of the causes foster non-equity growth strategies by hotel chains. Indeed, non-equity strategies are particularly appealing when hotel chains wish to exercise tight control over new markets, when they aim to speed up growth and/or when they wish to increase the size of the hotel chain to a significant extent.

Non-equity strategies are also especially advisable in scenarios where there are strong barriers to entry and/or big cultural gaps between the host country and hotel chain's country of origin. Indeed, in many places where governments impose strong barriers to entry on foreign companies, like Cuba, non-equity strategies are the only possibility that hotel chains have of penetrating these markets. Something similar also occurs in the case of countries that are culturally very different from the investor hotel's own country. In this case, several studies show that cultural gaps between a hotel chain's country of origin and the host country play a decisive role in the choice of expansion strategy. Nonetheless, in existing literature no conclusive significance has been found in the exact importance of this relationship. Traditional entry-mode literature holds that firms minimize the high information costs associated with operating in culturally unfamiliar countries by seeking collaborative modes (Franko, 1976; Stopford and Wells, 1972; Davidson, 1980; Gatignon and Anderson, 1988; Kought and Singh, 1988; Erramilli, 1991; Kim and Hwang, 1992; Agarwal, 1994).

On the one hand, a series of studies argue that the greater the cultural gap, the smaller the shareholding that will be held in the company and the greater the prevalence of collaborative alliances in growth strategies (Stopford and Wells, 1972; Franko, 1976; Kought and Singh, 1988; Erramilli, 1991; Kim and Hwang, 1992). Others authors claim, however, that when different cultures are involved, companies prefer to expand through ownership, thus imposing their own management methods (Davidson, 1980). In an unknown environment, a company does not fully trust local management methods and will prefer to run the hotel itself (Hymer, 1976).

From the data obtained from the panel of experts, as commented on above, this study has helped bring to light new evidence to corroborate literature that purports that the bigger the cultural gap between an investor chain's country and the host country, the more the hotel chain will opt for non-equity growth strategies.

Lastly, hotel chains will also mainly opt for non-equity formulae when they plan to strengthen their brand image and take advantage of the big economies of scale derived from running a large number of hotels, such as savings on purchases, marketing campaigns, etc. (Martorell, 2006).

On the other hand, it is more interesting to use an equity-based strategy when the hotel chain's goals are more closely focused on restructuring its investments, increasing staff training and taking advantage of possible economic liberalization processes, since all this will lead to a big increase in the hotel chain's turnover and, with this strategy, these profits will not have to be shared with other partners.

In the light of all the factors commented on above, non-equity growth strategies are the most widely used by the world's top hotel chains in terms of size and room portfolio when operating outside their country of origin. In fact, the top 10 most globalized hotel chains in terms of the number of countries in which they operate and the top 10 with the biggest room portfolios (Hotels, 2010) have both expanded through growth strategies that do not involve capital outlays.

References

Agarwal, S.: Socio-cultural Distance and the Choice of Joint Ventures: A Contingency Perspective. Journal of International Marketing 2(2), 63–80 (1994)

Agarwal, S., Ramaswami, S.: Choice of Foreign Market Entry Mode: Impact of Ownership, Location, and Internalization Factors. Journal of International Business Studies 23(1), 1–27 (1992)

Alexander, N., Lockwood, A.: Internationalisation: A comparison of the hotel and retail sectors. The Service Industries Journal 16, 458–473 (1996)

Altinay, L.: Factors influencing entry mode choices: empirical findings from an international hotel organisation. Journal of Hospitality & Leisure Marketing 12(3), 8–9 (2005)

Bhattacharyya, S., Lafontaine, F.: Double-Sided Moral Hazard and the Nature of Share Contracts. RAND Journal of Economics 26(4), 761–781 (1995)

Contractor, F.: Contractual and cooperative forms of international business: Towards a unified theory of modal choice. Management International Review 30(1), 31–54 (1990)

Contractor, F., Kundu, S.: Modal Choice in a world of alliances: Analyzing organisational forms in international hotel sector. Journal of International Business Studies 29(2), 325–358 (1998a)

Contractor, F., Kundu, S.: Franchising versus company-run operations: Modal choice in the global hotel sector. Journal of International Marketing 6(2), 28–53 (1998b)

Davidson, W.: The location of foreign direct investment activity: Country characteristics and experience effects. Journal of International Business Studies 12, 9–22 (1980)

Dunning, J., McQueen, M.: Transnational corporations in the international tourism. UNCTC, New York (1981)

Erramilli, M.: Entry Mode Choice in Service Industries. International Marketing Review 7(5), 50–62 (1990)

Erramilli, M.: The experience factor in foreign markets entry behavior of service firms. Journal of International Business Studies 22(5), 135–150 (1991)

Erramilli, M., Agarwal, S., Dev, C.: Choice between Non-Equity Entry Modes: An Organisational Capability Perspective. Journal of International Business Studies 33(2), 223–242 (2002)

Franko, L.: The European multinationals. Greylock Publishers, Stanford (1976)

Gatignon, H., Anderson, E.: The multinational corporation's degree of control over foreign subsidiaries: An empirical test of a transaction cost explanation. Journal of Law Economics, and Organisation 4(2), 305–336 (1988)

Gil, J.: La gestión interactiva de los recursos humanos en la incertidumbre. Ed. Centro de Estudios Ramón Areces, Madrid (1995)

Gil-Lafuente, A.M., Anselín, E.: Fuzzy logic in the strategic analysis: impact of the external factors over business. International Journal of Business Innovation and Research (2009)

Gil-Lafuente, A.M., Barcellos de Paula, L.: Una aplicación de la metodología de los efectos olvidados: los factores que contribuyen al crecimiento sostenible de la empresa. Cuadernos del Cimbage 12, 23–52 (2010)

Gil-Lafuente, A.M., Gil Aluja, J.: Fuzzylog. Patente (2009)

Gil-Lafuente, A.M., Luis-Bassa, C.: Identificación de los atributos contemplados por los clientes en una estrategia CRM utilizando el modelo de efectos olvidados. Cuadernos del Cimbage 13, 107–127 (2011)

Hymer, S.: The international operations of national firms: A study of direct foreign investment. The MIT Press, Cambridge (1976)

Ingram, H.: Book Review: Franchising Hospitality Services. International Journal of Contemporary Hospitality Management 13(5), 267–268 (2001)
Kaufmann, A., Gil, J.: Models per a la investigació d'efectes oblidats. Ed. Milladoiro, D.L. Vigo (1988)
Kim, C., Hwang, P.: Global Strategy and multinational entry mode choice. Journal of International Business Studies 23(1), 29–53 (1992)
Kogut, B., Singh, H.: The Effect of National Culture on the Choice of Entry Mode. Journal of International Business Studies 19, 411–432 (1988)
Lashley, C., Morrison, A.: In Search of Hospitality: Theoretical Perspectives and Debates. Butterworth-Heinemann, Oxford (2000)
Martorell, O.: The growth strategies of hotel chains. Best business practices by leading companies. The Haworth Hospitality Press (2006)
Martorell, O., Mulet, C., Rosselló, M.: Valuing growth strategy management by hotel chains based on the real options approach. Tourism Economics 14(3), 511–526 (2008)
Martorell, O., Mulet, C.: The franchise contract in hotel chains: a study of hotel chain growth and market concentrations. Tourism Economics 16(3), 493–516 (2010)
Okumus, F.: Implementation of Yield Management Practices in Service Organisation: Empirical Findings from a Major Hotel Group. The Service Industries Journal 24(6), 65–89 (2004)
Stanworth, J., Curran, J.: Colas, Burgers, Shakes and Shirkers: Towards a Sociological Model of Franchising in the Market Economy. Journal of Business Venturing 14(4), 323–344 (1999)
Stanworth, J., Stanworth, C., Watson, A., Purdy, D., Healeas, S.: Franchising as a Small Business Growth Strategy. International Small Business Journal 22, 539–559 (2004)
Stopford, J., Wells, L.: Managing the Multinational Enterprise. Organization of the Firm and Ownership of the Subsidiaries. Basic Books, Inc., New York (1972)

Tourism Expenditure of Airline Users: Impact on the Spanish Economy

M. Luisa Martí Selva, Consuelo Calafat Marzal, and Rosa Puertas Medina

Departamento Economía y Ciencias Sociales. Universidad Politécnica de Valencia
Camino de Vera s/n, 46022 Valencia Spain
{mlmarti,chelo,rpuertas}@esp.upv.es

Abstract. This paper aims to measure how the injection of money into the Spanish economy as a result of tourist spending contributes more or less to public income, job creation and new business opportunities depending on the air transport tourists choose. Input/Output (I/O) analysis is used to identify the Direct, Indirect and Induced Effect in both aggregate terms and also sector by sector.

Keywords: Low-cost airlines, Input/Output analysis, Tourism impacts.

1 Introduction

The European air transport market has become increasingly competitive since the European Commission presented the "Civil Aviation Memorandum No. 2" in 1984. The memorandum established the long term objectives of the sector and the strategy to gradually achieve them through three packages of deregulation measures in 1987, 1990 and 1992. The objectives permitted free access and definition of routes, flight regularity and capacity and fares, although the measures did not take effect in Spain until 1993 (Costas and Germá, 1997). As a result of the progress made in information systems, two different business models have been identified: traditional airlines that concentrate traffic at large logistics platforms to obtain significant improvements at the airports where they are based, and Low-Cost Airlines (LCA), which offer direct short and medium distance flights at highly competitive prices (Rodríguez et al, 2008).

LCA arrived in Europe on the back of the deregulation of American air traffic in 1971 and the success of Southwest Airlines. Their emergence and subsequent development is very closely related to the removal of intermediation from the supply chain through the introduction of internet sales and bookings, which saw traditional charter companies convert to LCA. Authors such as Miquel and De Borja (2006) believe that one of the keys to their development strategy has been the online marketing of their products (more than 90% sell tickets over the internet). However, other studies that have focused on the causes and reasons for the emergence of LCA have reached the conclusion that the main motives have been: the development of new business models, the deregulation of air transport, the

A.M. Gil-Lafuente et al. (Eds.): Soft Comput. in Manag. and Bus. Econ., STUDFUZZ 287, pp. 287–304.
springerlink.com

increase in competition and the prodigious development of the internet (Romero and Cortés, 2006 and Rey and Inglada, 2006).

The arrival of this type of company has revolutionised air transport in recent years, forcing traditional companies to change their business model. Deregulation has benefitted European travellers enormously due to the drastic reductions in the fares offered by LCA, which are up to 40%-60% cheaper than those offered by traditional companies (Piga and Filippi, 2001).

The United Kingdom, Spain, Germany and Italy account for 61% of all low-cost flights in Europe, according to the *Instituto de Estudios Turísticos*, (Institute of Tourism Studies in English, hereafter referred to as IET, 2011a). At present, the LCA with the greatest volume of traffic in Spain are Ryanair Ltd., EasyJet Airline Co. Ltd., Air Berlin, Channel Express, Transavia Holland Bv. and Vueling Airlines.

The market share of LCA recently surpassed that of traditional companies, recording ongoing and stable growth (Graphic 1).

Graphic 1 LCA share of travellers

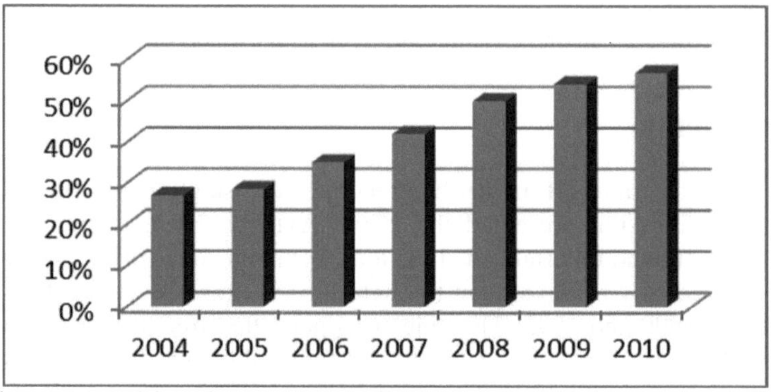

Source: own elaboration based on data from IET (several years).

In this scenario, two very different types of tourists have developed over the last few years: LCA users and those who fly with traditional companies. These two types of tourists have very different habits and their expenditure in the destination country has a different impact on the economy. In light of the minimal attention the latter has received from researchers, this paper conducts a comparative analysis of the economic impact of both types of passengers, in terms of expenditure, on the Spanish tourism industry. More specifically, the paper aims to measure how the amount of money that the tourism industry injects into the economy contributes more or less to public income, job creation and new business opportunities depending on the type of air transport chosen. A Leontief quantity model based on Input/Output analysis is used to identify the Direct, Indirect and Induced effect of tourism expenditure comparing the two modes of air transport.

The paper is organised as follows. Section 2 describes the current situation of the tourism industry, paying special attention to LCA and traditional airline passengers, as well as to the difference in user spending distribution. Section 3 presents a review of the literature and explains the methodology applied in the economic impact analysis. Section 4 defines the methodological aspects of the Input/Output analysis and in Section 5 the methodology is applied to passengers distinguishing between LCA and traditional airline users. Section 6 presents the results of the research, which will clarify the different economic effects caused by the differences in expenditure of the two groups. Finally, Section 7 presents the main conclusions of the research.

2 Tourism from the Perspective of LCA Versus Traditional Airlines and Differences in User Expenditure Distribution

The domestic air transport market has grown by 50% in Spain since 2000, contributing 7% to national GDP (Navío, 2011). The decrease in fares and increase in direct routes are some of the possible causes of this trend.

One of the features that sets Spanish tourism apart is its high level of seasonality, most flows taking place in summer (July to September). This trend is particularly noticeable in the case of foreign tourists arriving in Spain, although domestic tourism also displays this pattern (IET, 2011b). Another characteristic is the reason for travelling to Spain. Approximately 83% of the international tourists who visited Spain in 2010 did so for leisure and holidays. The majority of foreign tourists stayed in hotels and similar types of accommodation and opted for a not very organised trip. The latter is related to the new patterns of behaviour displayed by consumers when booking a trip, in that they have easier access to information technologies, seek more flexibility, are more demanding, etc.

As regards the amount of money international tourists spend, those who chose to fly to Spain register the highest average level of consumption, followed by the tourists who travelled by ship and in third place by those who came by road. This difference has increased since the emergence of LCA, which have absorbed almost 60% of international arrivals at Spanish Airports, recording a growth rate of close to 7% last year and confirming the conclusions reached in the study by Rey et al (2011).

According to Fernández (2008), the decrease in the price of air fares normally results in an increase in the number or regularity of trips, the amount of money spent during the trip and even the arrival of a new tourist who did not travel in the past, but can today as a result of the "increase" in their real income due to the reduction in air fares. However, the latest data published by IET suggest that LCA users spend almost 17% less than those who fly with traditional airlines. That is, every LCA user spends less than tourists who use traditional airlines (Table 1).

Table 1 Average expenditure per tourist by type of airline (€)

YEAR	TRADITIONAL	LOW-COST
2004	973.2	853.5
2005	1018.5	764.9
2006	931.7	793.8
2007	967.2	806.0
2008	966.7	799.6
2009	1010.0	805.5
2010	1041.4	821.1

Source: own elaboration with data from IET (several years).

However, there is limited information on the effect that this expenditure has on the various sectors of the tourism industry in the country. The main saving can be made in the choice of air transport. Costs can be reduced by 29.7% if LCA are chosen, although these passengers also spend less on accommodation, food and visits to museums, the theatre, leisure establishments (9.6%, 2.3% and 13.7%, respectively) than those who use traditional airlines.

Apart from the fact that LCA users spend less in total, they also spend their money on different things. The spending patterns of LCA and traditional airline users are detailed in Tables 2 and 3 below.

Transport accounts for the largest budget outlay of all users, representing almost 40% in the case of traditional airlines and between 31% and 34% in the case of LCA. This percentage has increased steadily until last year, when it decreased, particularly in the case of traditional airlines. LCA users spend a larger proportion of their budget on the rest of budget items, although it is worth recalling that they spend less in total.

Table 2 LCA user expenditure pattern

Year	Accommodation	Food	Restaurants	Transport	Extraordinary Expenses	Others
2004	22.40%	7.80%	18.90%	32.70%	4.60%	13.60%
2005	23.50%	7.30%	19.70%	33.80%	0.60%	15.10%
2006	26.80%	7.50%	18.20%	33.10%	-	14.50%
2007	28.60%	7.00%	16.70%	33.70%	0.30%	13.70%
2008	27.90%	7.20%	17.20%	33.00%	0.80%	13.90%
2009	27.40%	8.20%	17.50%	34.40%	-	12.60%
2010	28.49%	8.54%	17.79%	31.10%	0.00%	14.08%

Source: own elaboration with data from the IET (several years).

Table 3 Traditional airline user expenditure pattern

Year	Accommodation	Food	Restaurants	Transport	Extraordinary Expenses	Others
2004	24.30%	5.40%	15.00%	38.60%	2.60%	14.10%
2005	26.60%	4.30%	15.50%	37.50%	1.70%	14.40%
2006	25.20%	5.70%	15.40%	39.90%	-	13.80%
2007	26.40%	5.20%	15.00%	39.80%	0.50%	13.10%
2008	26.40%	4.70%	14.70%	40.90%	0.50%	12.80%
2009	25.60%	5.30%	14.90%	41.00%	-	13.30%
2010	28.16%	6.09%	16.26%	35.36%	0.00%	14.13%

Source: own elaboration with data from IET (several years).

Furthermore, LCA users no longer only go on business trips (conferences, fairs, meetings), but also on weekend trips, which is helping to reduce the seasonalisation of holiday periods. Low-cost companies have brought about a significant change in tourist habits and behaviour, which is forcing traditional airlines to adapt so as not to lose their share of the market. LCA flights are nowadays becoming increasingly less linked to cheap and low-quality tourism.

The latest report by IET (2011c) reveals that there are no significant differences between the socio-demographic characteristics of the tourists who choose LCA to fly to Spain and those who fly with traditional airlines. The largest group in both cases is aged between 25 and 44, has a university education, works and boasts an average level of income. Passenger behaviour is equally similar (reason, accommodation, activities, satisfaction…), the only differences being found in the degree of organisation of the trip. Only 26.4% of LCA users contracted a package deal, compared to 43% in the case of those who used a traditional airline. This is mainly due to the web pages provided by these companies being used more. In addition, this group of tourists accounts for 45.6% of the total expenditure of tourists who travelled to Spain by air, although LCA users spend less on average per person than traditional airline users (Table 1).

3 Review of the Literature and Explanation of the Methodology Applied

There are a wide variety of methods for analysing the economic impact of a final demand shock, such as tourism expenditure and its impact on industry. Which method to apply depends not only the objective of the analysis, but also on the information available, which is occasionally limited and not very accurate. The methods used most frequently are listed below:

- Econometric models estimate the importance of economic variables using regression techniques. They require a large amount of information. In fact, the more equations, the more information is required.
- Expenditure multiplier models based on macroeconomic variables obtain the secondary effects caused by a change in demand. They are based on the hypothesis that both public spending and also investment and exports are independent, whereas private consumption and imports depend directly on disposable income.
- Input/Output Models analyse the bilateral relationships between all economic sectors, determining the effects each has on the rest. They can obtain multipliers that reflect the disaggregated effects of variations in demand. This method requires an Input/Output (I/O) Table referring to a particular geographical region, as well as an expenditure framework that defines the activity under analysis.

This study employs Input/Output Models because the authors believe that they are the best suited to the objective of this paper. Hence, in reference to the context of the research, we are aware that multiple studies have been conducted in the field of tourism. The literature includes studies that focus on measuring tourism demand through expenditure and/or the number of tourists, most of which use time series data (Lim, 1997). Other studies approach tourist expenditure as the dependent variable in a regression model (Taylor et al, 1993). Rey et al (2011) builds a dynamic panel to analyse the effects of LCA on Spanish tourism, stating that the expansion of such airlines has been significant over the first decade of the 21st century and unaffected by the crisis. As regards the impact of taxes on airport fees, Sainz-Gonzalez et al (2011) were unable to find any differences between LCA and traditional companies, concluding that they both pass on such charges to consumers by increasing their fares.

From the perspective of a tourist destination, Ivars and Menor (2008) contributed a study on the repercussions of LCA in regard to the development of the hinterland of Alicante Airport. Likewise, Graham and Dennis (2010), Donzelli (2010) and Abda et al (2012) provide evidence on the effects of LCA on different geographical regions (Malta, Southern Italy and the USA), using a variety of methodologies. However, it is not advisable to generalise the results obtained in different studies, as different case studies can give rise to highly disparate conclusions.

In the same vein and focusing on the objective of this study, we have also explored the relationship between tourist activity and its contribution to economic growth, giving rise to two different, albeit complementary, methodological approaches: Input/Output models and computable general equilibrium models. Despite being used more frequently, the former tend to exaggerate the effect of growth in tourism on production, income and employment due to making not very realistic baseline assumptions. Meanwhile, general equilibrium models provide a better understanding of the nature of the impact in response to changes in economic policy (Brida, et al., 2008). So as to avoid incurring in past errors and to endow results with greater accuracy, the analysis presented here will only use the statistical information published by the *Instituto Nacional de Estadística* (Spanish National Institute of Statistics in English, hereafter referred to as INE) and IET, which enables us to compile an I/O Table of the national tourism industry, as well as the expenditure/tourist of LCA and traditional airline users.

One of the first applications of this methodology can be found in the research by Herderson (1955), where an Input/Output Model is used to analyse the Italian economy. In the field of maritime transport, the literature includes numerous papers, ranging from Hill (1975), which determines the impact of the Port of Baltimore (USA), to Martí et al (2009), which includes an application to the various ports in the Valencia Region (Spain). This methodology has also been employed to determine the effects of various recreational and sports competitions (Barker et al, 2002; Crompton et al, 2001; Kasimati, 2003) and, more recently in the field of tourism, to analyse the economic impact of hotels in the Spanish city of Seville (González, 2010).

The grounds for choosing the Input/Output methodology to analyse the impact of airline user expenditure on the tourism industry is the possibility of estimating the effects on typically touristic economic activities both at sector and aggregate level. However, as is the case with any other methodology, this technique is not exempt of criticism, which can occasionally be refuted by simply clarifying the objective of the analysis and the information available to perform it. This tool delves deeper into the productive fabric of the tourism industry, providing figures on production, employment and taxation, among other variables, which will shed light on the differences that exist between the two types of airlines and whether or not it is worth backing the development of one model of air transport or another.

4 Methodological Aspects of Input/Output Analysis

Input/Output analysis is one of the cornerstones for performing quantitative studies on how the activities of a facility or an event influence their economic environment. This methodology focuses mainly on determining economic impact and on the importance of decisions regarding large investments in transport, health and leisure infrastructures, or any other service that has a complex relationship with the rest of economic activities.

The fundamental tool used in this paper is the I/O Table, created on the basis of the Tourism Satellite Account (TSA), which provides information about the relationships in the tourism industry. The TSA is published by INE, the most recent version referring to 2007. The flowchart below summarises the stages of the empirical analysis performed.

The I/O Table created has a very similar structure to the traditional tables published by Statistics Institutes. It is divided into three large sections (Figure 2): the central section (Section 1), also called the inter-interdustry section, which in this case is made up of 12 sectors that represent the tourism industry: Hotels, Restaurants, Travel Agencies, Auxiliary transport services, Property rental services, Transport equipment hire, Market and Non-market cultural, recreational and sports services and also passenger road, rail, maritime and air transport. The Primary Inputs Section (Section 2) includes all the items that make up the value added of the foregoing sectors, while the Final Demand Section (Section 3) will provide information on the behaviour of consumption, investment and exports.

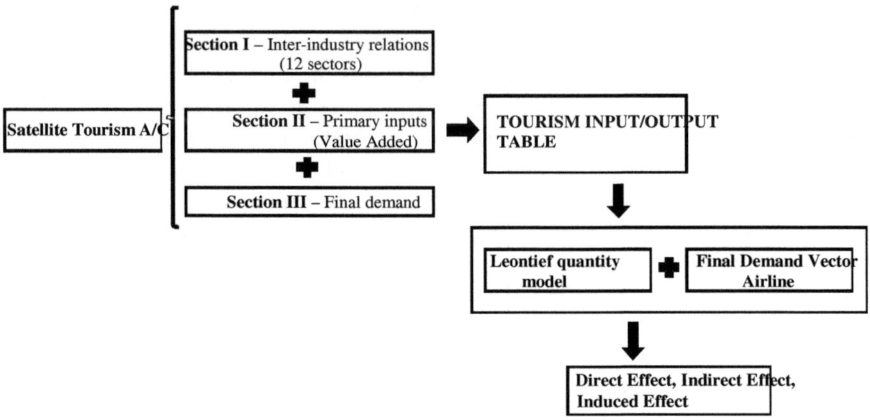

Source: own elaboration.

Fig. 1 Methodological flowchart of Input/Output Analysis

Section I: Inter-industry relationships

	sector 1	sector 2	......
sector 1			
sector 2			
......			

Section III: Final Demand

	Final demand	T. employers
sector 1		
sector 2		
......		

Section II: Primary inputs

	sector 1	sector 2	......
GVA			
T. resources			

Source: own elaboration.

Fig. 2 Structure of the Input/Output table

By applying Leontief's quantity model to the I/O Table representing tourism, we can calculate the response of the production system to an initial shock in final demand, such as the money spent by tourists who use either LCA or traditional airlines. In short, the fundamental notion is that an increase in demand is passed on, such that production not only increases by the amount required to meet the upturn in demand, but by more in order to satisfy the intermediate demand of the rest of industries so they can achieve their increase in production. This is a chain of subsequent input requirements on behalf of each branch of industry that must increase its production in order to supply the rest of industries.

The mathematical formula used to calculate how much each industry needs to produce to meet a given Final Demand is as follows:

$$[X_i] = [I - A_{ij}]^{-1} \cdot [Y_i] \quad (1)$$

whereby:

X: vector of Total Outputs (X_i), Total Inputs (X_j).
Y_i: vector of final demands.
$[I - A_{ij}]^{-1}$: Inverse Leontief matrix.
I: identity matrix.
A_{ij}: technical coefficients matrix, a_{ij} being

$$a_{ij} = \frac{X_{ij}}{X_j} \quad (2)$$

The coefficient a_{ij} is defined as the use industry j makes of products from industry i per unit of production. If this information is obtained for all industries, we have a technical coefficients matrix $[A_{ij}]$.

This model is useful because of the consistence and proportionality of the technical coefficients (a_{ij}). This makes it possible to assess the effects of an exogenous vector of Final Demand on production, income and employment. This model identifies several types of effects of a branch of economic activity on the rest of branches and vice-versa. More specifically, it can distinguish the Direct Effect, the Indirect Effect and the Induced Effect.

The Direct and Indirect Effect are calculated using model (1), described previously. If we develop the matrix $[I - A_{ij}]^{-1}$, we obtain the follow expression:

$$[I - A_{ij}]^{-1} = I + [A] + [A] \cdot [A] + [A] \cdot [A] \cdot [A] + \ldots + R \quad (3)$$

whereby R is the residual of a decreasing sequence.

By incorporating this development into the Leontief model, we will obtain:

$$[I - A_{ij}]^{-1} \cdot [Y] = [Y] + [A] \cdot [Y] + [A] \cdot [A] \cdot [Y] + [A] \cdot [A] \cdot [A] \cdot [Y] + \ldots + R \quad (4)$$

At the same time, it is broken down into two effects, namely the Direct and Indirect Effect.

The Direct Effect represents the direct production investment that industry stakeholders must make in order to satisfy a change in the final demand for its services.

$$E_D = [Y] + [A]\cdot [Y] \tag{5}$$

The Indirect Effect represents the impact stemming from the subsequent purchase and sale relationships that take place between the industries originally affected by the activity being studied and the rest of industries.

$$E_I = [A]\cdot [A]\cdot [Y] + [A]\cdot [A]\cdot [A]\cdot [Y] + \ldots + R \tag{6}$$

Indirect effects are generally larger because they arise from the demand generated by industries that have received a boost from the Direct Effect (when certain industries increase their purchases, they also demand from other industries and so on, generating a web of inter-industry relationships).

Finally, we address the calculation of the Induced Effect as that generated by the consumption and investment capacity of companies and stakeholders directly related to the activity of an industry. In order to perform this calculation, complementary assumptions must be made regarding household saving capacity (by assigning the rest to final consumption) and regarding investment capacity, both of savings from wages and also from gross operating surpluses. This is therefore the multiplier effect of the Direct and Indirect Effect on the economy.

In order to calculate the Induced Effect, the same Leontief model is used, but with a Final Demand vector enlarged by the demand for consumer and capital goods, generated by the income of the stakeholders that participate in the activities of the industry under study. The new model would be as follows:

$$[X_i] = [I - A_{ij}]^{-1}\cdot [Y_i^{R*}] \tag{7}$$

where, $[Y_i^{R*}]$ is the consumption vector generated by the income of the industry.

5 Application of the Methodology to Airline Users

The results of the analysis enable us to ascertain which activities in the tourism industry benefitted the most from the money spent by tourists during their stay in 2007 (the latest information available). We have distinguished between the tourists who used LCA and those who used traditional airlines in order to determine whether there are any quantitative differences between the two groups that provide objective data to help the regions under study to adapt their supply (hotels, restaurants, recreational areas) to the existing demand. A priori, the statistics referring to 2007 reveal that more travellers use traditional airlines (60%) than LCA (40%).

As indicated previously, we have used the I/O Table of tourism for 2007 published by INE. Out of the 12 industries representing tourism, it is worth highlighting Property rental services and Restaurants and similar establishments, as they have the most value added and largest total production (Table 4).

Table 4 Sector-by-sector distribution of Gross Value Added and Production (millions of €)

SECTORS	GVA mp	Production
Hotels and similar establishments	14,388	22,124
Property rental services	90,302	116,952
Restaurants and similar establishments	55,402	93,199
Passenger road transport	6,106	9,983
Passenger rail transport	1,186	2,042
Passenger maritime transport	317	809
Passenger air transport	3,398	9,801
Travel agency services	2,986	4,958
Auxiliary transport services	13,747	34,444
Transport equipment hire	2,380	4,413
Market cultural, recreational and sports services	17,601	29,473
Non-market cultural, recreational and sports services	6,131	11,459

Source: own elaboration with data from INE (2007).

In the case of restaurants, the results in terms of production and GVA are linked to a higher concentration of employment. Indeed, this industry accounts for more than 40% of total employment in the tourism industry, while the second largest group of employees belongs to "Cultural, recreational and sports services", with a share of 13.5% of the total (Graphic 2).

Graphic 2 Sector-by-sector distribution of employment in 2007 (in 1,000s)

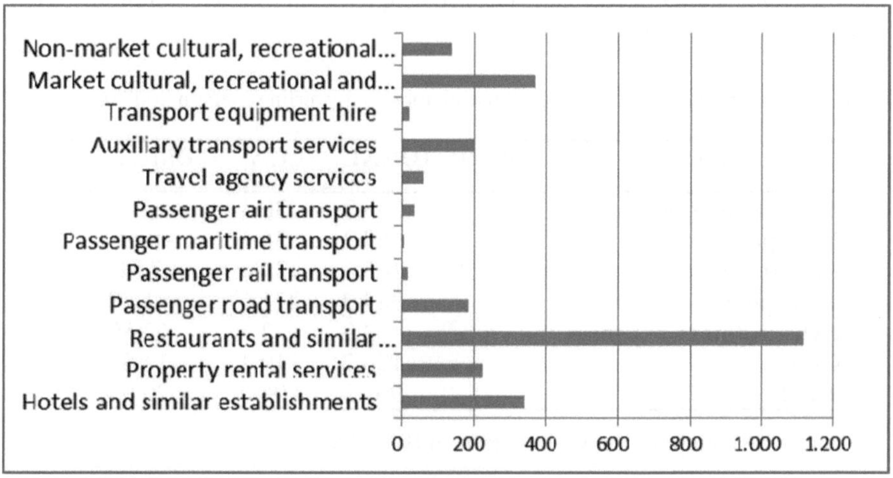

Source: own elaboration with data from INE (2007).

According to the data from the I/O Table, tourism consumption in 2007 amounted to a total of 93,198 million euros, distributed mainly among Restaurants, Hotels, Transport (including road, rail, maritime and air) and Property rental services (Graphic 3). The sectors that stand out the most are Accommodation, Restaurants and similar establishments and Transport, generating close to 78,000 million euros between them, while Travel agency services and Cultural, recreational and sports services had less of an impact on consumption.

Graphic 3 Distribution of tourism consumption in 2007 (millions of €)

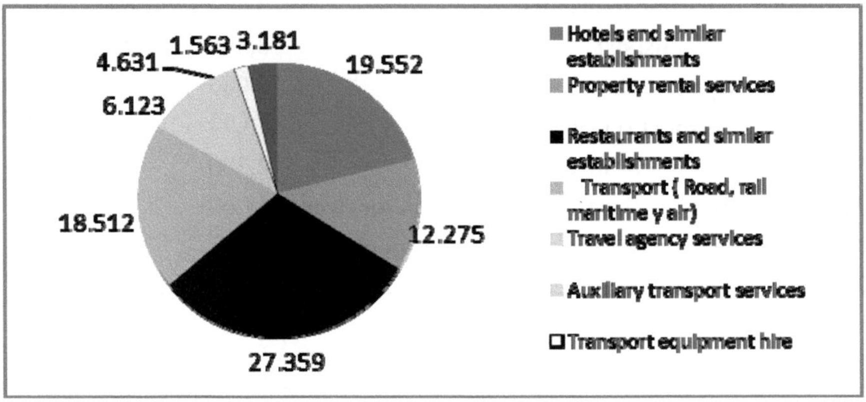

Source: own elaboration with data from INE (2007).

The vector of Final Demand used in the empirical analysis referred to the type of expenditure of travellers in Spain that use both LCA and traditional airlines (Table 5). This expenditure comprises five basic items: transport, accommodation, restaurants and similar establishments, travel agency services and cultural, recreational and sports services.

Table 5 Distribution of average expenditure on tourism by airline users in 2007 (€)

Items	TRADITIONAL	LCA	Difference
Accommodation	255.3	230.5	10.8%
Food purchases	50.3	56.4	-10.8%
Restaurants	145.1	134.6	7.8%
Transport	384.9	271.6	41.7%
Extraordinary expenses	3.9	2.4	62.5%
Other	126.7	110.4	14.8%
Total	967.2	806.0	20.0%

Source: own elaboration with data from IET (2007).

The total consumption detailed in Table 5 refers to travellers who did not contract a package deal[1], who represent 72.2% of the tourists who used LCA (23,900 people) and 58.8% of those who flew with traditional airlines (35,850 people). According to Fernandez (2008), it has been demonstrated that when low-cost tourists decide to consume, they do so to a lesser extent than traditional tourists. However, this poses the question of which items they spend less on. The information in Table 5 reveals that the largest share of LCA user total expenditure went towards transport, followed by accommodation. Meanwhile, the tourists who used traditional airlines proportionally spent somewhat more on the first item than LCA users, but slightly less on accommodation.

When we compare the average spending by industry per traditional airline passenger in regard to those who used LCA, air transport figures prominently. Low-cost tourists spend almost 42% less on this item than those who use traditional airlines. Meanwhile, traditional airline users spend 14.8% more than tourists who fly with LCA on Market cultural, recreational and sports services ("Others").

6 Economic Impact

Applying the proposed methodology, we first analysed the effects caused by the total volume of tourists, differentiating between them according to the type of air transport they used. Tables 6 and 7 show that the Direct Impact is more significant, that is, the money spent by tourists had a direct effect on the rest of sectors, increasing their demand and, consequently, GDP. In the case of LCA tourist expenditure, GDP rose by more than 10 million euros, compared to almost 19 million euros in the case of tourists who used traditional airlines. Furthermore, the distribution of GDP between wages and profits is observed to be very similar in both cases. Finally, we must highlight that tourist activity has created 556 jobs (205 jobs from the impact of LCA tourist expenditure and 351 from the impact of traditional airline users) as a consequence of the increase in demand generated in the industry.

Table 6 Economic impact of LCA travellers (1,000s of €)

	Direct	Indirect	Induced	Total
Wages	5,130	4,152	1,593	10,875
Business profits	5,509	4,159	2,086	11,754
Tax income	171	139	112	422
GDP	10,639	8,317	3,701	22,657
Total production	31,965	26,327	10,795	69,087
Employment (persons)	205	145	66	416

Source: own elaboration.

[1] The IET database does not take into account the tourists that have arrived in Spain on a package holiday, as the money spent on them distorts that spent on other expenditure items. This gives a distorted image of tourist expenditure on certain goods, such as the money spent on single and return journeys or group accommodation (Fernandez, 2008).

Table 7 Economic impact of traditional airline users (1,000s of €)

	Direct	Indirect	Induced	Total
Wages	9,269	7,811	2,870	19,950
Business profits	9,547	7,512	3,758	20,817
Tax income	286	243	202	730
GDP	18,817	15,334	6,669	40,820
Total production	57,804	49,646	19,449	126,899
Employment (persons)	351	257	119	727

Source: own elaboration.

As regards the indirect effect, we must underline that, as a whole, both LCA tourists and also traditional airline tourists boosted GDP by more than 23 million euros, increasing both wages and business profits (8.3 and 15.3 respectively). Likewise, they created no less than 402 jobs.

Finally, in order to ascertain the induced impact it is necessary to make an assumption regarding the impact of the income generated on consumption and investment. So as not to distort the results, we have conservatively assumed that only 20% of that income was channelled to domestic expenditure, giving rise to new demand in the rest of sectors that increases tourism activity. More specifically, a total production of 10.795 billion euros was generated in the case of LCA tourists, compared to 19.449 billion euros in the case of tourists who used traditional airlines.

The fourth column in Tables 6 and 7 shows the total impact of tourist expenditure. We can clearly see how the expenditure of tourists who travelled with traditional airlines has a much larger effect in absolute terms than that of LCA tourists. This shows that traditional airlines currently still generate more activity in the tourist industry than LCA, as their turnover remains higher.

Nevertheless, if we analyse the unit contribution of each tourist to the rest of sectors, their behaviour is not so different (Table 8) and they distribute their spending similarly. However, the difference between them (last column) provides objective data in support of the fact that the tourists who use traditional airlines have more of an impact on the productive framework of the tourism industry (more than 20% in terms of GDP).

One of the advantages of the Input/Output methodology is the possibility of breaking down effects to sector level, thereby enriching the results of the analysis. In order to determine which sectors of the tourism industry benefit the most, the disaggregated results are detailed below. This information makes it possible to assess the activities most favoured by tourism expenditure (Table 9).

A comparative analysis in absolute terms reveals similarities in the sector-by-sector distribution of the Direct Effect of the expenditure of the two types of tourists. Seven out of the 12 sectors considered increase their production, air transport itself recording the largest rise, followed by the activity of Restaurants and similar establishments and Hotels and similar establishments.

Table 8 Total economic impact of each user of traditional airlines and LCAs (€)

	Total E.I. TRADITIONAL	Total E.I. LCA	Difference (%)
Wages	556	455	22.2
Gross operating surplus	581	492	18.1
Tax income	20	18	11.1
GDP	1139	948	20.1
Total production	3.540	2.891	22.4

Source: own elaboration.

Table 9 Sector-by-sector distribution of the economic impact of LCA and traditional airline user expenditure (1,000s of €)

	DIRECT EFFECT		INDIRECT EFFECT		INDUCED EFFECT	
Sector	**LCA**	**TRAD.**	**LCA**	**TRAD.**	**LCA**	**TRAD.**
Hotels and similar establishments	4,236	7,030	1,478	2,453	659	1,187
Property rental services	73	135	187	371	551	992
Restaurants and similar establishments	5,672	8,851	4,496	7,122	1,787	3,220
Passenger road transport	-	-	-	-	271	488
Passenger rail transport	-	-	-	-	133	239
Passenger maritime transport	-	-	-	-	48	86
Passenger air transport	7,868	16,777	8,009	17,079	1,036	1,867
Travel agency services	95	203	280	596	334	602
Auxiliary transport services	384	816	1,425	3,033	1,073	1,934
Transport equipment hire	-	-	7	15	263	473
Market cultural, recreational and sports services	2,829	4,890	1,995	3,412	631	1,137
Non-market cultural, recreational and sports services	-	-	-	-	219	394
Total	**21,156**	**38,702**	**17,877**	**34,081**	**7,004**	**12,619**

Source: own elaboration.

The Indirect effect resembles the Direct Effect in that it has an impact on the same sectors of the tourism industry, although passenger air transport benefits even more in this case. Therefore, inter-industry demands follow the same pattern both in the first relationships between expenditure and the rest of sectors and also in subsequent relationships, although the same does not apply to the Induced effect. As we can see in Table 9, the increase in income boosts the production of 12 branches of the tourism industry, although to a lesser extent than the previous effects (Induced effect). In this case, Restaurants and similar establishments is the sector that benefits the most.

7 Conclusions

Studying the economic impact of the money spent by the tourists who fly to Spain is of particular interest as they represent a significant source of income and employment for our country. In this sense, the economic repercussions of their expenditure will boost the main macroeconomic variables of the Spanish economy.

The empirical research in this paper differentiated between tourists who used LCA and those who flew with traditional airlines in order to ascertain whether the savings on fares resulted in substantial differences in their demand, allowing us to establish an "ad-hoc" profile for this type of traveller. The use of low-cost companies is becoming increasingly widespread among the great majority of the population, whose social characteristics are also becoming increasingly varied. For this reason, it is initially difficult to establish a stereotype to define them.

There are various methodologies for measuring economic impact on the productive fabric of a country. However, Input/Output models are the only ones that enable us to bilaterally analyse the relationships between all economic sectors, determining the effects of each on the others. Furthermore, unlike other methodologies, the Leontief model makes it possible to determine the different effects at sector level.

The results confirm our initial idea that the total impact of the tourists who arrive in our country on traditional airlines is greater than in the case of those who fly with LCA. This is mainly due to the fact that, according to the latest information available, their turnover remains higher than that of LCA, although this trend is expected to change. This notion already existed in the literature and the research undertaken in this paper confirms that the consumption of traditional airline users has a greater direct, indirect and induced impact on GDP, which is spread almost equally between an increase in wages and business profits.

In addition, after analysing unit expenditure per traveller, we observe that the initial savings tourists make by travelling with LCA does not result in an increase in expenditure, or, therefore, in the productive activity of the destination country. Moreover, the sector-by-sector analysis reveals that passenger air transport itself, Restaurants and similar establishments and Hotels are the branches of the tourism industry that most benefit from tourist expenditure, with similar consequences being observed for the Indirect Effect.

References

Abda, M.B., Belobaba, P., Swelbar, W.: Impacts of LCA growth on domestic traffic and fares at largest US airports. Journal of Air Trasport Management 18, 21–25 (2012)

Barker, M., Page, S.J., Meyer, D.: Evaluating the impact of the 2000 America's Cup on Auckland, New Zealand. Event Management 7, 79–92 (2002)

Brida, J.G., Pereyra, J.S., Such, M.J., Zapata, S.: La contribución del turismo al crecimiento económico. Cuadernos de Turismo 22, 35–46 (2008)

Costas, A., Germá, B.: Los beneficios de la liberalización de los mercados de productos. Colección de Estudios Económicos La Caixa (1997)

Crompton, J., Lee, S., Shuster, T.J.: A guide for undertaking economic impact studies: The Springfest example. Journal of Travel Resarch 40, 79–87 (2001)

Donzelli, M.: The effect of low-cost air transportation on the local economy: Evidence from Southern Italy. Journal of Air Transport Management 16, 121–126 (2010)

Fernández, E.: Análisis del gasto de los turistas que acceden a España por avión. Documento de Trabajo. Universitat Autónoma de Barcelona (2008)

Gonzalez, M.: Impacto económico de los hoteles: Aplicación a la ciudad de Sevilla. PASOS. Revista de Turismo y Patrimonio Cultural 2, 319–338 (2010)

Graham, A., Dennis, N.: The impact of low cost airline operations to Malta. Journal of Air Transport Management 16, 127–136 (2010)

Herderson, P.H.: El método del factor-producto: Una aplicación del mismo a la economía italiana. Moneda y Crédito 54, 3–25 (1955)

Hill, J.: The economic impact of the Port of Baltimore and Maryland. Division of transport, business and public policy. College of Business and Management. University of Maryland (1975)

INE. Cuenta Satélite de Turismo de España (2010)

Instituto de Estudios Turísticos (2011a), `http://www.iet.turismoencifras.es`

Instituto de Estudios Turísticos, Balance del Turismo Secretaria de Estado de Turismo y Comercio. Madrid(Abril 2011b)

Instituto de Estudios Turísticos, Turismo, tráfico aéreo y compañías aéreas de bajo coste en el año 2010, Secretaria de Estado de Turismo y Comercio, Madrid (2011c)

Ivars, J., Menor, M.: El impacto de las compañías aéreas de bajo coste en la actividad turística del área de influencia del aeropuerto de Alicante. Estudios Turísticos 175-176, 89–104 (2008)

Kasimati, E.: Economic aspects and the summer Olympics: a review of related research. International Journal of Tourist Research 5, 433–444 (2003)

Lim, C.: Review of International Tourism Demand Models. Annals of Tourism Research 24, 835–849 (1997)

Martí, M., Puertas, R., Fernández, J.I.: Metodología para el análisis de impacto portuario: Aplicación a los puertos de Gandía, Sagunto y Valencia. Fundación Valenciaport (2009)

Miquel, J., De Borja, L.: Modelos de intermediación en el marco de un nuevo paradigma de la intermediación turística. In: VI Congreso Turitec, Málaga (2006)

Navío, F.: El cambio del sector aéreo español en los últimos treinta años. Transporte aéreo y turismo, XXX enero-febrero, 38–44 (2011)

Piga, C., Filippi, N.: Booking and flying with low cost airlines. Business School, Nottttingham University, Nottingham, UK (2001)

Rey, B., Inglada, V.: Evolución reciente de las compañías de bajo coste en Europa. Una referencia al caso de España. Economistas Número Extraordinario, 100–107 (2006)

Rey, B., Myro, R., Galera, A.: Effect of low-cost airlines on tourism in Spain. A dynamic panel data model. Journal of Air Transport Management 17, 163–167 (2011)

Rodríguez, B., Vargas, A.M., Montes, M.P.: Las compañías de bajo coste y sus usuarios. El caso de España. In: Congreso Asociación Europea de Dirección y Economía de la Empresa, Salamanca (2008)

Romero, J.J., Cortés, F.J.: Consideraciones y perspectivas de la estrategia de bajo coste. Boletin ICE Económico 2871, 28–36 (2006)

Sainz-Gonzalez, R., Nuñez-Sánchez, R., Coto-Millán, P.: The impact of airport fees on fares for the leisure air travel market: The case of Spain. Journal of Air Transport Management 17, 158–162 (2011)

Taylor, D.T., Fletcher, R.R., Clabaugh, T.: A comparison of characteristics, regional expenditures and economic impact of visitors to historical sites with other recreational visitors. Journal of Travel research 32, 30–49 (1993)

Tourism and Learn Spanish in Historic Cities: A Case Study in Córdoba

Inmaculada Piédrola Órtiz, Carlos Artacho Ruiz, and Eduardo J. Villaseca Molina

Area of Business Organisation, University of Córdoba,
Av. Medina Azahara, s/n 14071 Córdoba, Spain
{cu9piori,gs1arruc,direccion.corporacion}@uco.es

Abstract. Spanish is the official language of communication in 21 countries. This study aims to connect the increasing internationalisation of Spanish with the increasing demand for cultural products such as tourism, cinema and publications. It also analyses the impact that the teaching of Spanish may have on tourism and the economy. To this end, data were collected from questionnaires, personal interviews and from field observations. The objective of these investigations was to emphasize the importance of language tourism can have for a world heritage city such as Córdoba. At the same time we aim to describe the profile of the typical language tourist who chooses the city of Córdoba and its university as a destination and as a place of study. The data were collected in Córdoba during 2010. The results of this study show that the language tourist stays for a longer period of time than the conventional tourist, and that the language tourist is very satisfied with his or her stay, both with the city of Córdoba itself and with the university.

Keywords: Internationalization of Spanish, communication language, language tourism, Córdoba.

1 The Importance of the Spanish Language

According to Alonso y Gutiérrez (2006), the consumption of natural resources implies a process of exhaustion. The contrary can be said of a language: it does not waste away nor can it become exhausted, it can only be enriched. Its value increases in step with its increased use (Jiménez Jiménez, 2006; Jiménez Redondo, 2006), the main economic characteristics of a language are illustrated in Figure 1.

Spanish is one of the most widely used languages in international communication, and currently holds second place. This can be attributed to various factors, the most outstanding of which are its demographic importance and its relative homogeneity (López Morales, 2006). If one can say that the primary objective of a language is to promote communication, then it would seem obvious that the value of the language increases in step with potential channels of communication and the quantity of communications made in this language (Dalmazzone, 2000)

A.M. Gil-Lafuente et al. (Eds.): Soft Comput. in Manag. and Bus. Econ., STUDFUZZ 287, pp. 305–318.
springerlink.com

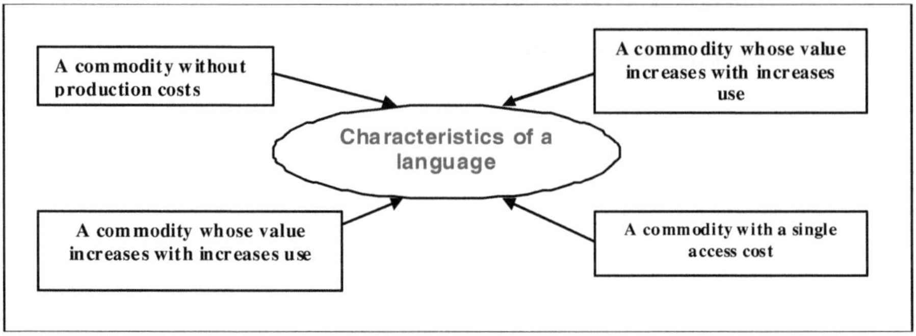

Source: Alonso Rodríguez (2006).

Fig. 1. Economic characteristics of language

herefore it can be said that Spanish, as a means of communication, is connected to the number of people who speak it. This point applies both to the citizens of the 22 countries where Spanish is the native language, and to the approximately 40 million Spanish-speaking legal residents and some 10 million illegal native Spanish speakers living in the United States.

Its economic importance is associated with its use in the world of commerce and business. The economic potential of the Spanish language is therefore influenced by a series of factors: the potential market of Spanish language learners, its use as a second language, the economic development of Spanish-speaking countries and the internationalisation and connectedness of its businesses (Durán Herrera, 2006).

2 Globalisation and Internationalisation of Spanish

Globalisation is understood as a process of increased and increasing integration of the world economy, of information, of communication and of culture. This in turn affects the interactions and relations that guarantee the internal cohesion of each language community (Simón Ruiz, 2010). The process of economic and cultural globalisation has been an unavoidable fact for at least the past 20 years, and is causing a cultural homogenization which is causing the disappearance of minority languages and cultures, as can be seen today in many Native American groups.

Jakob Marschak, the acknowledged father of language economics, classified this process as the most developed communication system among human organisations. To put this in other words, it acts as a means of exchange, or as a single currency whose use reduces the costs of transaction (Jiménez and Narbona, 2007).

A number of languages have become prominent in a world becoming increasingly globalised and integrated. English has become global (Garrido Medina, 2010). Spanish is among the other important languages such as French, Portuguese,

Chinese, Japanese, German, Italian and Russian. Indeed, Spanish is gaining international importance as a second language in comparison to French or to other widely spoken languages (for example in Brazil or the United States).

According to Alonso and Jiménez (2010), the number of Spanish speakers will soon reach that of English, and is increasing faster than the number of native Chinese speakers, which at present has by far the largest number of native speakers. Only Arabic is increasing significantly faster. Spanish is spoken by 440 million people, whose combined purchasing power represents approximately 9% of the world gross domestic product (GDP). This is a powerful argument for interrelation with Spanish-speaking countries. This could be advantageous for Spain, which now has a mature business and economic structure, and requires a more active projection of itself towards the wider world.

From an international perspective, it is worth noting that Spanish is the third most widely spoken second language, behind English and only a little behind French (which is the official language in 27 countries), and ahead of Hindi, Russian and Chinese.

If Spanish is to play a part in the world of the future, it will be because of its attraction as a second language, due to the number of its native speakers and also to its economic importance and its cultural prestige.

Following Ruiz Zambrana (2009) and García Delgado, et al. (2010), we can summarise the internationalising phenomenon of Spanish in the following points:

- The high birth rate in Latin America. (Molina Sánchez, 2006),.
- The advance of economic, political and cultural globalisation, which is now going further and wider than in any previous stage in history. The increasing openness of the economies of Latin America, and the increasing international commercial integration of countries such as Spain, Mexico, Chile, Argentina and Colombia.
- The extension of the information society, whose principal help – what is known and how to transmit what is known – is language. The dominance if the internet illustrates this perfectly. Its foundation stone is the idea of free exchange of information and knowledge.
- The interest shown in Spanish, for both cultural and economic objectives, in countries where Spanish is not the first language (Instituto Cervantes, 2006).
- The use and maintenance of Spanish as an official language in key international organisations.

As well as these explanations, further factors need to be taken into account:

- Spanish is a language with a long and important literary history, both in its original European homeland and its geographical extensions across the world.

- It is a relatively uniform language, both in its pronunciation and in its vocabulary.
- It is geographically compact: most Spanish-speaking countries border other Spanish-speaking countries.

3 The Economics of the Spanish Language

Although it is practically impossible to quantify the economic value of a language, Martín Municio (2003), in projections to the year 2004, included all activities related to language and estimated its value to be equivalent to 15% of its Gross Domestic Product (GDP), or €150bn. Several authors (Alonso Rodríguez, 2007 and 2009, García, Alonso and Jiménez, 2010), following on from these figures, with newer and wider analyses of the value of Spanish, and with projections up to the year 2010 have confirmed these figures, although with important qualifications.

Two complementary techniques were used to carry out these analyses. The first was based on the selection of products and their later market valuation. The second technique was more innovative, and was based on the selection of employees within a company who perform tasks for which language is an essential and primary commodity. Combining the results of these two analyses reveals the importance of Spanish in a more precise and profound manner, in terms of employment and of profit.

There was an increase of one per cent in the value of the Spanish language to Spanish GNP between the years 2000 and 2007, with its percentage share of the economy increasing from 14.6% to 15.6%. This rise expressed in monetary terms shows an increase from €92bn to €164bn across the same time period.

The statistics regarding employment are clearer still. While there were almost 2.6m jobs related with language in the Spanish economy in 2000, some 900,000 additional new jobs were created, bringing the total to almost 3.5m jobs in 2007. In percentage terms, there was an increase from 15% to 16.2%.

At the International Congress on ´Spanish as a cultural, economic and tourism asset and resource´, which was held in Salamanca in November 2008, Sanz Sáiz (2008) drew attention to a study made by the University of Granada on the impact of language tourism in the city of Granada. The study was centred on students who chose to learn Spanish at the University of Granada. It was found that 10% of the GDP of the province of Granada was due to the high number of foreign students registered at the university – foreigners and natives. This figure is comparable to that generated by the construction industry for the same province.

3.1 The Potential of the Spanish Language as a Resource for Tourism

It is important to point out the learning of Spanish is a basic necessity and a means of integration for Latin America, and it therefore possesses a considerable potential for development.

According to Turespaña in 2008, the definition of tourism made by the World Organisation of Tourism (WOT) in 1991 was: *'the activities made by people during their travels and stays in places that are different and distinct form their normal surroundings, for a period of time of less than one year, and with the objective of being immersed in a language that is different from that of their normal surroundings'.*

It is well known to all that one of the principal strengths of the Spanish economy lies in its tourism industry (Güemes Barrios, 2001). The Spanish tourism industry is often characterised as that of sun and beach holidays, but there are various other sectors of the industry now emerging, one of which is language tourism.

The increase of this sector has important positive economic effects, and not only on employment, but also in all other industries related to tourism, such as transport, hospitality and the hotel industry, to name just a few.

Foreign tourists who visit our country with the objective of learning Spanish demand services similar to those asked for by more conventional tourists. The majority of language tourists are young and come from age groups 14 to 30, although a significant number come from the more mature age groups of over 55. They combine educational activities along with enjoyment of Spanish cuisine and the local culture (Montero *et al*, 2010).

The public sector has an important role in the development of policies of promotion and quality which are necessary for the development of Spanish as a foreign language (SFL).

According to García Delgado *et al.* (2010), the Erasmus Programme should be included inside the ambit of SFL, given that it generates a very important total of real and potential students of the Spanish language. More than 25,000 students come to Spain annually, with an average stay of 6.5 months per student, and spend €135m. This places Erasmus students in the highest position of all SLE students.

The countries which provide the largest number of language tourists are Germany, the United States and France, and it is predicted that European Union countries such as Poland, the Czech Republic and Hungary will provide students interested in learning Spanish in the coming years. Other countries which could become important clients in the future are the emergent economies or so-called BRICs (Brazil, Russia, India and China).

Language tourism is not something that first originated in Spain. Countries such as the United Kingdom, France and Germany have long included this type of tourism in their promotional activities, where it is considered a subsector of cultural tourism (Huete, 2008). Nevertheless, the number of language tourists coming to Spain has risen from 130,000 in the year 2000 to 267,000 in the year 2007, according to figures provided by Turespaña (2008) and FEDELE (2009). In other words, the number of language tourists coming to Spain has risen by 9% annually. The reasons for this increase are due to the important publicity campaigns made over the Internet, and the value of the internet as a channel of information.

Language tourism brings approximately 200,000 visitors to Spain annually, and whose daily average spending is higher than that of a conventional tourist.

The average annual spending of language tourists is around €300m, and the average length of stay is long – three to four weeks. The typical language tourist is marked by both the length of their stay and their high spending.

Although language tourists arrive all year round, this type of tourism tends to peak in the spring and summer months. This shows that language tourism, like conventional tourism, is seasonal in its behaviour, although the seasonal differences are less marked with language tourism (Baralo Ottonello, 2007).

One fundamental characteristic of language tourism is that the demand is usually concentrated in city destinations, especially those destinations that possess a rich heritage or culture. Because of this, the image of the destination (Taboada de Zuñiga, 2010) s a decisive factor in the mind of the language tourist when making a decision about exactly where to go.

4 Teaching Structure of SFL

There are some difficulties in estimating the number of SFL courses offered in Spain. Firstly, it is difficult to estimate the total number of companies and institutions that offer courses of Spanish for foreigners (Carrera y Gómez, 2007) A study made by Turespaña (2008), to which we have previously referred, sought to determine the total offer of courses, and to establish which segments offered the best guarantees of completing the courses successfully.

The analysis of courses on offer must start by establishing structure of sectors based on classification criteria. These classification criteria must permit the grouping of teaching centres in homogenous units, according to their courses. Figure 2 shows the current distribution in Spain according to the type of centre.

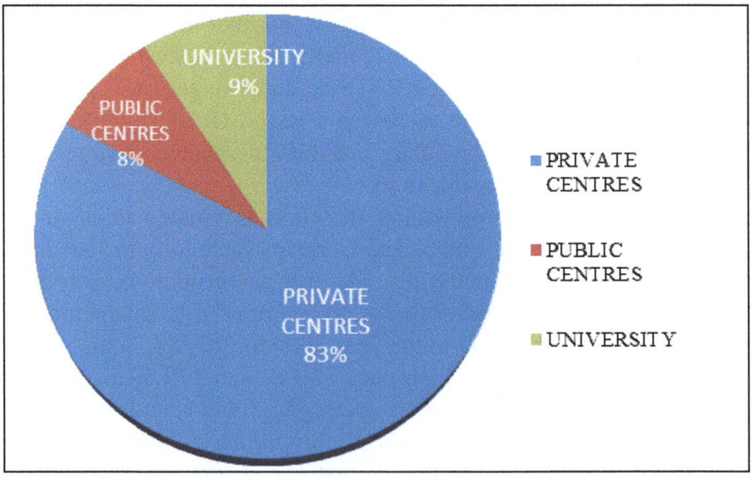

Source: Turespaña (2008) y FEDELE (2009).

Fig. 2 Distribution of academic offer by type of centre

Teaching centres available in Spain can be classified as *public centres* (public universities and official language schools) or as *private centres* (private universities and private language centres of Spanish for foreigners).

Of the 627 centres counted, only 163 were dedicated exclusively to the Spanish language sector, which conforms to what is called *reference offer*. These centres develop the best business practices and can count on all the guarantees necessary to compete successfully, as they consider that the tourist experience is goes far beyond the teaching experience. Because of this, and as an additional complement to their academic services, they offer additional activities and assistance to their students.

On the other hand, while characterization is useful for identification, it is insufficient if one focuses the analysis of the supply from the concept of the language product – the supply placed in the market. In other words, it does not exactly reflect the reality of the practice from a business point of view.

There are various elements which shape the types of education that are offered by the education centres, some of the most important being:

- Demand which is motivated by academic goals.
- Demand which is directed towards the improvement of skills and professional opportunities.
- Demand which is motivated by an interest in Spanish culture.

A business proposal, like that which is established in the analysis of supply, also means the adaptation of supply to demand. Also, one must not forget the educational experience of the language tourist takes place within a wider framework of experience – accommodation, leisure, transport, personal security and so on.

5 Methods of Analysis

The underlying motivation for this study is the lack of an exhaustive examination of the language tourism sector in Córdoba. Such a study could serve as the foundation for developing marketing strategies that could maintain and promote existing demand.

The analysis was divided into various steps. An initial review was made of the scientific literature available on the matter, which was followed by data collection by means of a series of questionnaires, interviews and field observations. Data were gathered in a continuous and ongoing manner during 2010.

The questionnaire (see Table 1) was designed to define, analyze and evaluate the situation of the language tourist who chooses Córdoba as his destination city and the university as his chosen teaching centre.

The questionnaire contained 22 items which dealt with the sociodemographic profile of the language tourist and other topics such as:

- The motivation for choosing Córdoba as a destination for language tourism.
- Assessment of the tourism resources in Córdoba.
- Best aspects of the stay in Córdoba
- The design of the Spanish language course.

Table 1 Technical description of the questionnaire

Geographical area	Córdoba city.
Frame of reference	Language tourists who choose the University of Córdoba as their Spanish language teaching centre.
Sample group	628 students registering during 2010.
Type of questionnaire	Semi-structured questionnaire (opened and closed questions).
Pre-test	20 questionnaires were given to different students to confirm that they understood the questions.
Sample group treatment	Sample group was subdivided according to nationality.
Data collection period	January to December 2010.
Valid responses	221 valid responses
Sample error	5.3%. (confidence level of 95%).

The questionnaires were batch-processed using SPSS and distributed through teachers of Spanish as a foreign language, course coordinators of Master´s degree programmes, and by e-mail. The final number of valid questionnaires received was 221, with a sampling error of 5.3%

6 Results

6.1 Sociodemographic Profile of the Students

Table 2 shows a sociodemographic breakdown of the students who choose Córdoba for their language stay. The majority are currently students, either completing their secondary or university education. They have an interest in travel and adventure and with an age distribution from 17 to 30.

The largest group of students come from the United States, followed at some considerable distance by Germany, Italy, France and China. The students are mainly following courses in business studies, sciences and humanities. There is a very clear preponderance of women in the sample group, which is something that contrasts with previous studies.

There was no correlation of gender with age among the respondents (contingency coefficient 0.442, P = 0.079), nor with the level of education (cc = 0.234, P = 0.189) nor with the country of origin (cc 0.429, P = 0.254).

Table 2 Sociodemographic characteristics of the sample group

Variable		%	Variable		%
Gender	Male	30.8%	**Educational level attained**	Primary	1.4%
	Female	69.2%		Secondary	37.9%
	—	—		Bachelor´s	51.0%
	—	—		Postgraduate	8.6%
	—	—		Doctorate	0.5%
Age group	17-20 years	30.2%	**Country of origin**	Germany	13.1%
	21-25 years	55.5%		China	5.4%
	26-30 years	10.9%		United States	43.9%
	over 30 years	3.4%		France	5.4%
	—	—		Italy	9%
	—	—		Morocco	1.4%
	—	—		Poland	5%
	—	—		United Kingdom	2.3%
	—	—		Tunisia	2.7%
	—	—		Others	11.8%

The students form Germany, the United States and China were noteworthy for their good command of English, and to some degree Spanish and French as well. In general, 41% of the students had good command of a second language, firstly English (26.9%), then French (19.5%), then Spanish (12.3%) and finally Italian (2.3%).

6.2 The Principal Reasons for Doing a Language Course in Spain

As can be seen in Figure 3, one of the principal motivations for coming to Spain to do a Spanish language course is their interest in its culture. This is followed by the awareness that Spanish is becoming increasingly useful as a second language, and the possible benefits it may have on future career prospects, and the hospitality. Other minor considerations were the price level, to improve pronunciation and the climate.

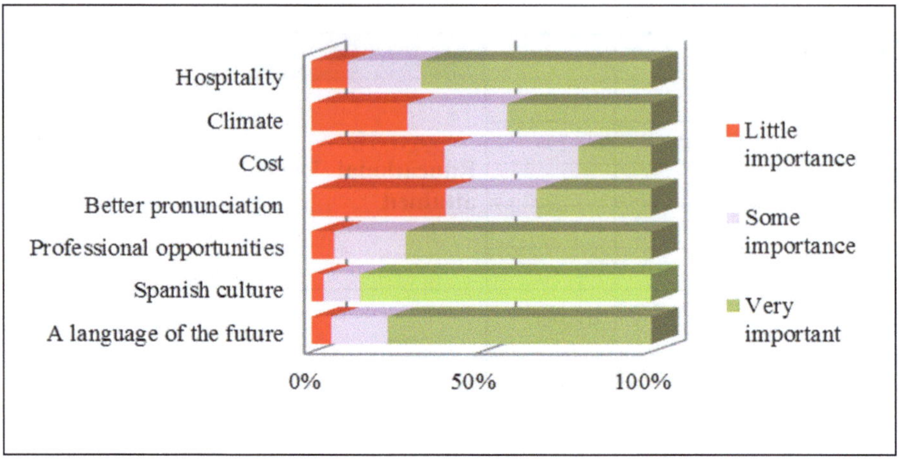

Fig. 3 The principal reasons for doing a language course in Spain

6.3 *Design of the Courses*

The ideal number of hours for a Spanish course would be five to 15 hours per week, according to 51.4% of respondents. The ideal duration of a stay is from three to six months for 49.6% of respondents, while 39.3% thought that a stay of more than six months would be better. There were no significant differences according to the country of origin of the respondents.

These data confirm the findings of Güemes Barrios (2001), in that the seasonality of language tourism is less than that of conventional tourism. Furthermore, the level of Spanish expressed by the language tourist is advanced in 12.1% of respondents and intermediate in 68.2% of respondents.

6.4 *Principal Reasons for Choosing Córdoba for Language Tourism*

One of the principal goals of this study was to evaluate the reasons for choosing Córdoba as a destination for language tourism. The different aspects related with the choice of destination are presented in Figure 5. The data are expressed using a five-point Likert scale where 1 represents little importance and 5 means extremely important.

Figure 5 illustrates the means by which students obtained information in order to choose a destination. It is clear that the main source of information about their destination cities was the teaching centre – 66.97%. At a great distance behind was the internet, with 15.38%. In third place were comments and recommendations given by family and colleagues, at 10.41%. It cannot be doubted that ´word of mouth` recommendations will correlate with the level of tourism for a given locality.

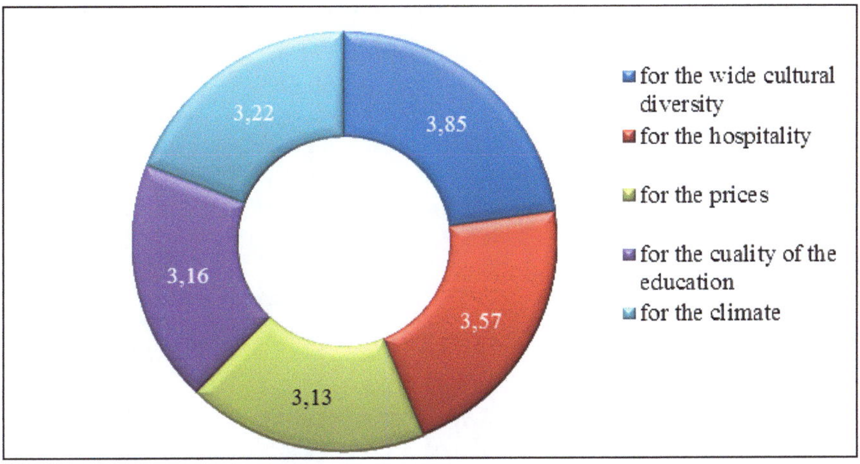

Fig. 4 Evaluation of the reasons for choosing Córdoba as a destination for language tourism

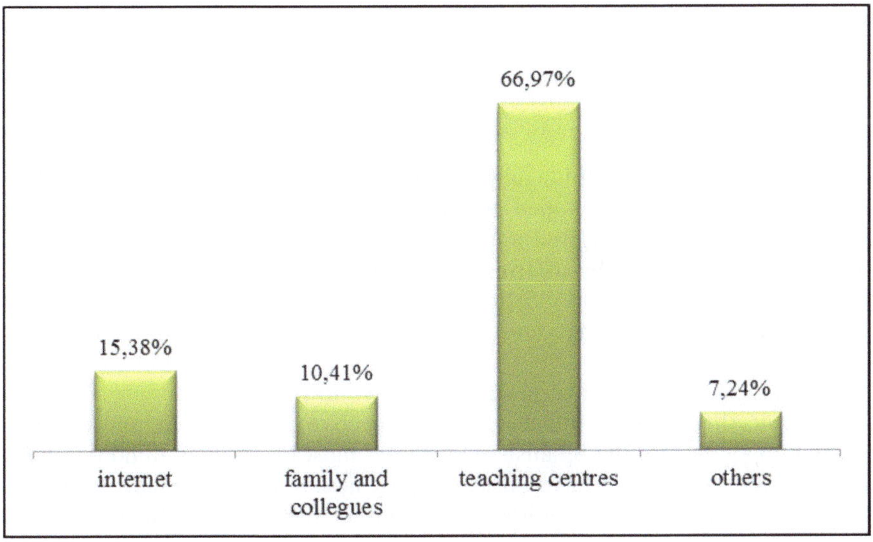

Fig. 5 Sources of information for choosing a study destination

6.5 *Level of Satisfaction with the Language Teaching at the University*

Finally, it can be said that the there is a high level of satisfaction with the teaching of Spanish at the University. The results are presented in Figure 6 using a five-point Likert scale, where 1 indicates dissatisfied and 5 indicates very satisfied. The average evaluation was 4.15 points.

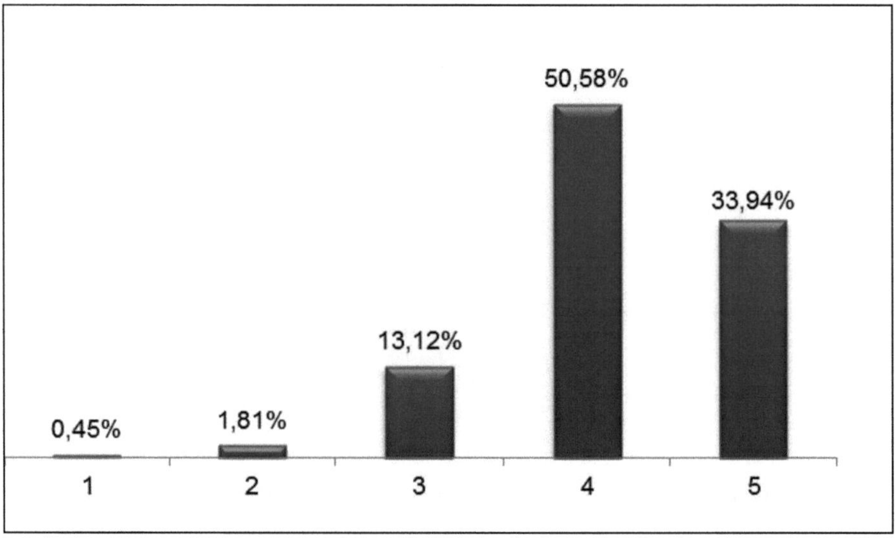

Fig. 6 Level of satisfaction with the Spanish teaching at the University of Córdoba

7 Conclusions

We should think of Spanish language tourism as a sector with great potential, but which still needs to be developed and matured. We need to improve the services which are already on offer, and not just in terms of academic quality. Language tourism can be enhanced by offering additional and complementary services such as help with finding accommodation, guided tours to heritage sites and natural parks. Enriching our language tourism model will make it more attractive in foreign markets.

Universities and local governments will have to play an important and active role in projecting both the image of their respective cities abroad, and of the increasing importance of the Spanish language. It must be borne in mind the impact that language tourism can have on the employment market. Language tourism is already recognized – albeit indirectly - as one of the so-called 'new sources of employment´.

Returning to the present study, we can affirm that Córdoba possesses the correct combination of necessary elements for it to become an attractive destination for language tourists, although the city still remains a long way behind its competitor cities in Andalusia – Seville, Málaga and Granada. Because of this, it is essential to develop a clear strategy which includes clearly fixed objectives, products that have been carefully thought over, and the selection of priority markets. The range of tourism products offered in the future will have to be wide, and should range from products that fully inclusive and highly structured, to those that are more flexible and offer greater independence to the learner tourist.

References

Alonso Rodríguez, J.A.: Naturaleza económica de la lengua. Working Document. Fundación Telefónica e Instituto Complutense De Estudios Internacionales (Icei), Madrid (2006)

Baralo Ottonello, M.: Enseñanza De Español yY Turismo: Las Estancias Lingüística. Revista Mosaico (20), 32–36 (2007)

Carrera, M., Bonete, R., Muñoz, R.: El Programa Erasmus en el marco del valor económico de la enseñanza del español como lengua extranjera. In: La Economía de la Enseñanza del español como Lengua Extranjera. Oportunidades y Retos. Instituto Complutense de Estudios Internacionales, Fundación Telefónica, Madrid, pp. 39–82 (2007)

Consorcio De Turismo De Córdoba: Plan Estratégico de Turismo de Córdoba. Ayuntamiento de Córdoba (2009), `http://www.turismodecordoba.Org/plan-estrategico.Cfm`

Dalmazzone, S.: Economics of Language: A Network externalities approach. In: Breton, A. (ed.) Exploring the Economics of Language, pp. 63–87. London (2000)

Durán Herrera, J.J.: La relevancia del idioma español en el proceso de generación de empresas multinacionales. In: actas del I Congreso Internacional de la Lengua Española: Activo cultural con valor económico creciente (2006)

Fedele (Federación De Escuelas De Enseñanza De Español Para Extranjeros): Plan Estratégico de Turismo Idiomático 2009-2012. Servicio de Publicaciones, Madrid (2009)

García Delgado, J.L.: Collaboration. In: Alonso, J.A., Jiménez, J.C. (eds.) Economía Del Español. Una introducción, Ariel y Fundación Telefónica, Barcelona (2007)

García Delgado, J.L., Alonso, J.A., Jiménez, J.C.: Valor Económico del Español: Una Síntesis. In: El Español: Lengua Global. La Economía. Santillana/Instituto Cervantes, pp. 17–44 (2010)

Garrido Medina, J.: Lengua y Globalización: Inglés Global y Español Pluricéntrico. Historia y Comunicación Social 15, 51–66 (2010)

Güemes Barrios, J.J.: El Español como Recurso Turístico: El Turismo Idiomático. In: Actas Del II Congreso De La Lengua. Valladolid, Cvc (2001)

Hernández, J.M., Campón, A.M.: Evolución Del Turismo Idiomático en España: Un sector estratégico en auge. Estudios Turísticos 186, 45–68 (2010)

Huete, R.: Tourism Studies in Spain. Journal of teaching in Travel &Tourism 77, 73–92 (2008)

Instituto Cervantes: Enciclopedia del Español en el Mundo. Círculo De Lectores, Barcelona (2006)

Jiménez Jiménez, J.C.: Cuánto Vale El Español Anuario del Instituto Cervantes 2006-2007, pp. 459–462. Plaza & Janés, Barcelona (2006)

Jiménez Redondo, J.C.: La Economía de la Lengua: Una visión de Conjunto. Instituto Complutense de Estudios Internacionales. Fundación Telefónica, Madrid (2006)

Jiménez, J.C., Narbona, A.: El Español como Instrumento de la Internacionalización Empresarial. In: el Español: Lengua Global. La Economía (2010)

López Morales, H.: El futuro del español. En Instituto Cervantes, Enciclopedia del español en el Mundo, Barcelona, Egedsa, pp. 476–491 (2006)

Marschak, J.: Economics Of Language. Behavioral Science 10, 35–40 (1965)

Martín Municio, A.: El valor Económico de la lengua española. Espasa-Calpe, Madrid (2003)

Molina Sánchez, C.A.: El valor de la lengua. In: Enciclopedia del español en el Mundo, pp. 17–20. Instituto Cervantes (2006)
Molinero, Mar, C.: The Politics of Language in the Spanish-Speaking world from colonisation to Globalisation Routledge (2000)
Montero, J.M., Fernández, G., Higueras, M., García, C.: El Turismo Idiomático en España. Una Panorámica. In: Investigaciones, Métodos Y Análisis De Turismo, pp. 55–66. Septem Ediciones, Oviedo (2010)
Moreno Fernández, F.: La Diversidad Lingüística de Hispanoamérica: Implicaciones Políticas y Sociales. Análisis Del Real Instituto Elcano (Ari) 38, 4–8 (2006)
Otero, M.S., Mccoshan, A.: Survey of the Socio-Economic Background Of Erasmus Students. Dg Eac 01/05 Final Report, Ecotec Research and Consulting Limited (2006)
Ruiz Zambrana, J.: La Situación actual de la Lengua Española en el Mundo. In: Contribuciones A Las Ciencias Sociales (September 2009), `http://Www.Eumed.Net/Rev/Cccss/05/Jrz.Htm` (Fecha De Acceso August 3, 2011)
Sanz Sáiz, I.: Beneficios Económicos derivados de los Programas De Movilidad De Estudiantes. In: Actas del Congreso Internacional sobre el español como valor y recurso cultural, Turístico y Económico 24 y 26 de noviembre, Salamanca (2008)
Siguan, M.: La Europa de las Lenguas. Alianza, Madrid (2005)
Simón Ruiz, C.: Español Actual: Globalización e Interculturalidad. Decires, Revista Del Centro De Enseñanza para Extranjeros 12(14), 75–89 (2010)
Taboada De Zúñiga Romero, P.: Una aproximación al Turismo Idiomático En España. El caso particular de las ciudades históricas. Actas de las III Jornadas De Investigación en Turismo: nuevas perspectivas para el turismo para la próxima década, Sevilla (2010)
Tamames Gómez, R.: La Dimensión Económica del Español en el mundo. Editorial Venecia, Madrid (2009)
Turespaña: Estudios de Productos Turísticos. Turismo Idiomático. Egraf, Madrid (2008)

Sustainable Stakeholder Relationship Patterns: An Analysis Using a Case Study in the Spanish Hotel Sector

Ana Gessa-Perera and María del Amor Jiménez Jiménez

Department of Financial Economics, Accountancy and Operations Management,
University of Huelva
Plaza La Merced 11, 21001, Huelva, Spain
{gessa,amor.jimenez}@decd.uhu.es

Abstract. Relationships with stakeholders are gaining greater importance on companies' corporate social responsibility agendas. The purpose of this study is to examine these relationships in greater detail using a case study of the Spanish hotel sector. Analysing the findings we have been able to observe the changing patterns of relationships with stakeholders, both with respect to the numbers of stakeholders involved, which have risen considerably, and the type of relationship, which has moved on from being a merely informative unilateral communication to one of dialogue and communication that in many cases favours stakeholder engagement and involvement in company management.

Keywords: stakeholders, corporate social responsibility, stakeholder engagement, sustainability.

1 Introduction

The changes that have taken place in the business context in recent decades have driven the development of a new focus that includes economic, social and environmental criteria in business management, thus responding to the expectations of organisations' various stakeholders (SH). This entails the recognition and voluntary incorporation of social and environmental concerns into commercial transactions and into relationships with their interlocutors with the aim of improving their competitive positions and added value (Post et al., 2002).

Relationships with SH has therefore become one of the most representative aspects of this new business management focus and the basis for a number of studies (Clarkson, 1995; Donaldson and Preston, 1995; Mitchell et al., 1997; Jones and Wicks, 1999; Hart and Sharma, 2004) since Freeman (1984) set out the stakeholder theory and the role of stakeholders in organisations, according to which the 'reason for being' for most organisations is to serve as a tool for satisfying their stakeholders' expectations, with the author himself identifying stakeholders as *any*

A.M. Gil-Lafuente et al. (Eds.): Soft Comput. in Manag. and Bus. Econ., STUDFUZZ 287, pp. 319–334.
springerlink.com

group or individual who can affect or is affected by the achievement of the organisation's objectives.

The purpose of this study is to analyse the management of these SH relationships in the Spanish hotel sector and to identify relationship conduct patterns with stakeholders. Thus in the following section we set out the theoretical framework on which our empirical study is founded. Next, the methodology and findings are presented. Finally, the conclusions are given.

2 Stakeholder Relationships under the Sustainable Focus of Stakeholder Theory

Having accepted the importance of stakeholders in the responsibility assumed by firms for their setting in the economic, social and environmental sphere, we aim to use a dynamic and holistic focus to analyse the relationships companies have with different stakeholder groups that determine their participation in business management. For our analytical scheme we have adopted and adapted the Deegan and Unerman (2006) model, which we consider suitable for explaining the stakeholder participation process in the company, and seek to answer a logical sequence of inter-related issues (see Figure 1). We specifically seek to answer the following questions: Why do they form relationships? Who do they form relationships with? For what reason do they form relationships? and How do they form relationships?

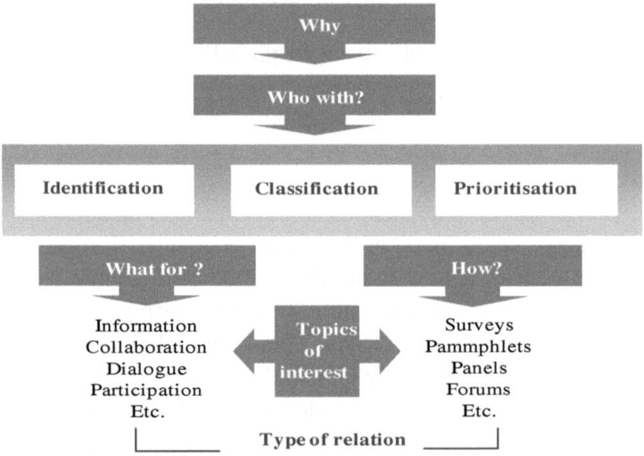

Fig. 1 SH relationships model

2.1 Why Do They Form Relationships?

The reasons which might lead companies to allow SH to engage in their management underpin the instrumental and normative aspects of Stakeholder Theory that, according to Donaldson and Preston (1995), justify their implementation, basing the management and administration of SH on strategic and moral or ethical criteria, respectively. This can therefore legitimise their actions and become a source of sustainable competitive advantage via trust, reputation and innovation (Schroeder and Kilian, 2007; Rodríguez et al., 2002), or for generating innovations and disruptive business models (Hart and Sharma, 2004).

Despite recognising different motives to justify SH participation in business management, Noland and Phillips (2010) group them into two categories, one strategic and the other moral.

Under the strategic focus of CSR, SH participation can be understood to be beneficial both for SH and for the company itself (a win-win situation) with a consequent re-think of the objectives to be achieved, and social and environmental goals included alongside economic objectives (companies must be economically viable, socially integrated and respectful of the environment).

Meanwhile, the following moral reasons have been identified, inter alia: the establishment by democratic means and with the participation of SH (including the less powerful or influential) of the actions that the company takes that can be considered socially acceptable (Lewis and Unerman, 1999); SH involvement in decision-making (Van Buren, 2001); an attempt to effectively satisfy SH needs (Phillips, 1997, Evan and Freeman, 1979); and improved company responsibility (Gray et al., 1996; Owen et al., 2000).

2.2 Who Do They Form Relationships with?

Responding to the question posed in this section entails determining the criteria for establishing which SH the company should interact with. This requires determining who the SH are, defining them, classifying them and prioritising them.

The similar sound and semantics of the terms *stakeholders* and *shareholders* and the fact that new interests are at stake in organisations may be behind the origin of the term *stakeholders*. Authors such as Goodpaster (1983) and Freeman (1984) attribute the expression's emergence to a play on words intended to demonstrate to company management that there were large numbers of interests at risk –i.e., at stake- not only those of the shareholders, owners and employees.

Although many contributions have been made to this topic despite its short existence and simplicity, Freeman's (1984) definition of SH has become a benchmark for the literature on stakeholders, and this is why many later contributions are adaptations of the various focuses from which the analysis is being conducted (economic, social, ethical, commercial, etc.).

Freeman's proposal implicitly includes the two-way nature of the relationship between the company and the various groups as well as the contribution made to business objectives. This has enabled us to understand how an organisation works, interpreted from angle of the plurality of agents (with changing and, occasionally, conflicting interests) that intervene and, as Freeman (1984) himself states: They can have an endless number of forms and classifications and they will, moreover, be determined to a great extent by the company's characteristics and dimensions.

There is no one list of SH, not even for a single company, and the main collectives usually considered among SH are the shareholders/investors/owners, suppliers, customers, employees and society (Spiller, 2000; Papasolomou et al., 2005), who can be categorised according to different criteria.

A number of proposals have been made for identifying and classifying the various SH since Evan and Freeman (1979) distinguished between SH who are vital (for the existence or survival of the company) and those who, although not vital, affect or can be affected by its activity. Amongst the most cited and referenced we can highlight the classifications proposed by Savage et al. (1991), Clarkson (1995) and Mitchell et al. (1997). Savage et al. (1991) group SH into four categories depending on their potential for cooperation or support and their potential for threat: those with the highest levels of support and threat (mixed blessings), those with the lowest levels of support and threat (marginals), the most cooperative and least risky (supportive) and the most risky and least cooperative (nonsupportive). Clarkson (1995) distinguishes between primary (essential for survival) and secondary (while not essential they can affect or be affected by the organisation) SH on the basis of their levels of risk. The Mitchell et al. (1997) contribution determines the importance of SH depending on the influence, urgency and legitimacy of their interests. These authors distinguish between latent, expectant and definitive SH depending on the concurrence of one, two or all three of these attributes, respectively. This also defines how important they are (low, medium or high) for company management (more attributes, more important).

However, SH identification and typology is still acknowledged as an SH focus line of research. Different proposals have emerged in keeping with those mentioned above which include new elements and criteria. A number of these are presented in Table 1.

The different classification criteria noted enable, or should enable, organisations to prioritise their SH, who are adaptable and, on occasion, unique to each organisation, depending on the moment in time. Granda and Trujillo (2011), to propose a set of criteria on which to base the decision for prioritising relationships with SH: the level or capacity of influence or dependence (current and future); expectations, and degree of interest in engagement and willingness to participate; typology of pre-existing relationships with SH; knowledge of the organisation and link to the ultimate goal of the dialogue process; SH type (public, internal, social, corporative, etc.); the geographical aspect of the process and the social context.

In this same line of prioritisation, Olcese et al. (2008) classify SH as criticals, basics or complementaries depending on their impact and relevance for the company. The first of these have a key economic impact, strengthen or influence company reputation, grant or limit licences or access or create the future of the sector. The basics have an average impact on business results, might partially affect company reputation, but in some way impact on key company processes. Finally, the complementaries are those who have a minimal economic impact, a very slight influence on company reputation and could supply complementary services or products.

Table 1 SH Typology

SH Types	Criteria	References
Nuclear and satellite	Capacity for influence and urgency.	García-Echevarría (1982)
Socioeconomically attractive, socially vulnerable, economically vulnerable and socioeconomically vulnerable	Level of social and economic support	Whitehead et al. (1989)
Primary and secondary	Capacity to influence chance of survival	Clarkson (1995) Donaldson and Preston (1995)
Social and non-social (primary and secondary)	Type of relation (direct/primary or indirect/secondary) and capacity for defence (own or otherwise) of their interests	Wheeler and Sillanpäa (1997)
Organisational, community, regulatory and mediating	Capacity to influence environmental behaviour	Henriques and Sadorsky (1999)
Regulatory, primary external Primary internal and secondary	Type of relation	Buysse and Verbeke (2003)
Latent, expectant and definitive	Power, legitimacy and urgency	Vos (2003) Currie et al. (2009)
Internal and external	Responsibility and relationship with surroundings	Guerras and Navas (2007)
Organisational, policy and social	Environmental pressure	Diez et al. (2008)
Structural, managerial and complementary	Type of relation	Olcese et al. (2008)

This prioritisation phase will enable the business organisation to identify the type of relationship it should have with each of its SH and thus answer the "*what do they form relationships for?*" and "*how do they form relationships?*" questions that are the objective of the following phase.

2.3 What Do They Form Relationships for and How Do They Form Them?

Under the dynamic and holistic focus that we proposed in the preceding section and following the above-described process, the next step is to determine the strategies that will be adopted with each of the SH, the criterion that will be used to include their interests, and the way that they can be included. This will enable the type of relationship that is formed with each of the interested parties (what do they form relationships for?) and the mechanisms used for this (how do they form their relationships?) to be characterised. These two questions are closely linked, which is why we have opted to analyse them jointly in this section, as they are at the very core of the company's engagement with the stakeholder.

The controversy that arose around SH participation in organisations' management has not been an obstacle to attempts being made to characterise and classify the type of relationship that they have with the SH that they have identified according to different criteria. For its interest to business practice, we highlight the proposal made by the Institute of Social and Ethical AccountAbility[1], which sets out different levels of relationship depending on the type of communication established and the objective to be achieved (monitoring, information, contractual, consultation, convening, collaboration and delegation). Depending on the degree to which SH participate in company decisions, it also distinguishes between a unilateral relationship (the company takes it upon itself to inform its stakeholders about concerns), bilateral with no participation or verification (the company asks and the SH answer) and bilateral with participation in the decision-making.

Olcese et al. (2008) distinguish between *essential* and *basic* relationships (attention given to queries and complaints, satisfaction studies, promotion, etc.) gauged by the impacts that they have, and the more *complex* relationships that require greater engagement and integration with SH (benchmarking, improvement projects, partnerships, etc.). Also, depending on the actions engaged in with the various SH, the same authors distinguish between different types of relationship: information, exploration and analysis, commercial and post-commercial actions, development actions and advanced actions.

Savage et al. (1991) distinguish four types of strategies for relating to SH depending on the potential for threat or for collaboration and support that each SH has with regard to the company objective: involvement (a high degree of collaboration and a low threat level), control (low cooperation and threat), defence (high threat level and low collaboration level) and collaboration (high collaboration and threat).

The reach and content of SH participation can also vary greatly irrespective of the type of relationship (Viviani, 2006; Harris, 2007). It can be temporary and disappear when the extraordinary action with which it is linked comes to an end, or be permanent. It can also affect a specific area of the company or the company as a whole.

The company will use different types of dialogue mechanisms or tools depending on the type of relationship it has with an SH and the relevance that it has for the company. In keeping with the classification proposed by the Institute of Social and Ethical AccountAbility, Granda and Trujillo (2011) group these in three categories which go from a marginal and non-interventional focus to one of full inclusion in decision-making (see Table 2). They distinguish between unilateral mechanisms, which are generally applicable to the less relevant SH or those that have a lesser capacity for engagement; bilateral mechanisms, which envisage information-sharing with the most relevant SH (from active information to queries); and inclusion tools that enable SH to be integrated into company decision-making processes.

This classification will enable us to conclude the proposed process through which companies put their relationships with SH into practice by responding to the last question: How do they form relationships with SH?

---

[1] Has prepared the "The Stakeholder Engagement Manual" in collaboration with the United Nations Environment Programme and Stakeholder Research Associates Canada Inc, to find answers to the large number and wide range of questions in the area of SH relationships.

Table 2 SH relationship mechanisms

Type of mechanism	Stakeholder relevance	Mechanisms	Relationship with stakeholders
Unilateral	–	Reports/Statements/Proceedings Pamphlets Briefings Meetings/Seminars/Conferences	Unilateral
Bilateral	↓	Interviews/Questionnaires Periodic meetings Satisfaction studies Suggestion and ideas box/board	Bilateral with no participation or verification
Integrating	+	Monographic workshops Stakeholder panels Communication and dialogue portal/platform Projects Arrangements/agreements/partnerships Work teams	Bilateral with direct or indirect participation

3 Case Study

3.1 Methodology Design

Case study methodology was considered suitable for this study. Although it does not allow statistical generalisation, it does permit contextual generalisation and enables a deep analysis to be made of a complex research phenomenon (Yin, 1989; Patton, 1990 and Maxwell, 1996).

Cases were selected by theoretical, not statistical, sampling and we sought to choose those that could provide the greatest opportunity for learning (Denzin and Lincoln, 1994). Apart from accessibility to different information sources[2], another reason for selecting the cases presented below was that they fitted in well with the research objective, i.e., to explain how hotel companies relate to SH. These are companies in the Spanish hotel sector that engage in different relationships with SH for the development of their strategies (as their CSR reports show[3]) and which are widely-recognised in environmental and social matters. Their engagement with

[2] Questionnaires and personal interviews with the two hotel company CSR directors were used for our study as well as other documentary sources (CSR reports and other company documents on different media platforms).

[3] The experts agree that one of the main signs that demonstrate that management models and policies with CSR content are implemented in companies is precisely that they publish CSR or Sustainability reports in which they, the companies, give an account of the main objectives and advances in the issue.

sustainability can also be seen in the effort that they make to take on a number of commitments, adhering to various CSR-linked initiatives[4], and are currently a benchmark for sustainable management in the business world.

3.2 Description of Cases

The chosen hotel groups are two large Spanish chains in both cases with over thirty years' experience, over 300 hotel establishments all around the world, a capacity in excess of 50,000 rooms and a workforce that exceeds 15,000 employees (see Table 3).

Table 3 Business profile of hotel chains

	Case 1	Case 2
Year of start up	1956	1978
Type of establishment	Urban/Holiday	Urban/Holiday
No. Employees/(nationalities)	35,728/(109)	18,294/(118)
No. Hotels	310	397
No. Countries	27	16
No. Rooms	78,598	58,687
No. Customers	23,000,000	18,000,000
Operating regime (% hotels)		
Owned/Rented/Managed/Franchise	22/28½/40/9½	22/57/21/0
Published CSR reports	2008-2010	2006-2010

Both companies take part in different CSR-linked initiatives and both use the Global Reporting Initiative Guide (GRI) as a reference when writing up their CSR reports. Special mention should also be given to the fact that they both belong to the United Nations World Pact, whose principles shape the framework of their CSR strategies and management.

The two hotel groups are widely acknowledged for their environmental and social work both in the tourist and general business sectors and have been presented with various awards and distinctions. They have also taken part in a range of projects that prioritise the sustainable behaviour of the tourist sector.

3.3 Analysis of Findings

We now go on to find out the degree to which relationships with SH are implemented and formalised with the aim of determining the patterns of sustainable SH

[4] The Global Reporting Initiative Guide (GRI) and the United Nations World Pact, among others.

relationships in the two cases and responding to the questions posed in the previously presented theoretical framework which underpin our study's empirical content.

3.3.1 Why Do They Form Relationships?

In both cases, under the CSR focus the hotel chain executives identify their SH relationship as the differentiating feature of the responsible, dynamic and plural business model that currently characterises the sustainable management of their respective chains, which is moving further and further away from the classic shareholder and owner-centric business model. Their objectives are not solely economic but adapt to the new demands and changes in the environment. This is what the two company executives told us. Case 1 "*the relationship with stakeholders is an immense source of opportunities. It's what makes us a more sustainable, more stable and secure company, and one that is more attractive for anyone who approaches it*". The SH relationship is equally essential for Case 2 "*because it's impossible to run your business and your activity without including the SH*"... as "*we have no capital without shareholders, no services without employees, no sales without customers...*".

So for both groups, under the CSR focus good relations and communication with SH (envisaged in their missions) have an obvious strategic angle associated with the opportunities for obtaining multiple economic, social and environmental benefits (see Graph 1).

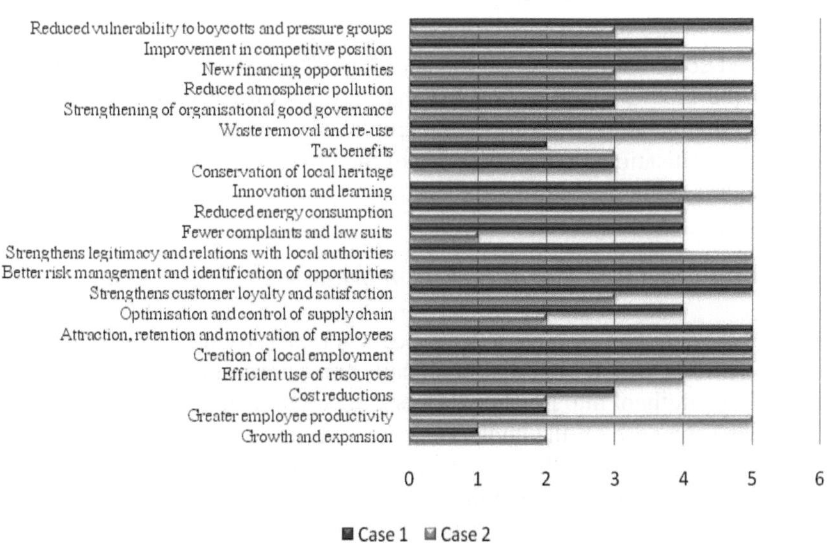

Graph 1 Benefits of relationships with SH

Although the greatest percentage (68-70%) of benefits for the two hotel groups is still acknowledged to be economic, they currently have to make room for environmental and social benefits, too (see Graph 2). This confirms that the hotel chains have reconsidered their objectives to include environmental and social goals and, consequently, also the focus under which the two companies enter into their SH relationships. They thus adopt a proactive and anticipatory attitude in order to minimise risks and take advantage of the underlying strategic opportunities in SH relationships with the objective of achieving an improvement in their competitive position that enables them to strengthen their legitimacy, reputation, credibility and prestige.

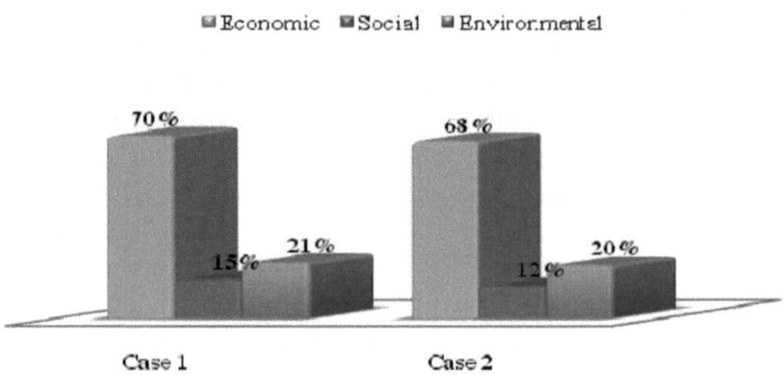

Graph 2 Types of benefits associated with the relationship with SH

The stated considerations reveal the addition of new stakeholders (society, suppliers and the environment) to the list that has traditionally been considered in businesses (shareholders, customers and employees). Their identification, classification and prioritisation are the objective of the next phase of the proposed model, which is addressed in the following section.

3.3.2 Who Do They Form Relationships with?

Although both companies have a set process for identifying and managing relationships with SH which envisage practically the same aspects (see Table 4), the way that they put them into practice differs to a certain degree.

For Case 2 they are all the audiences with which the company relates and that are part of its business value ecosystem. They are its employees, customers, shareholders, suppliers, the community and the environment.

Identifying these SH was one of the main tasks included in this hotel group's 2007 Master Plan, which numbered among its objectives the establishment of a CSR management system to facilitate dialogue with all the company's SH and

Table 4 Management of relationships with SH

Aspects	Case 1	Case 2
Identification of SH needs, expectations and capacity to influence	√	√
Risk identification and assessment	√	√
Periodic review and updating of process	√	√
Prioritisation of SH		√
Action plans	√	√
Requirements demanded of SH		√
Identification of most relevant topics for each SH	√	√

their involvement. As the company's CSR Director told us *"the first thing that we had to do was to identify all the main SH, the state that they were in and to prioritise the actions directed at them"*. This task was undertaken in partnership with an outside organisation which focused on doing a diagnostic analysis of the initial situation. The company identified the main SH in accordance with the criteria of power, urgency and legitimacy.

After identifying the main SH the company mapped the so-called *stakeholder initiative priorities* which it would use to determine the framework of relations and policies, programmes and responsible value initiatives for each audience. This procedure permitted the company to set out priority lines of action to enable active communication and an open dialogue with the SH which was key to finding out their needs and expectations, telling them what is being done and promoting their values, engaging and involving them in the company's responsibility culture with a view to determining the levels of engagement and relationship (can be improved, good or excellent) with each of the identified audiences, as well as establishing the priority with which actions would be undertaken (first, second or third degree in descending order of importance) with each of them,

Case 1 also identifies the same SH. For this an analysis matrix was devised according to the degree of influence and urgency that SH have on the company's corporate reputation. Although for the company "*all SH groups are the same and on the same level when it comes to being listened to and the company responding to their needs*" and so it has not implemented any mechanism to establish a scale of priority to rate them although it does recognise that "*the priority is the customer and the search for a balance in conformity among all the remaining stakeholder groups*".

3.3.3 What Do They Form Relationships for and How Do They Form Them?

The mechanisms for dialogue and communication used by the two hotel groups to form relationships with their various audiences are the main indicator for characterising the relationship and its strength, and thus responding to the two questions of "what for?" and "how?"

The analysis reveals the wide range of mechanisms that the hotel chains in the study use to maintain an active relationship with their SH, adapting to the characteristics and needs of each (Graph 3).

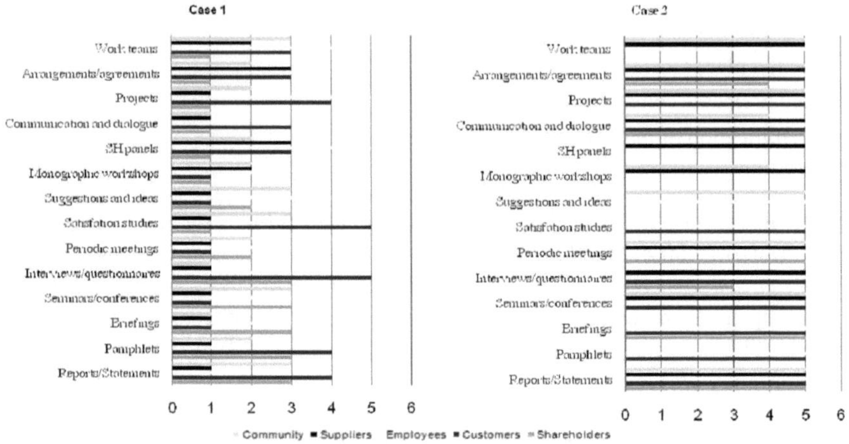

Graph 3 Extent to which SH relationship mechanisms are used

The two companies use different resources to keep their SH informed. Reports, statements and briefings stand out for their frequency and periodicity. Of the methods used to ascertain SH demands and expectations and their degree of satisfaction with their needs, the interviews and periodic meetings that are had with almost all SH stand out, whilst surveys are basically used to find out how satisfied customers and employees are. The remaining mechanisms used (monographic workshops, panels, platform, portal, projects, work teams, etc,) not only foster SH participation in company management, but are characterised by producing interaction both between the different SH groups and the individual SH within them.

Incorporating advances in information and communication technology (ICTs) into the SH relationship process has also helped to keep the information that is conveyed or received through specific communication channels (web, social networks, e-mail, etc.) continually updated in both chains and enables them to maintain a constant flow of instant and up-to-date information in any direction (company-SH; SH-company; SH-SH). This therefore enables information to be more widespread and to reach a greater number of SH.

Graph 4 presents the type and strength of the hotel chains' relationships with each of their SH according to the proposed classification in Table 2 and the amount of use (frequency and periodicity) made of the mechanisms available to the companies for relating with their SH.

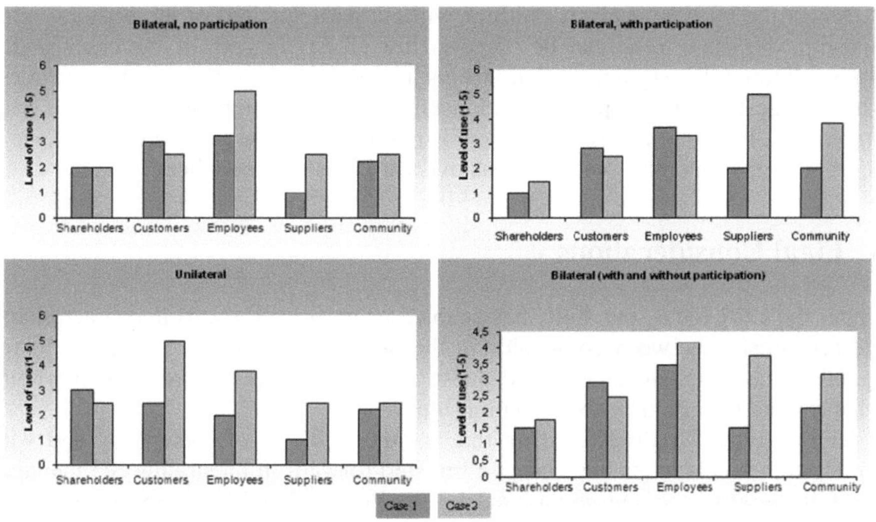

Graph 4 Type of SH relationship

The degree to which use is made of the different information mechanisms reveals the size of the unilateral relationship that the two chains have with various SH and which is characterised by being merely advisory (the company informs the SH). So, using reports, statements, pamphlets, briefings, talks etc. the hotel groups keep their various SH informed about any aspects that they consider to be relevant for each (e.g., shareholders, about the current share value, customers about new services or special offers, etc.). This type of relationship is on virtually the same level for the two chains' shareholders and community, although the relationship that Case 2 has with the remainder of its SH (customers, employees and suppliers) is stronger than for Case 1. In both cases, however, the frequency and periodicity with which the various mechanisms are used follows the same trend (from greater to lesser relation, respectively).

In other regards, the extent to which the mechanisms are used in bilateral relationships in the two cases under analysis confirms this trend with respect to shareholders, customers and employees although unlike what happens in unilateral relationships, Case 2 does not always do so more frequently and with greater periodicity than Case 1. In both cases, the most active relationship is with employees, who cooperate by taking part in projects, in work teams, on panels, etc.

Despite both companies committing to a solid and long-term relationship with suppliers, Case 2 has ramped up their participation (3.75 out of 5 compared to 1.5 for Case 1) by putting a range of projects into operation of which the Sustainable Club stands out for its innovation, relevance and impact. Its goal is to drive dialogue and participation between the company and suppliers who are committed to the development of sustainable products and services so that they might join forces in a search for new ideas and efficient products that enable eco-efficient and responsible solutions to be found, the most relevant projects that they are working on to be publicised and any achievements to be disseminated.

As for the by-type analysis (with or without participation) of the bilateral relationship, the same trend can be observed for all SH in both of the cases under study, although the strongest relationships are not the same in the two companies. The strongest non-participation relationship in the two companies is the one with employees. In relationships where there is participation, the most intense in Case 1 is, as before, with employees, whereas in Case 2 it is with suppliers.

4 Final Considerations

Under the CSR focus, our study has enabled us to move forward in characterising the relationship between companies in the hotel sector and their SH, and to respond to the series of inter-related questions that were posed: why, with whom, what for and how do they form relationships?

The analytical findings confirm that for the first two of these (why and who with?) both hotel chains have taken a big step forward in integrating SH management at the strategic and operational levels. Proof of this is the extent to which relations with SH have been put in place and formalised in their respective companies. This has led to them using a broad variety of mechanisms to facilitate communication and dialogue with each of their audiences and enable the reciprocal exchange of information that is essential for steps to be taken towards sustainable relationship models. This last aspect is linked to the remaining two questions (what for and how?), for which we highlight the following:

The use of communication and dialogue channels specifically adapted to the needs of some particular SH.

The predominance of bilateral relationships verifies that their engagement and relationships with their SH have taken a new direction in making them direct or indirect participants in the decision-making process.

ICT development has made a positive contribution to the implementation of new relationships through the growing use of specific communication channels (web, social networks, etc.).

On the basis of the above, and being mindful of the limitations of the research methodology used in this study, the findings confirm the importance that SH take on in the new management models under the CSR focus. As the people responsible for AccountAbility and Utopies stated, *basic relationship models* (a reactive or defensive attitude when faced with social or environmental situations) have been abandoned to make way for *advanced* or *improved models* (a proactive and anticipatory attitude that minimises risks and exploits the strategic opportunities that underlie relationships with SH) that enable SH to participate and engage in company decisions.

The findings of this study will serve as a basis for future research which will enable us to go further into the SH relationship implementation and formalisation process in the sector under analysis, and to draw conclusions from a more representative sample bearing in mind that, although the research methodology used is not without its limitations, it is valid for the objectives that were set at the beginning of this study to be achieved.

References

Buysee, K., Verbeke, A.: Proactive environmental strategies: A stakeholder management perspective. Strategic Management Journal 24(3), 453–470 (2003)

Clarkson, M.B.E.: A stakeholder framework for analyzing and evaluating corporate social performance. Academy of Management Review 20(1), 92–117 (1995)

Currie, R.R., Seaton, S., Wesley, F.: Determining stakeholders for feasibility analysis. Annals of Tourism Research 36, 41–63 (2009)

Deegan, C., Unerman, J.: Financial accounting theory: European edition. MacGraw Hill, Maidenhead (2006)

Denzin, N.K., Lincoln, Y.S.: Handbook of Qualitative Research. Sage Publications, Thousand Oaks (1994)

Diez, F., Medrano, M., Diez de Castro, E.: Los grupos de interés y la presión medioambiental. Cuadernos de Gestión 8(2), 81–96 (2008)

Donaldson, T., Preston, L.E.: The stakeholder theory of the corporation: concepts, evidence, and implications. The Academy of Management Review 20(1), 65–91 (1995)

Evan, W.M., Freeman, R.E.: A stakeholder theory of the modern corporation: Kantian capitalism. In: Beauchamp, T.L., Bowie, N.E. (eds.) Ethical Theory and Business. Prentice-Hall, Englewood Cliffs (1979)

Freeman, R.: Strategic Management: a Stakeholder Approach. Pitman, Boston (1984)

García-Echevarría, S.: Responsabilidad social y balance social de la empresa. Fundación Mapfre, Madrid (1982)

Goodpaster, K.E.: The concept of corporate responsibility. Journal of Business Ethics 2, 1–22 (1983)

Granda, G., Trujillo, R.: La gestión de los grupos de interés (stakeholders) en la estrategia de las organizaciones. Economía Industrial 381, 71–76 (2011)

Gray, R., Owen, D., Adams, C.: Accounting and Accountability: Changes and challenges in corporate social and environmental reporting. Prentice-Hall, London (1996)

Guerras, L.A., Navas, J.E.: La Dirección Estratégica de la Empresa. Teoría y Aplicaciones. 4ª Edición, Thompson Civitas (2007)

Harris, R.: Perchè il dialogo è diverso. In: Freeman, R.E., Rusconi, G., Dorigatti, M. (eds.) Teoria degli Stakeholder, Franco Angeli, Milano, pp. 217–223 (2007)

Hart, S.L., Sharma, S.: Engaging fringe stakeholders for competitive imagination. Academy of Management Executive 18(1), 7–18 (2004)

Henriques, I., Sadorsky, P.: The relationship between environmental commitment and managerial perceptions of stakeholder importance. Academy of Management Journal 42(1), 87–99 (1999)

Jones, T.M., Wicks, A.C.: Convergent Stakeholder Theory. Academy of Management Review 24(2), 206–221 (1999)

Lewis, L., Unerman, J.: Ethical relativism: A reason for differences in corporate social report-ing? Critical Perspectives on Accounting 10(4), 521–547 (1999)

Maxwell, J.A.: Qualitative Research Design: An Interactive Approach. Sage Publications, Thousand Oaks (1996)

Mitchell, R.K., Agle, B.R., Wood, D.J.: Toward a theory of stakeholder identification and salience: Defining the principle of who and what really counts. Academy of Management Review 22(4), 853–886 (1997)

Noland, J., Phillips, R.: Stakeholder engagement, discourse ethics and strategic management. International Journal of Management Review 12(1), 39–49 (2010)

Olcese, A., Rodríguez, M.A., Alfaro, J.: Manual de la empresa responsable y sostenible. In: Manual de la empresa responsable y sostenible. McGraw-Hill, New York (2008)
Owen, D., Swift, T., Humphrey, C., Bowerman, M.: The New Social Audits: Accountability, Managerial Capture or the Agenda of Social Champions? European Accounting Review 9(1), 81–98 (2000)
Papasolomou, I., Krambia, M., Katsiloudes, M.: Corporate social responsibility: The way forward? Maybe not! European Business Review 17(3), 263–279 (2005)
Patton, M.Q.: Qualitative Evaluation and Research Methods, 2nd edn. Sage Publications, California (1990)
Phillips, R.: Stakeholder Theory and a Principle of Fairness. Business Ethics Quarterly 7(1), 51–66 (1997)
Post, J.E., Preston, L.E., Sachs, S.: Managing the extended enterprise: the new stakeholder view. California Management Review 45(1), 6–28 (2002)
Rodríguez, M.A., Ricart, J.E., Sánchez, P.: Sustainable development and the sustainability of competitive advantage: A dynamic and sustainable view of the firm. Creative & Innovation Management 11(3), 135–146 (2002)
Savage, G.T., Nix, T.W., Whitehead, C.J., Blair, J.D.: Strategies for assessing and managing or-ganisational stakeholders. Academy of Management Executive 5(2), 61–75 (1991)
Schroeder, K., Kilian, B.: El efecto de las prácticas de RSE en los ingresos de los negocios. Banco Interamericano de Desarrollo, Washington (2007)
Spiller, R.: Ethical business and investment: a model for business and society. Journal of Business Ethics 27(1-2), 149–160 (2000)
Van Buren, H.: If Fairness is the Problem, is Consent the Solution? Integrating ISCT and Stake-holder Theory. Business Ethics Quarterly 11(3), 481–499 (2001)
Viviani, M.: Il coinvolgimento degli stakeholder nelle organizzazioni socialmente responsabili. Maggioli Editore, Rimini (2006)
Vos, J.F.J.: Corporate social responsibility and the identification of stakeholders. Corporate Social Responsibility and Environmental Management 10, 141–152 (2003)
Wheeler, D., Sillanpäa, M.: The stakeholder corporation. Pitman, London (1997)
Whitehead, C., Blair, J., Smith, R., Nix, T., Savage, G.: Stakeholder supportiveness and strategic vulnerability: Implications for competitive strategy in the HMO industry. Health Care Management Review 14(3), 65–76 (1989)
Yin, R.K.: Case Study Research. Design and Methods. Applied Social Research Methods Series. Sage Publications, London (1989)

Hotel Environmental Impact Management: A Case Study in Cádiz Province

Mª Teresa Fernández-Alles and Ramón Cuadrado-Marqués

Department of Marketing and Communication, University of Cádiz.
Avda. Duque de Nájera 8, 11002 Cádiz, Spain
{teresa.alles,ramon.cuadrado}@uca.es

Abstract. The high consumption of natural resources and energy made by hotel companies, together with the corresponding waste generation, has placed the hotel sector in an ideal application field for environmental social responsibility policies. So much so that the environmental performance of hotel companies has received increasing attention, which has led to an increasing number of hotels designing and implementing socially responsible measures focused on environmental protection.

The present work is focused on the study, from a social responsibility approach, of the implications of policies and measures for environmental management in the hotel sector, exposing not only the various operational areas in which they arise, but also the possible motivations of management to start these programs. To complete the work, we analyse, using the case method, the environmental policies carried out in one of the leading hotels of the province of Cadiz, Spain, for pioneering the application of the same, the Hotel Playa Victoria, owned by Palafox hotel chain.

Keywords: Corporate social responsibility, Environmental management, Hotel, Tourism.

1 Introduction

Corporate Social Responsibility, understood as "a concept whereby companies integrate social and environmental concerns in their business operations and in their interaction with their stakeholders on a voluntary basis" (Commission of the European Communities, 2001) is currently the subject of growing demand from all stakeholders, reflected in increasingly frequent treatment of this theme in various media, both Scientific and mainstream publications. Thus, corporate social responsibility is, today, one of the most important issues for the various stakeholders, either customers, shareholders or public institutions.

Based on the concept of corporate social responsibility, the purpose of this study is to analyse the practices carried out in the field of one of the most important sectors of Spanish economy, given its contribution to GDP, balance of payments and the job creation (World Travel and Tourism Council, 2011), as is the tourism sector, focusing on one of its main pillars, the hotel subsector, which shares with the sector in which it fits its high economic, social, and environmental impact (Bohdanowicz & Zientara, 2009; Shah, 2011).

A.M. Gil-Lafuente et al. (Eds.): Soft Comput. in Manag. and Bus. Econ., STUDFUZZ 287, pp. 335–346.
springerlink.com

In this way, those hotel policies that are aimed either to minimize the impact on the environment or to contribute to nature preservation and regeneration will be studied in this paper. According to this goal, firstly, a review of the scientific literature on corporate social responsibility in the hospitality industry, and environmental practices will be conducted, with the purpose of knowing about the state of the art on sustainable and environmental management in the lodging industry and highlighting future research paths, enrichening this field of knowledge, being a key to fulfill the multiple objectives of organizations, customers and society. Secondly, it will be complemented by the exhibition of the results obtained from a real case study in which the environmental policies that are taking place today in one of the most prominent hotels in the province of Cadiz are the object of the analysis, being chosen for its commitment to society, in general, and the environment in particular.

2 Corporate Social Responsibility in the Hotel Sector

Corporate social responsibility is a topic that has gained increasing attention from the scientific community, since its origins in the 30's of XX century. Thus, it is worth mentioning Bowen's "Social Responsibilities of the Businessman" (1953), which generated an intense debate on the subject, and also the contributions of Carroll (1979, 1991 and 1999), who modelled the corporate social responsibility to facilitate its implementation in companies, and others, such as Kotler & Lee (2005), or Porter & Kramer (2006), who have significantly contributed to enrich the state of the art by giving it a strategic focus.

Specifically, in the hotel sector, it is noticeable the existence of a large number of studies aimed to the relationship between hotels and the environment, considered the latter as the local environment in which hotel companies operate, in many cases, and as the impact on the environment globally in all of them. Also, the environment factor is gaining weight in the destination choice for tourists (Mensah, 2006). Stakeholder pressure, is definitely leading the industry to develop effective environmental policies, and allocating resources to keep them running (Shah, 2011).

This fact has given rise to many discussions about the operational areas of integration of corporate social responsibility policies, its convenience, or the communication of those measures, despite the high level of legislation about many of the aspects that may manifest socially responsible behaviour in this sector, meaning that many initiatives are considered just legal, leaving little room for ethical and discretionary actions (Carroll, 1979).

From the point of view of hotel management, there is a fairly widespread misconception of corporate social responsibility and eco-tourism, even equalising the terms. When regarding the impact of the activity of tourist accommodation on environment, this confusion gets even stronger (Ayuso, 2006). This statement is especially true in the case of city hotels, where the connection between the hotel business and the environment, understood as the natural environment, becomes blurred (Bohdanowicz, 2005). In this context, cognitive processes, coupled with CSR knowledge, largely determine the involvement of the hotel managerial staff in social responsibility programs, so that those most qualified and experienced managers, from a CSR point of view, will be more prone to adopt these policies.

Thus, in the midst of this situation, the motivation for taking action or launching socially responsible programs does not seem to be caused, in most cases, by the knowledge and belief on these policies, but rather by a set of factors Miller gathered, in order of importance, in the following classification (2001): Sectorial structure, law enforcement, public image, cost savings, and ethical considerations.

Several authors have studied the low relevance of moral obligation and motivation for environmental action by the hotel managers, considering the profitability or cost savings as the main motivating factor, being necessary to develop a value-creating insight in order to spread these practices against the vision above (Gurney & Humphreys, 2006). Returning to the question of the profitability of these measures, there is evidence of the profits derived from their application:

- In the short run, on average, there's a share price increase as a result of the greater trustfulness (Nicolau, 2008), and the better economic outcome, regardless of the size of the company, provided that there are high levels of investment in social responsibility policies (García & Armas, 2007), reinforcing confidence and strategy-designing flexibility with increasing levels of corporate social responsibility (Lee & Park, 2009).
- In the long run, the market value of the company, defined as long-term profitability, will be greater than if they had not engaged socially responsible policies, for the same reasons given in the previous point (Lee & Park, 2009).

Consequently, and given the arguments presented throughout this section, it can be asserted that the formulation of strategies that integrate social responsibility in hotel management is plainly convenient, with particular attention to those actions and operational capabilities that have the closest relationship to the environment, given the growing awareness of the target group on the environmental impact of hotels.

3 Environmental Actions in the Hotel Sector

In the past 20 years, the global hotel industry has increased its involvement with the environment, responding to widespread and growing demand for more environmentally friendly products and services with it. Research on social sustainability and social responsibility has radically increased during the past 10 years (Yoo, Lee & Bai, 2011), reflecting this growing concern for the environment.

Arguably, almost from its inception, the main thrust of environmental policies by major hotel chains has been the creation of the International Hotels and Environment Initiative (IHEI) in 1992, replaced in 2004 by the International Tourism Partnership (ITP), but also there are efforts by other organizations such as the United Nations' World Tourism Organization (WTO) or the International Hotel & Restaurant Association (IH&RA).

3.1 Application Areas

In the study of the environmental impact created by a hotel, there are several steps to be taken into account beyond its activity period and after it. The stages of site selection, design and construction of the hotel, the operational phase, and finally, the phase that follows the cessation of the hotel business in the building, should be adopted regarding environmental criteria to minimize the impact.

Since the first phase will only affect those hotels that are still under design, it is preferable to focus on the operational phase, so every hotel could help to protect the environment by making changes in the way of managing the different operational areas. In environmental management, there many distinct areas in which to act, by its direct relationship with the environment, being energy, water use, and waste procedures (International Tourism Partnership, 2010), bio-diversity or eco-system protection, noise pollution, green building design, food, environmental education and environmental partnership (Shah, 2011; Hsieh 2011).

It is noted that, contrary to the widespread idea that environmental management requires large investments (Bohdanowicz, 2005b), many of the proposed actions involve only a little change in the way of organizing and managing daily operations, without resorting to strong capital investment (which it may be desirable or even necessary (Bouquet, Crane, & Deutsch, 2009; Tarí et al, 2010) at higher levels of environmental performance).

However, from the point of view of a socially responsible management, it is not enough to reduce the impact of the facility itself through the implementation of specific policies on energy and water consumption and waste production. It is necessary to adopt a holistic approach, integrated into the strategy of the hotel or chain, so that is reflected in codes of conduct published by them. Accordingly, hotels' framework and policies, staff training and raising its awareness and community relations must consider the environmental impacts and the influences on environmental performance caused by the hotel managerial staff, hotel's departments, customers and the community surrounding the hotel (ITP, 2010). Finally, it is necessary to review the environmental policy of the hotel purveyors, as, by rewarding the consumption of goods and services that come from environmentally friendly companies, the hotel can contribute indirectly to the preservation of the environment (ITP, 2010).

3.2 Hotel, Environment, and Customers

There are several arguments that attempt to justify the environmental programs of hotels from different approaches. From a resource based view of the firm, by undertaking environmental or social programs, hotels can gain competitive advantages, both for saving costs and for the "development of some valuable capabilities, such as stakeholder integration, continuous innovation or higher-order learning" (Garay & Font, 2012). Apart of obtaining competitive advantages by going green, the hotel industry can become a channel for social change, due to its high visibility (Ryan, 2002; Kang et al, 2012). Furthermore, there are both legal and economic reasons, but equally important is customers' perception of the hotel commitment to both society and environment.

In the past 20 years, environmental awareness has been arisen among hotel customers (Becker-Olsen et al, 2006), so their exigencies regarding the hotel services can easily involve environmental issues. On the other hand, a mass of customers, both potential and real, does not take into account the environmental performance (or social) of the hotel when making a reservation. Therefore, the right communication strategy differs in one case and in another. Communications must aim to capture the prospect's conscious attention in the first case, and in the second case, try to change the behaviour of actual customers, bringing them closer to socially responsible practices (Dinan & Sargeant, 2000). Apart of targeting customers the right way, hoteliers could be concerned by the return of investment in green initiatives, or if they can charge a premium price to the customer. Findings are contradictory, since Manaktola & Jauhari (2007) found that Indian hotel customers weren't prone to pay a plus, while Kang et al (2012) state that US travellers are likely to pay it, as they're more interested in environmental issues. Furthermore, the higher the grade of the hotel, the more beneficial these initiatives can result, "since customers for these establishments show, on average, a higher level of willingness" to pay the eco-premium price (Kang et al, 2012).

Regarding what measures are most requested by customers, it seems that there is no clear differentiation between different types of measures (reduce, reuse or recycling), but the most significant match in its tangibility to the client. These measures will perform their best the greater the visibility of these measures, and the more customer implication they demand (Manaktola & Jauhari, 2007).

Finally, regarding how the measures should be communicated to customers, there is evidence on the suitability of materials especially designed for this purpose, issuing cards and posters about environmental commitment including the company logo, against default designs for specific programs outside the hotel (as many of the ones commonly used for the reuse of towels). If you want to promote the response to these programs, indicate that social causes are to be used the funds derived from savings made, or if management wants to retain those savings, to communicate to what extent the personal contribution to help reach the appropriate achievement environment, although the hotel's image perceived to be worse in the first case (Shang, Basil, & Wymer, 2010).

Anyway, most authors agree that, apart from discussions on the effectiveness, performance and other issues of environmental and social policy of the hotels, the customer has much power over the future of the hospitality industry either by boycotting products of little responsibility, or by the generation of competitive advantages in marketing for companies committed to the environment (Miller, 2001).

3.3 Hotel Environmental Management in Spain

Hitherto, a general overview of the state of the art in environmental management in the global hospitality industry has been depicted. Focusing on Spain, it can be affirmed that, in many cases, Spanish hotels are governed by the same motivations and policies that hotels located in other countries. However, there are some differences that make Spain a different scenario with some particularities in terms of environmental management.

Comparing studies in the Spanish hotel sector with those from other regions, it can be concluded that the motivations that drive Spanish hoteliers are essentially the same as those of their foreign counterparts, coinciding with the prevailing confusion and lack of awareness, which explains the low level of implementation of environmental management practices (Ayuso, 2006). However, there is a positive reinforcement in the managerial staff of those hotels that have focused attention on environmental performance, over its management, as it does not guarantee higher levels of profitability, whereas the correlation between environmental performance and higher levels of profitability is positive (García & Armas, 2007).

The main difference to other local research, lies in the fact that Spanish managers prioritize those measures that result in competitive advantages, regardless of whether they have a direct or indirect effect on the outcome, although it is difficult to determine the influence of managers' awareness (Ayuso, 2006), compared to the results of other studies that indicated cost savings as the primary incentive (Stipanuk, 2001). However, Cespedes et al (2003) argue that the economic legitimacy of environmental measures have a greater say in decision-making of Spanish hoteliers that social legitimacy or the pressure of lobbies. The cognitive processes that Ayuso exposes in her work (2006) could explain this apparent discrepancy.

Regarding the operational practices employed, there are no differences between regions, excepting the different levels of development in the hotel industry of each country. However, it was observed that the number of Spanish hotel companies that employ codes of conduct as regulators of its environmental performance is very small compared to other analysed regions (Ayuso 2006).

4 Research Methodology

Case study methodology has been chosen to illustrate the current model of environmental management in hospitality industry. To meet this objective, the Hotel Playa Victoria, part of the Palafox chain, was selected for its high level of awareness, both environmental and social. The Palafox hotel chain is based in Saragossa (Spain), and directly manages 6 hotels, 5 of them in Saragossa, and the selected one in Cadiz, the southernmost province in continental Europe. The hotel, apart from its linkage to a chain highly sensitized to the impact of its business, perfectly exemplifies the integration of environmental policies in the strategy of the hotel since its construction in 1994. It is a special case both for its location, being a city hotel alongside one of the most awarded urban beaches in Europe, and for its architectural design, which has matched modern materials, such as steel and glass, with local traditional materials such as oyster stone, in a building designed to minimize power consumption and the usage of natural resources, having been awarded with the Principe Felipe Prize for Business Excellence in the section of energy saving and efficiency in 1995. This hotel is, therefore, a reference in its region in terms of environmental management and quality, and one out of four hotels with 4 stars certification currently serving in the city of Cadiz.

Since the objective of this research is to obtain an overview of the current situation of environmental management in hotels in the province of Cadiz from the perspective of corporate social responsibility, the case study methodology is the appropriate tool for in-depth understanding of the object of study, since the results of the implementation and management of such policies do not have a clear and unique set of consequences (Baxter & Jack, 2008). Moreover, although the selection of Hotel Playa Victoria does not meet the criteria of representation, this case presents a hotel with an environmental commitment and management quality above average, becoming a model for benchmarking for other establishments in the region. To acquire the necessary information, various sources have been used, such as in-depth interview with the Quality Manager of the hotel, the information available on the website of the hotel chain which it belongs to, the internal magazine concerning the environment, customer satisfaction questionnaires, and screen captures from the hotel ERP. Moreover, there was available information on environmental policy of the other hotels in town competing in the same category (Fernández Alles & Cuadrado Marqués, 2010), supporting the selection of this hotel. However, despite of applying the most suitable research strategy to achieve the goals of this work, as the case study method is the tool of choice, statistical generalisation cannot be applied to the obtained results. Nevertheless, analytical generalisation can be applied to any hotel meeting the standards of the Hotel Playa Victoria, or to any company with a strong commitment to the environment.

5 Findings

Although six primary stakeholder groups can be differentiated (Hult et al, 2011), the results of this research have been sorted into three main stakeholder groups. Third parties, involving suppliers, community and government, internal stakeholders standing for the employees, and customers.

5.1 Relationships with Third Parties

Hotel management contributes to environmental performance through establishing relationships with various external agents, either with a contractual relationship or without it. This contribution, regardless of raising management awareness on social responsibility, is encouraged from the top managerial level of the chain, through the policy statement of integrated quality and environmental management in Palafox chain, which calls for responsible individuals to involve and engage suppliers and customers in environmental management. They understand this commitment as a shared responsibility, in which establishments must meet afunction of sensitization. So, first of all, the suppliers of food, hotel housekeeping, as well as outsourced laundry, are shortlisted after assessing their fitness to ISO 9001 standards for continuous quality management, which includes a series of basic measures relating to environmental performance of certified companies or organizations, or to the ISO 14001 on environmental management that specifically addresses the problematic analysed in this research. Secondly, as is clear from the analysis of ERP and control system of the hotel, the collaboration with the Consejería de Medio Ambiente de la Junta de Andalucía (public agency for environmental protection in Andalusia, Spain) helps ensure proper treatment of all hazardous waste, by working with public certified waste processing companies, minimizing the impact on the environment. In addition, the hotel director intensively collaborates with the Department of Beaches of Cadiz Town Hall, in protective measures such as raising awareness and divulgation of environmental heritage.

5.2 Internal Relationships

Internal communication of CSR policies can be done in several ways, trying to involve the staff in the development and improving stages of these policies. About this issue, several actions must be highlighted on an individual basis:

Environmental and Quality Management Statement: Posted in a highly visible place for both customers and staff of the hotel, makes constant reference to the reduction of the hotel's impact on the environment, as well as the importance of these actions in relation to the sustainability of economic activity. In addition, explicitly incorporates the figure of the internal customer, at the same level of the external one, declaring the need for continuous training and coaching. Higher degree of financial performance can be obtained from jointly applying environmental and quality management, although size and resources can moderate this influence (Tarí et al, 2010).

Palafox's Annual Conference for Managers: The chain organises every year a conference in which every hotel manager in Palafox hotels can convey the key innovations and events to their colleagues. One of the most interesting issues at the conference is the exposition of the environmental measures that exceed the standards of the chain, so they can be benchmarked by other managers and, therefore, turned into standards.

Palafox's Environmental Magazine: Publication edited by the company to showcase the main contributions to environment preservation and related news, and to promote the local natural richness. This magazine is available for any guest of the hotel.

Operations Handbook: Designed to standardise all the processes, it covers environmental management. The environmental management chapter is aimed to all the hotel staff, with the goal of wide-spreading the policies regarding this issue along the organisation. Apart from the main part of chapter, there are environmental specific procedures and tasks, aimed only to those individuals involved in each specific procedure.

Involving the Staff in the Environmental Policy: Each department is responsible of writing down an environmental leitmotiv every few months, so they achieve two objectives:

- Keeping a high level of awareness over environmental issues among the staff.
- Reminding that every individual contributes to the overall environmental performance of the hotel.

Furthermore, the staff is responsible for organising the World Environment Day at the hotel. By offering several informative activities they raise awareness of both customers and partners. They also promote water savings by organising a contest of water consumption reduction ideas.

5.3 Relationships with Customers

The hotel managerial staff does not only communicate its environmental commitment as a sort of advertisement, but they ask customers for sensitization and cooperation. Cooperation shows in many ways, like the common linens and towels reuse programs, but some measures differentiate this hotel from the competitors:

- Permanent Environmental Stand: It is placed in front of the reception desk, offering leaflets and additional info about environmental practices at home, at the beach, and at the hotel, becoming the center of World Environment Day celebrations and green laces give-away.
- Recyclable ashtrays give-away: Hotel guest are offered conical ashtrays either at the environmental stand, reception desk or hammocks control, preventing them to pollute the adjacent beach with ashes, cigarettes, and other waste.

Hotel managerial staff tries also to know about the environmental performance in terms of customer satisfaction and of reservation decision-making by a closed-answer question in which the customer assesses the hotel environmental management, and another open-answer question to find hiring decision criteria. The main reservation decision criterion for Playa Victoria Hotel's customers is location, but the Quality Manager of the hotel pointed out that focusing on German customers only, a 5% of them decided to reserve a room in this hotel for its environmental policy or environmental certification.

6 Concluding Remarks and Future Directions

Centred on environmental management, the main objective of this research was to depict the way a salient hotel managerial staff manages its environmental performance and impact, analysing both its external and internal stakeholder relationships.

This case study is focused on a certain hotel facility for two main reasons: First, this establishment is a reference in its sector and region, with high quality standards, and second, it has a long track record in environmental management from the very moment of its architectural design process.

The Playa Victoria Hotel, at the sight of the present findings, can be called a socially responsible hotel, due to the fact that its managerial staff actively takes into account the influence of the different stakeholders linked to the hotel activity and its economical, social, or environmental performance.

The stakeholders show their demands over the hotel activities. The hotel staff answers to these demands, reflecting the social and environmental commitment of the company, and also informs the customers about this performance, combining stakeholder management and market orientation (Ferrell et al, 2010).

This hotel has got a sound positioning, both in customers and stakeholders minds, shifting its environmental management into a sustainable competitive advantage. The managerial staff, under a proactive and public relations approach, has improved the environmental performance of the hotel, leaving the company in a favourable competitive position. This situation wouldn't have been possible if the managers had adopted a "posthumous" approach (Jahdi & Acikdilli, 2009), just reacting to environmental pressures in order to limit damage and loses coming of the fact of not managing the environmental impact of the hotel. Furthermore, the customers perceive the voluntariness of the hotel policies and actions as an evidence of their commitment (Manaktola & Jauhari, 2007; Kang et al, 2012) reinforcing the green positioning of the hotel in customers' minds.

If the hotel policies are imposed as a legal standard, this hotel will be in a privileged position, because they won't have to face new investment to accomplish with the law requirements. Additionally, when the hotel is exceeding the legal standards about environmental protection, it is communicating a well differentiated image of the company, making a profit of such positioning when selling vacancies to market segments strongly influenced by environmental awareness. This is the case for the German customers segment in Playa Victoria Hotel, representing a relevant share of customers in peak season.

From a stakeholder management approach, it can be considered that the managerial staff of the hotel could improve, even more, their stakeholders' relations. This could be achieved by applying any systematic procedure, specifically aimed to data-mining stakeholders' demands, as they currently do regarding customers. Following the same approach, we can highlight the high level of information provided to each worker, with the goal of meeting the environmental objectives. By using this internal marketing process, each individual knows how to contribute to the company's goal, so they can fix personal objectives.

Finally, we wanted to make reference to environmental communication strategies aiming the customers. These communications try to achieve two goals, sensitizing the customer, and involving the customer in the hotel's environmental commitment, in order to gain their collaboration.

Organising sensitization conferences, the permanent environmental stand, and the disposable ashtrays give-away contribute to reinforce the credibility of the hotel's environmental policies and its sustainable positioning.

At the sight of results of literature review, and the relevance of sustainable and environmental management both for the lodging industry and for society, future research in this field should address issues regarding environmental impact measuring, and its tangible outcomes in terms of customer loyalty and brand differentiation, since hotel managers often reflect a lack of knowledge about environmental management benefits and its impacts on their hotels' bottom line, or how they can capitalize on their environmental commitment to strengthen the brand and to gain competitive advantages via differentiation as a green hotel.

References

Ayuso, S.: Adoption of Voluntary Environmental Tools for Sustainable Tourism: Analysing the Experience of Spanish Hotels. Corporate Social Responsibility and Environmental Management 13, 207–220 (2006)

Baxter, P., Jack, S.: Qualitative Case Study Methodology: Study Design and Implementation for Novice Researchers. The Qualitative Report 13(4), 544–559 (2008)

Becker-Olsen, K.L., Cudmore, B.A., Hill, R.P.: The impact of perceived corporate social responsibility on consumer behaviour. Journal of Business Research 59, 46–53 (2006)

Bohdanowicz, P.: "European Hoteliers' Environmental Attitudes: Greening the Business". Cornell Hotel and Restaurant Administration Quarterly 2, 188–204 (2005)

Bohdanowicz, P.: Environmental awareness and initiatives in the Swedish and Polish hotel industry – survey results. International Journal of Hospitality Management 4, 662–682 (2005b)

Bohdanowicz, P., Zientara, P.: Hotel companies' contribution to improving the quality of life of local communities and the well-being of their employees. Tourism and Hospitality Research 9(2), 147–158 (2009)

Bouquet, C., Crane, A., Deutsch, Y.: The Trouble with Being Average 3, 79–80 (2009)

Bowen, H.R.: Social responsibilities of the businessman. Harper & Row, New York (1953)

Carroll, A.B.: A three-dimensional conceptual model of corporate performance. Academy of Management Review 4(4), 497–505 (1979)

Carroll, A.B.: The pyramid of corporate social responsibility: Toward the moral management of organizational stakeholders. Business Horizons 34, 39–48 (1991)

Carroll, A.B.: Corporate Social Responsibility: evolution of a definitional construct. Business and Society 38(3), 268–291 (1999)

Cespedes, J., De Burgos, J., Alvarez, M.J.: Stakeholders' environmental influence. An empirical analysis in the Spanish hotel industry. Scandinavian Journal of Management 19, 333–358 (2003)

Comission of the European Communities (2001): Libro Verde: Fomentar un marco europeo para la responsabilidad social de las empresas(2001) `http://eur-lex.europa.eu/LexUriServ/LexUriServ.do?uri=COM:2001:0366:FIN:ES:PDF` (browsed in July 2011)

Dinan, C., Sargeant, A.: Social Marketing and Sustainable Tourism – is there a Match? International Journal of Tourism Research 2(1), 1–14 (2000)

Fernández Alles, M.T., Cuadrado Marqués, R.: La responsabilidad social corporativa en el sector hotelero: estudio empírico en la ciudad de Cádiz. In: Proceedings of the XX Jornadas Luso Espanholas de Gestão Científica, Setúbal, Portugal (2010)

Ferrell, O.C., González-Padrón, T.L., Hult, T.M., Maignan, I.: From Market Orientation to Stakeholder Orientation. Journal of Public Policy & Marketing 29(1), 93–96 (2010)

Garay, L., Font, X.: Doing good to do well? Corporate social responsibility reasons, practices and impacts in small and medium accommodation enterprises. International Journal of Hospitality Management (article in Press, 2012)

García-Rodríguez, F.J., Armas, Y.D.: Relation between social- environmental responsibility and performance in hotel firms. International Journal of Hospitality Management (26) 824–839 (2007)

Guadamillas, F., Doñate, M.J.: Responsabilidad social corporativa, conocimiento e innovación: hacia un nuevo modelo de dirección de empresas. Revista Europea de Dirección y Economía de la Empresa 17(3), 11–26 (2008)

Gurney, P.M., Humphreys, M.: Consuming Responsibility: The Search for Value at Laskarina Holidays. Journal of Business Ethics (64), 83–100 (2006)

Hart, W.: The three Rs. Cornell Hotel and Restaurant Administration Quarterly 34(5), 18 (1993)

Hotel Playa Victoria (2011), `http://www.palafoxhoteles.com/hotel.aspx?id=24` (browsed in August 2011)

Hult, T., Mena, J., Ferrell, O.C., Ferrell, L.: Stakeholder marketing: a definition and conceptual framework. Academy of Marketing Science Review 1(1), 44–65 (2011)

Instituto para la Calidad Turística Española (2011), http://www.icte.es/ (August 2011)

International Tourism Partnership, El camino verde: estándares mínimos de un hotel sostenible (2010), http://www.tourismpartnership.org/downloads/El%20camino%20verde.pdf (browsed in August 2011)

Jahdi, K.S., Acikdilli, G.: Marketing Communications and Corporate Social Responsibility (CSR): Marriage of Convenience or Shotgun Wedding? Journal of Business Ethics (88), 103–113 (2009)

Kang, K.H., Stein, L., Heo, C.Y., Lee, S.: Consumers willingness to pay for green initiatives of the hotel industry. International Journal of Hospitality Management (in press, 2012)

Kotler, P., Lee, N.: Corporate Social Responsibility: Doing the Most Good for Your Company and Your Cause. John Wiley & Sons, Inc., United States (2005)

Lee, S., Park, S.-Y.: Do socially responsable activities help hotels and casinos achieve their financial goals? International Journal of Hospitality Management (28), 105–112 (2009)

Manaktola, K., Jauhari, V.: Exploring consumer attitude and behaviour towards green practices in the lodging industry in India. International Journal of Contemporary Hospitality Management 19(5), 364–377 (2007)

Mensah, I.: Environmental management practices among hotels in the greater Accra region. International Journal of Hospitality Management (25), 414–431 (2006)

Miller, G.: Corporate responsibility in the UK tourism industry. Tourism Management (22), 589–598 (2001)

Nicolau, J.L.: Corporate Social Responsibility: Worth-Creating Activities. Annals of Tourism Research 35(4), 990–1006 (2008)

Porter, M.E., Kramer, M.R.: The Link Between Competitive Advantage and Corporate Social Responsibility. Harvard Business Review 84(12), 78–92 (2006)

Ryan, C.: Equity management, power sharing and sustainability-issues of the new tourism. Tourism Management 23(1), 17–26 (2002)

Shah, K.U.: Strategic organizational drivers of corporate environmental responsibility in the Caribbean hotel industry. Policy Sci. (44), 321–344 (2011)

Shang, J., Basil, D.Z., Wymer, W.: Using social marketing to enhance hotel reuse programs. Journal of Business Research (63), 166–172 (2010)

Stipanuk, D.M.: Energy Management in 2001 and Beyond: Operational options that reduce use and cost. Cornell Hotel and Restaurant Administration Quarterly 42(3), 57–70 (2001)

Tarí, J.J., Claver-Cortés, E., Pereira-Moliner, J., Molina-Azorín, J.F.: Levels of quality and environmental management in the hotel industry: Their joint influence on firm performance. International Journal of Hospitality Management (29), 500–510 (2010)

Yoo, M., Lee, S., Bai, B.: Hospitality marketing research from 2000 to 2009. Topics, methods, and trends. International Journal of Contemporary Hospitality Management 23(4), 517–532 (2011)

Hsieh, Y.C.: Hotel companies' environmental policies and practices: a content analysis of their web pages. International Journal of Contemporary Hospitality Management 24(1), 97–121 (2011)

World Travel and "Tourism Council Travel & Tourism Economic Impact: Spain" (2011), http://www.wttc.org/site_media/uploads/downloads/spain.pdf (browsed in November 2011)

How to Identify Regional Specialization Measurement of Clusters in Tourism Industry?

Cristina Estevão[1] and João J.M. Ferreira[2]

[1] School of Management of Idanha-a-Nova,
Polytechnic Institute of Castelo Branco and NECE – Research Unit of Business Science,
UBI, Largo do Município, 6060-163, Idanha-a-Nova, Portugal
kristina.estevao@hotmail.com
[2] Management and Economics Department,
University of Beira Interior (UBI) and NECE – Research of Business Science,
UBI, Pólo IV – Edifício Ernesto Cruz, 6200-209 Covilhã, Portugal
jjmf@ubi.pt

Abstract. Clusters theory and their application have been subject to several theoretical and practical approaches within the framework of regional economic development throughout the last decade and worldwide. Many studies have gone about identifying clusters in diverse contexts, nevertheless fundamentally based upon qualitative or quantitative (rarely both) approaches tending to overlook fundamental methodological aspects inherent to the operational functioning of clusters. Furthermore, very few have focused on the tourism sector with recourse to regional specialisation measurements. This research aims to contribute towards narrowing the empirical research methodological shortcomings through the proposal and application of a quantitative methodology capable of robustly identifying and locating tourism industry clusters in Portugal. As results we have identified a specific number of clusters distributed across ten activity tourism sectors related to accommodation, restaurants, entertainment and events with recourse to regional specialisation measurements.

Keywords: clusters, concentration, locational coefficient, cluster index, tourism regions.

1 Introduction

The concept of clusters is not new. Interest first emerged at the beginning of the 19th century through pioneering studies and especially the works by Ricardo (1817), Von Thünen (1826) and Launhardt (1882). The question of specialised industrial location took on particular significant with the seminal study by Marshall (1890). Many subsequent studies have traced their roots to the writings of Marshall (Hoover, 1937; 1948; Becattini, 1979; Brusco, 1982; Dore, 1983; Piore and Sabel, 1984; Solinas, 1988; and Arthur, 1994).

A.M. Gil-Lafuente et al. (Eds.): Soft Comput. in Manag. and Bus. Econ., STUDFUZZ 287, pp. 347–360.
springerlink.com

The cluster concept is itself problematic and their definition ambiguous and as such has gradually taken on a series of rather different meanings (Malmberg, 2003; Malmberg and Power, 2006; Waxell and Malmberg, 2007). The ambiguity primarily derives from the cluster definition oscillating between industrial and geographic definitions (Waxell and Malmberg, 2007). Clusters have been defined by some as a set of companies located within close geographic proximity of each other (Swann and Prevezer, 1996; Rosenfeld, 1997; Porter, 1998; Cooke and Morgan, 1998; Crouch and Farrell, 2001; Cooke, 2001), or located in a specific area (Swann and Prevezer, 1996; Cortright, 2006), or producing a similar product or service (Rosenfeld, 1997), by others, as a groups of interrelated industries (Simmie and Sennet, 1999; Porter, 2000, 2003), without overlooking the importance of industries (Porter, 1998), the synergies established between companies located in clusters (Roselfed, 1997) and possible mutual competition (Feser, 1998; Bergamn and Feser, 1999; Ketels and Memedovic, 2008), as well as catalysers for competitiveness policies (Shakya, 2009).

And while some maintain that clusters are a synonym for competitiveness (Porter, 1990, 2002; Rocha, 2004; Shakya, 2009), innovation (Baptista and Swann, 1998; Nordin, 2003; Hospers et al., 2009; Business Europe, 2009) economic performance (Porter, 2003; Folta et al., 2006; Pe'er & Vertinsky, 2006; Gilbert et al., 2007; Porter et al., 2007; Wennberg and Lindqvist, 2008; COM, 2008; Gugler and Keller, 2009; Delgado et al., 2011) and entrepreneurship (Glaeser et al., 2009; Delgado et al., 2010), other hold that clusters display various drawbacks especially in terms of productive over-specialisation, technological apathy, institutional and industrial lock-ins, environmental influences, local congestion, pressures resulting in rising labour and property costs (Martin and Sunley, 2002) and as well as the fact that new companies are adversely impacted when locating in an economic cluster (Wennberg and Lindqvist, 2008).

Independently of the meaning attributed to clusters, in fact, there has been a vast range of studies on their identification (Helmstädter, 1996; Glassmann e Voelzkow, 2001; Brenner, 2003), but very few have focused on the tourism sector with recourse to regional specialisation measurements. Furthermore, many of those studies identifying tourism clusters are based upon simplistic observations or conceptions overlooking factors fundamental to the existence of clusters.

Hence, taking into consideration the two research failings identified – the lack of quantitative and objective methodologies for cluster identification and the overall lack of studies on tourism sector clusters in the majority of countries and in Portugal in particular – this research hereby strives to contribute with a methodology procedures able to identify tourism clusters in Portugal. The different tourism activities and regions in Portugal serve as the framework for the application of this study.

We shall firstly provide a review of state of the art in cluster research before moving onto describe our methodology alongside a description of the data and variables incorporated into the study. We then proceed to discuss the results and close with conclusions.

2 Literature Review

The interest in agglomeration and geographic distribution of economic activities dates back to the 19th century and the early part of the 20th century (Ricardo, 1817; Von Thünnen, 1826; Launhardt, 1882; Marshall, 1890, Weber, 1909). In the last two decades, research has led to important developments in the field of industrial location and distribution driving the emergence of new concepts such as new industrial spaces (Scott, 1988), innovative structures (Aydalot, 1986; Maillat, 1991), Neo-Marshallian nodes (Amin and Thrift, 1994), intelligent regions (Asheim, 1995), local production systems (Crouch et al., 2001), not to mention innovative systems (Lundvall, 1994; Cooke et al., 1997; Cooke and Heidenreich, 1998). This growing interest in agglomerations and the geographic distribution of economic activities led to the founding of a new geographic economics, which has also proposed new models of location (Krugman, 1991; Krugman and Fujita, 2004).

According to Martin and Sunley (2002), there remains great controversy around the concept of clusters. They maintain that while it is simple to identify clusters in space, however, this is no longer the case as regards their actual definition proving far more ambiguous and obscure and failing to gain any form of unanimity and sometimes resulting in a more anecdotal and less academically robust identification of clusters. Meanwhile, Maskell and Kebir (2005) hold that the lack of rigour in cluster definition leads to the concept getting applied to a broad spectrum of situations and by a broad range of actors, ranging from academics, consultants and politicians.

Porter (1990) stated that clusters are the geographic concentrations of companies and institutions involved in a specific sector of activity and where interrelationships reinforce competitive advantages. From the perspective of Doeringer and Terkla (1995), a cluster consists of geographic proximity between its component members that generate an agglomeration of economies of scales and scopes through specialisation and the internal division of labour. Meanwhile Swann and Prevezer (1996) define clusters more simply as groups of companies within a particular industry in a given geographic space. Rosenfeld (1997) points to clusters being used to represent concentrations of companies so that they may thus produce synergies out of their geographic proximity and their interdependence. Feser (1998) highlights how economic clusters do not refer only to industries and institutions but rather to highly competitive industries and institutions and how this competitiveness should be reflected in the ongoing relationships.

Cooke (2001) portrayed the definition of clusters as based upon three fundamental pillars: the first is geography, that is, the clusters are driven by their proximity and frequently concentrated within a region in a major country, and sometimes in a city. The second pillar is the creation of value, hence, clusters include companies from different sectors that are mutually related with others in terms of the production of goods and services valued by clients. The third pillar is the business environment, with clusters impacted by others with specific conditions in terms of the business environment resulting from each of their actions as well as cooperation between companies, government agencies, universities and other institutions participating in national and regional innovation systems.

According to Crouch and Farrell (2001), the clusters are a trend where companies in the same sector of activity tend to locate in close proximity even while they do not hold any particularly important presence in the surrounding area. The clusters are furthermore considered synonymous with competitiveness given they make a positive contribution towards innovation processes in facilitating relationships with other institutions and enabling a better understanding of consumers, concentrating the knowledge and information necessary to technological development (Porter, 2002).

The clusters are, additionally, a natural manifestation of specialist knowledge, competences, infrastructures and support to industries in raising productivity as the main determinant in maintaining high levels of prosperity in a location. A combination of relationships with suppliers, shared labour markets, rivalries, spread of knowledge and learning effects all shape the economic environment faced by companies in clusters (Ketels and Memedovic, 2008).

Shakya (2009) points to another dimension to clusters as interconnected systems involving both the public and private sectors and adds that approaches to clusters should be deployed beyond the scope of economies of scale and the common or garden analysis by sector with the objective of encouraging the involvement of a diversified group of interested parties through which they may develop an underlying shared comprehension as to public policy questions and be able to act in conjunction with them. The development of such shared platforms with strong stakeholder participation from both the public and private sectors is very often crucial and represent the departure point for a broader reaching process of economic reform in developing countries. The cluster initiatives may thus serve as catalysers for competitive policies.

According to Porter (1998), clusters cover a range of industries associated with other entities playing key roles in competition. Porter proposes that competitiveness in modern economies depends on productivity and not on access to inputs or companies on an individual scale. This productivity stems from the way companies compete and not how they behave individually. Clusters impact on this competition across three aspects: (i) boosting the productivity of companies in a specific region, (ii) pointing the companies in the direction of the innovation that will bring productivity gains in the future, and (iii) fostering the founding of new businesses, which in turn expand and reinforce the clusters in themselves.

Porter (1990) proposes an instrument for the analysis of national competitive advantages - Porter diamond, which features the existence of interconnected companies and activities, hence, clusters. For this author, the cluster formation process rises in intensity the greater the extent of geographic concentration of companies involved in this process. Furthermore, for countries hosting these processes (cluster formation), the more rapidly they occur, the more rapidly they attain success.

3 Methodology

No specific methodology for the identification and mapping of clusters has gained consensus whether in terms of the core variables to be measured or in terms of the procedures by which the geographic limits of clusters should be determined

(Martin and Sunley, 2002). Nevertheless, Sternberg and Litzenberger (2004) argue that among the various results and methodologies put forward for this task, there are two broad differences in the approaches: top-down and bottom-up. Making recourse to a top-down approach involves first researching up to what point industry is spatially concentrated prior to attempting to localise regional clusters. The spatial distribution of an industry is not uniform and therefore a certain spatial level of company concentration is a precondition for the formation of clusters (Martin and Sunley, 2002, Sternberg and Litzenberg, 2004).

This concentration is very often described through measurements that detail the extent of the spatial division of labour or, more simply, industrial specialisation (Amiti, 1998, 1997; Helmstädter, 1996; Kim, 1995; Krugman, 1991). The most common tool to this end is the location coefficient (Kim, 1995), initially defined by Hoover (1936), that captures the degree of specialisation of a region in a particular industry. It is used to establish the locational Gini coefficient thereby measuring the distribution of an industry throughout the sub-regions of the area under analysis (Kim, 1995).

The location coefficient and therefore the locational Gini measure regional specialisation as deviation of the distribution of the total industrial employment, considering overall employment as the referring variable. However, should we consider the cluster definitions presented – in which clusters are perceived as companies in close proximity to each other – the spatial dimension is neglected by measurements of specialisation (Helmstädter, 1996; Glassmann and Voelzkow, 2001; Brenner, 2003). There is no value providing any indication as to the scale of the region analysed or the magnitude of the proximity between companies. Nevertheless, Sternberg and Litzenberger (2004) maintain that the reference value should be considered by area and not by employment or by the inhabitants of a region. Where there is a reference value other than that of regional area, equal spatial distribution of industry cannot be theoretically accepted (Roos, 2002).

The Gini coefficient (Gstd) is an appropriate instrument for measuring concentration (Devereux, et al., 1999), but does however have to be weighted according to the region's size. Hence, the Gstd weighted with the area of the region (Gα) is proposed as a measurement of spatial industrial concentration.

Furthermore, in order to calculate regional clusters, a Cluster Index (CI) was calculated as suggested by Sternberg and Litzenberger (2004). It is defined as the product of the relative industry density, the relative industrial stock, and the relative size of the establishment. These authors state that its flexibility, simplicity of calculation and the availability of the data necessary combine to ensure its ease of use and operational deployment. This CI may take on values ranging from zero to infinite but which, at the minimum, should be greater than one (average value) in order to potentially identify a cluster. To ensure there is at least one cluster in a specific region, the CI should return a value of over 4.00 (Sternberg and Litzenberger, 2004). This value is attained where two of the three CI components (relative industrial density, relative industrial stock, and relative size of the establishment) are twice as high for the sub-region as the average of the total region (for the third component being the average, hence, one).

However, it is recognised that this value may return an arbitrary result. According to Keeble and Nachum (2002), it is possible that a cluster covers only part of a region or covers a series of regions. The cluster area therefore needs not only to be sufficiently large to gain critical mass but also sufficiently small to enable a sense of community to exist.

The CAE economic activity codes in accordance with the activities characteristic of the WOT et al. (2001) satellite account as detailed a five digit level of disaggregation, employment and the number of establishments engaged in each activity as listed in the Directorate-General of Studies, Statistics and Planning of the Portuguese Ministry of Labour and Social Solidarity (DGEEP-MTSS, 'Direcção-Geral de Estudos, Estatística e Planeamento do Ministério do Trabalho e Solidariedade Social') data base. Additionally, we incorporated the area of each counties studied in conjunction with its resident population with these figures supplied by the Portuguese National Statistics Institute (INE, "Instituto Nacional de Estatistica"). All data refers to the year of 2009.

4 Identification and Location of Tourism Clusters

The first phase in empirical application was the calculation of the locational Gini coefficient (Kim, 1995) so as to determine the peaks in tourism sector activity concentrated across the 308 counties making up the regional areas and tourism development poles on both the Portuguese mainland and its two archipelagos before subsequently constructing the CI for identifying regional clusters. The next phase in identifying and mapping the clusters saw the inclusion of only those regions returning a CI in excess of 4. We might eventually set limits for the value of each respective activity (Brenner, 2003), however, we sought to identify clusters with comparable characteristics in relation to the rest of the region. Thus, the ten activities containing the greatest number of clusters were selected and, in order to avoid an over-concentration effect, due to the lower number of companies in relation to the number of regions, industrial sectors with less than 308 establishment were excluded from our analysis.

For the ten tourism sector activities subject to analysis with a CI of greater than 4.00, a total of 505 clusters were identified and distributed as follows: 55202 – Rural Tourism (89 clusters), 55112 – Pensions Standard Accommodation with Restaurants (64 clusters), 55124 – Other Hotel Establishments without Restaurants (47 clusters), 56305 – Clubs and Pubs with Event Facilities (52 clusters), 56106 – Residential Food Delivery Services (49 clusters), 56304 – Other Clubs and Pubs without Stage Facilities (45 clusters), 56104 – Traditional Restaurants (44 clusters), 56302 – Bars (44 clusters), 56102 - Restaurants with Counter Service (36 clusters) and 93294 – Other Non Fixed Pleasure and Recreational Activities (41 clusters). The majority of activities fall within the scope of the hotel and accommodation sector, followed by catering and finally by entertainment and nightlife.

As an activity, rural tourism particularly stands out and encountered in the majority of counties deemed rural according to the Kayser criterion (1990), hence counties home to fewer than 5,000 inhabitants and applicable to around 29% of

Portuguese counties, followed by pensions standard accommodation with restaurants and pubs and clubs with event facilities and recorded in 21% and 17% of the total counties, respectively.

So as to measure the correlation between the concentration of activities and the number of clusters, we applied the Pearson (r) correlation coefficient. We found an insignificant level of correlation (r=0.033) similar to the study undertaken by Sternberg and Litzenberger (2004). We may furthermore state that the average of $G\alpha$ is higher (0.672) than the average weighted by employment (0.298), hence demonstrating that small and medium sized companies are geographically more concentrated than large companies and corporations. This result stems from the numbers of micro, small and medium sized companies in general in Portugal and in the tourism sector in particular where they make up 99% of all companies and account for some 75% of employment and thus generating a significant influence on the concentration of activities.

Highlights the fact that around 36% of Portugal, corresponding to a total of 111 counties does not include any clusters. With one, two and three clusters, we have 86, 48 and 18 counties, respectively. Only one counties hosts all the selected activities.

We furthermore analysed the correlation between the number of clusters and the population, size and population density of counties. We found that there was a modest degree of correlation (r=0.545) between the number of clusters and the population, hence, the higher the level of population, the higher the number of clusters. As regards size and cluster numbers, there is moderately negative correlation (r=-0.305), implying that the larger the counties size, the lower the number of clusters. In relation to population density, deployed as the indicator combining population and size, there is strong correlation between cluster numbers and population density (r=0.665) and hence, the greater the population density, the greater the number of clusters. We carried out similar analysis contrasting cluster numbers with the number employed (0.509), the number of companies (0.609) and the number of employees per company – the average number of employees (0.65). The cluster numbers correlate moderately with employee numbers and strongly with the number of companies and average number of employees per company. All of these correlations were positive and therefore the greater the number of jobs, companies and works and employees per company, the greater the number of clusters.

Observing figure 1, we would highlight that the largest single number of tourism clusters are located in the city of Oporto, represented by a total of ten activities with this counties covering one of the smallest areas (41.30 km2) in the country while representing one of the most densely populated counties (210,558 inhabitants), contributing around 4.70 % of employment in these activities on a nationwide basis.

Displaying nine clusters, we have the counties of Almada, Funchal, Lisboa and Vila Real de Santo António, with eight clusters are the counties of Albufeira, Cascais, Lagoa, Nazaré, Portimão, S. João da Madeira and Vila do Bispo, while Amadora, Machico, Matosinhos, Odivelas, Oeiras, Ponta Delgada and Santa Cruz attain seven clusters, while the counties of Braga, Coimbra, Espinho, Faro, and

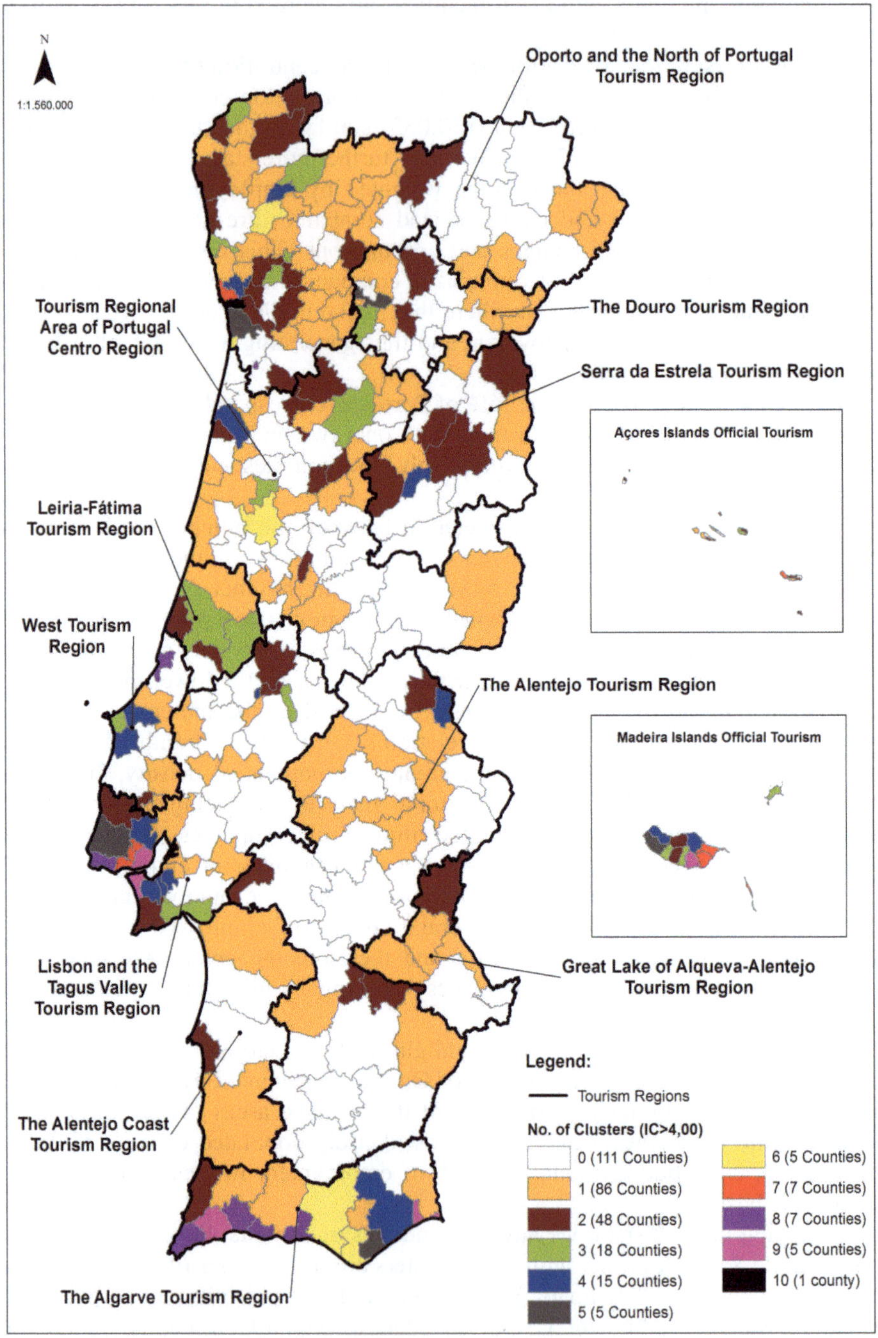

Fig. 1 Location of Clusters Identified in Portugal

Loulé manage six clusters and Calheta (the Azores), Olhão, Peso da Régua, Sintra and Vila Nova de Gaia each host five clusters. Returning zero, one, two, three and four clusters are around 90% of the counties of Portugal and those endowed with larger geographic areas.

Applying the analysis to the level of tourism region, we find that the region displaying the greatest number of clusters in relation to its total size is the Algarve with a total of 77 clusters, with a particular emphasis on coastal areas. The results also highlight the coastal zone of the Lisbon and the Tagus Valley region due to its diversity of clusters, especially Lisbon and Almada (both with 9), Cascais (8), Amadora, Odivelas and Oeiras (7) and Sintra (5), forming the only tourism cluster agglomeration. These results may be interpreted and justified by the tourism image and reputation that these regions hold internationally. Regarding the tourism region of Oporto and the North of Portugal, despite containing the greatest number of clusters explained by the fact the region holds the largest number of counties in the country, the results throw the coastal counties of Oporto (10), Matosinhos, (7), Espinho and Braga (6) and Vila Nova de Gaia (5) into the spotlight. In the Douro tourism region, the only highlight is Peso da Régua, counties renowned for its production of the fortified wine, port, and the impressive beauty of its vineyards, classified by UNESCO as World Heritage in 2001.

In the Centro region, the majority of counties do not contain any clusters and with a large number of counties home to between one and three clusters. Leading in this tourism region is Coimbra (6), a county with centuries of academic traditions and Aveiro (4) known as the Venice of Portugal, given the trips taken out onto its lagoon on the traditional local vessel – moliceiros. In the Serra da Estrela tourism region, only the county of Manteigas (4) turns in a significant result due to its location in the heart of the Serra da Estrela tourism destination with the remaining counties hosting only very low numbers of clusters. This must in no small part be due to being a region lacking in business investment in the tourism sector despite otherwise being a region rich in natural and historical resources. In the West tourism region, Nazaré county stands out from other counties given it displays eight clusters, once again, a coastally located county. In relation to the Alentejo tourism region, Marvão stands out with four clusters and a county capitalising not only on its border location with Spain but also with deep historical and natural roots and currently a candidate to World Heritage status.

The number of clusters in the remaining tourism regions in mainland Portugal does not have a great deal of weight in relation to the others and never exceeding a total of three clusters. In relation to Madeira, with a total number of 44 clusters, we highlight the counties of Funchal (9), Machico and Santa Cruz (7). The Azores contain 37 clusters, with a highlight being the county of Ponta Delgada (7), the archipelago's capital.

5 Conclusions

Following the results of the research, we find that the most concentrated activities do not necessarily form the most clusters and the larger the size of the region studied, the lesser are their numbers. We may also conclude that a greater population density generates a greater number of clusters and the larger the number of jobs, companies or employees by company, the greater the number of clusters. Given the relationships identified in ascertaining the number of clusters, it is correspondingly important not to overlook these variables when determining the presence of clusters. Through recourse to the cluster index, it proved possible to portray the potential clusters already existing in the national economy in the tourism sector. We thus confirm that clusters, beyond being susceptible to empirical identification, also vary considerably according to the activity, location and dimension (Martin and Sunley, 2002).

We may furthermore argue that there are shared synergies between counties returning the most clusters as they gain mutual productive advantages based upon their proximity in terms of the overall tourism regions. We thus verify, for example, that the fact Oporto county represents the local government region with the single largest number of tourism clusters, as well as the fact it is a competitive region with high economic growth, reflects one core assumption to the concept of cluster: promoting greater regional economic distinction and differentiation. The same happens with the Lisbon region and its respective counties as well as the Algarve tourism region. Indeed, the Algarve is a region that economically depends, and to an almost exclusive extent, on tourism with activities related to this sector enabling continuity in its economic growth in conjunction with its competitiveness as a regional tourism pole.

One of the major problems facing Portugal is the underdevelopment of inland regions and this is reflected in the low numbers of clusters found away from coastal centres. Clearly, the entities responsible should ensure the terms and infrastructures necessary to fostering a generally positive environment for tourism businesses are in place. Rendering support to structures and efforts to nurture clusters may add up to the difference between the relative success and failure of a tourism region and similarly much may be learned from the clusters developed in other national tourism regions.

Although this study provides important insights into the identification, operational approach and mapping of clusters, the same also suffers certain limitations. These limitations, in turn, do open up the door to future research.

Given the results attained, it would be of relevance to apply this cluster identification methodology to a geographic area of greater scope, such as Europe for example and compare these results (clusters) with those identified by the European Cluster Observatory. Another proposal for future research would be to empirically relate the clusters identified with economic performance and local/regional entrepreneurship.

References

Amin, A., Thrift, N.: Globalization, Institutions and Regional Development in Europe. Oxford University Press, Oxford (1994)

Amiti, M.: Specialisation Patterns in Europe. CEP-London School of Economics, London (1997)

Amiti, M.: New Trade Theories and Industrial Location in the EU: a Survey of Evidence. Oxford Review of Economic Policy 14(2), 45–53 (1998)

Arthur, W.: Increasing Returns and Path Dependence in the Economy. University of Michigan Press, Ann Arbor (1994)

Asheim, P.: Industrial Districts as Learning Regions: A Condition for Prosperity? STEP Report, vol. (3). STEP Group, Oslo (1995)

Aydalot, P.: Milieux Innovateurs en Europe. GREMI, Paris (1986)

Baptista, R., Swann, P.: Do Firms in Clusters Innovate More? Research Policy 27, 525–540 (1998)

Becattini, G.: Dal 'Settore' Industriale al 'Distretto' Industriale. Alcune Considerazioni Sull'unitá di Indagine dell'economia Industriale. In: Goodman, E., Bamford, J. (eds.) Rivista di Economia e Politica Industriale, 1st edn., pp. 7–21. Routledge, Small Firms and Industrial Districts in Italy London (1979)

Bergamn, E., Feser, E.: Industrial and Regional Clusters: Concepts and Comparative Applications, University of West Virginia (1999), `http://www.rri.wvu.edu/WebBook/Bergman-Feser/contents.htm` (acced em January 22, 2009)

Brenner, T.: An Identification of Local Industrial Clusters in Germany, Papers on Economics and Evolution, vol. 04. Max Planck Institute for Research into Economic Systems, Evolutionary Economics Group, Jena (2003)

Brusco, S.: The Emilian Model, Productive Decentralization and Social Integration. Cambridge Journal of Economics 6(1), 167–184 (1982)

Bussiness Europe, Unite and Innovate! European Clusters for Recovery (acedido em October, 2009), `http://www.clusterobservatory.eu/library/100068.pdf` (April 14, 2010)

COM, "The Concept of Clusters and Cluster Policies and Their Role for Competitiveness and Innovation: Main Statistical Results and Lessons Learned", Europe INNOVA/PRO INNO Europe Paper No.9 (2008)

Cooke, P.: Regional Innovation Systems, Clusters and the Knowledge Economy. Industrial and Corporate Change 10(4), 945–974 (2001)

Cooke, P., Uranga, M., Etxebarria, G.: Regional Innovation Systems: Institutional and Organizational Dimensions. Research Policy 26, 475–491 (1997)

Cooke, P., Heidenreich, M.: Regional Innovation Sistems – The Role of Governaces in Globalized World, 1st edn. UCL Press, London (1998)

Cooke, P., Morgan, K.: The Associational Economy: Firms, Regions and Innovation. OUP, Oxford (1998)

Cortright, J.: Making Sense Of Clusters: Regional Competitiveness and Economic Development, a discussion paper The Brookings Institution Metropolitan Policy Program Summary of Publications (2006)

Crouch, C., Le Galés, P., Trogilia, C., Voelzkou, H.: Local Production System in Europe: Rise or Demise? Oxford University Press, Oxford (2001)

Crouch, C., Farrell, H.: Great Britain: Falling Through the Holes in the Network Concept. In: Crouch, C., Le Galés, P., Trogilia, C., Voelzkou, H. (eds.) Local Production System in Europe: Rise or Demise?, pp. 161–211. Oxford University Press, Oxford (2001)

Delgado, M., Porter, M., Stern, S.: Clusters, Convergence, and Economic Performance, US Cluster Mapping Project Dataset (2011)

Delgado, M., Porter, M., Stern, S.: Clusters and Entrepreneurship. Journal of Economic Geography, 1–24 (May 2010)

Devereux, M., Griffith, R., Simpson, H.: The Geographic Distribution of Production Activity in the UK, IFS Working Paper 26/99. Institute for Fiscal Studies, London (1999)
Doeringer, P., Terkla, D.: Business Strategy and Cross-industry Clusters. Economic Development Quarterly 9, 225–237 (1995)
Dore, R.: Goodwill and the Spirit of Market Capitalism. British Journal of Sociology 34(24), 459–482 (1983)
Feser, E.: Old and New Theories of Industry Clusters. In: Steiner, M. (ed.) Cluster and Regional Socialisation: On Geography. Tecnology and Networked, pp. 18–40. Pion, Londres (1998)
Folta, T., Cooper, C., Baik, Y.: Geographic Cluster Size and Firm Performance. Journal of Business Venturing 21, 217–242 (2006)
Gilbert, B., McDougall, P., Audretsch, D.: Clusters, Knowledge Spillovers and New Venture Performance: An Empirical Examination. Journal of Business Venturing 23, 405–422 (2007)
Glaeser, E., Kerr, W., Ponzeto, G.: Clusters of Entrepreneurship, Working Paper 10-019. Harvard Business School, Boston (2009)
Glassmann, U., Voelzkow, H.: The Governance of Local Economies in Germany. In: Crouch, C., et al. (eds.) Local Production Systems in Europe. Rise or Demise?, pp. 79–116. University Press, Oxford (2001)
Gugler, P., Keller, M.: The Economic Performance of Swiss Regions, Indicator of Economic Performance, Composition of Cantonal Economies and Clusters of Traded Industries, Center for Competitiveness University of Fribourg Switzerland (December 2009)
Helmstädter, H.: Regionale Struktur und Entwicklung der Industriebeschäftigung: Konzentration oder Dekonzentration? Seminarbericht37. Gesellschaft Fur Regionalforschung e.V., Heidelberg (1996)
Hospers, G., Desrochers, P., Sautet, F.: The Next Silicon Valley? On the Relationship Between Geographical Clustering and Public Policy. Int. Entrep. Manag. J. 5, 285–299 (2009)
Hoover, E.: The Measurement of Industrial Localization. Review of Economics and Statistics 18, 162–171 (1936)
Hoover, E.: Location Theory and the Shoe and Leather Industries. Harvard University Press, Cambridge (1937)
Hoover, E.: The Location of Economic Activity. McGraw-Hill, New York (1948)
Kayser, B.: La Renaissance Rurale, Sociologie des Campagnes du Monde Occidental. Edition Armand Colin, France (1990)
Keeble, D., Nachum, L.: Why do Business Service Firms Cluster? Small Consultancies, Clustering and Decentralization in London and Southern England. Transactions of the Institute of British Geographers 27(1), 67–90 (2002)
Ketels, C., Memedovic, O.: From Clusters to Cluster-based Economic Development. International Journal of Technological Learning, Innovation and Development 1(3), 375–392 (2008)
Kim, S.: Expansion of Markets and the Geographic Distribution of Economic Activities: the Trends in U.S. Regional Manufacturing Structure, 1860-1987. The Quarterly Journal of Economics, 881–907 (November 1995)
Krugman, P.: Geography and Trade. Leuven University Press, Leuven (1991)
Krugman, P.: Competitiveness: A Dangerous Obsession, Chapter 1, and Myths and Realities of US Competitiveness. In: Pop Internationalism. ch. 6, pp. 3–24, 87–104. MIT Press, Cambridge (1996)

Krugman, P., Fujita, M.: The New Economic Geography: Past, Present and the Future. Regional Science 83, 139–164 (2004)

Launhardt, W.: Die Bestimmung des zweckmaÈûigsten Standorts einer gewerblichen Anlage ("Determining the optimal location of an industrial site"). Zeitschrift des Vereins Deutscher Ingenieure 26, 105–116 (1882)

Lundvall, B.: The Learning Economy: Challenges to Economic Theory and Policy. Paper presented at the EAEPE Conference, Copenhagen (October 1994)

Maillat, D.: The Innovation Process and the Role of the Milieu. In: Bergmann, E., Maier, G., Tödtling, F. (eds.) Regions Reconsidered: Economic Networks, Innovation and Local Development in Industrialised Countries, pp. 103–117. Mansell, London (1991)

Malmberg, A.: Beyond the Cluster – Local Milieus and Global Connections. In: Peck, J., Yeung, C. (eds.) Remaking the Global Economy: Economic-Geographical Perspectives, pp. 145–159. Sage, London (2003)

Malmberg, A., Power, D.: True Clusters. A Severe Case of Conceptual Headach. In: Asheim, B.T., Cooke, P., Martin, R. (eds.) Clusters and Regional Development: Critical Reflections and Explorations. Routledge, London (2006)

Marshall, A.: Principles of Economics. Macmillan, London (1890)

Martin, R., Sunley, P.: Deconstructing Clusters: Chaotic Concept or Policy Panacea? Journal of Economic Geographic 3, 5–35 (2002)

Maskell, P., Kebir, L.: What Qualifies as a Cluster Theory?. DRUID Working Paper No. 05-09, Danish Research Unit for Industrial Dynamics (2005)

Nordin, S.: Tourism Clustering and Innovation - Path to Economic Growth and Development, No.14, ETOUR, Ostersund (2003)

Pe'er, A., Vertinsky, I.: The Determinants of Survival of De Novo Entrants in Clusters and Dispersal. Tuck School of Business, Working Paper (2006)

Piore, M., Sabel, C.: The Second Industrial Divide. Basic Books, New York (1984)

Porter, M.: The Competitive Advantage of Nations. Free Pass, New York (1990)

Porter, M.: Clusters and the New Economics of Competition. Harvard Business Review 76(6), 77–90 (1998)

Porter, M.: Location, Competition and Economic Development: Local Clusters in a Global Economy. Economic Development Quarterly 14(1), 7–20 (2000)

Porter, M.: Regional Foundations of Competitiveness and Implications for Government Policy. Paper Presented to Department of Trade and Industry Workshop (April 2002)

Porter, M.: The Economic Performance of Regions. Regional Studies 37(6/7), 549–578 (2003)

Porter, M., Ketels, C., Delgado, M.: The Microeconomic Foundations of Prosperity: Findings From the Business Competitiveness Index. Global Competitiveness Report 2007-2008. Palgrave Macmillan, London (2007)

Rocha, O.: Entrepreneurship and Development: the Role of Clusters. Small Business Economics 23, 363–400 (2004)

Roos, M.: Ökonomische Agglomerationstheorien—Die Neue Ökonomische Geographie im Kontext. Josef Eul, Lohmar (2002)

Rosenfeld, S.: Bringing Business Clusters into the Mainstream of Economic Development. European Planning Studies 5(1), 3–23 (1997)

Ricardo, D.: On the Principles of Political Economy and Taxation. John Murray, London (1817)

Scott, A.: New Industrial Spaces: Flexible Production Organization and Regional Development in North America and Western Europe. Pion, London (1988)

Shakya, M.: Competitiveness Assessment of Tourism in Sierra Leone, Policy Research Working Paper, Poverty Reduction and Economic Management Network (October 2009)

Simmie, J., Sennett, J.: Inovation in the London Metropolitan Region. In: Hart, D., Simmie, J., Wood, P., Sennett, J. (eds.) Innovative Clusters and Competitive Cities in the UK and Europe. Oxford Brookes School of Planning (1999)

Solinas, G.: Productive Structure and Competitiveness in the Italian Footwear Industry. Paper Presented at the 10th Conference of the International Working Party on Labour Market Segmentation. University of Porto, Portugal (1988)

Sternberg, R., Litzenberger, T.: Regional Clusters in Germany—their Geography and their Relevance for Entrepreneurial Activities. European Planning Studies 12(6), 767–791 (2004)

Swann, G., Prevezer, M.: A Comparison of Dynamics of Industrial Clustering in Computing and Biotechnology. Research Policy 25, 1139–1157 (1996)

Von Thünen, J.: Der Isolierte Staat in Beziehung auf Landwirtschaft und Nationalökonomie. Teil, vol. 1. Friedrich Perthes, Hamburg(1826); translated by Wartenberg, C.: Von Thunen's Isolated State. Pergamon Press, Oxford (1996)

Waxell, A., Malmberg, A.: What is Global and What is Local in Knowledge-Generating Interaction? The Case of the Biotech Cluster in Uppsala, Sweden. Entrepreneurship & Regional Development 19(2), 137–159 (2007)

Weber, A.: Über den Standort der Industrien (On the Location of Industries). Mohr Verlag, Tübingen (1909)

Wennberg, K., Lindqvist, G.: How do Entrepreneurs in Clusters Contribute to Economic Growth? SSE/EFI Working Paper Series in Business Administration No. 2008:3 (2008)

Innovation in Tourist Management through Critical Success Factors: A Fuzzy Map

Luis Camilo Ortigueira and Dinaidys Gómez-Selemeneva

Department of Business Administration, Pablo de Olavide University,
Carretera de Utrera, km1, 41013 Sevilla, Spain
{Lcortsan,digomsel}@upo.es

Abstract. This article presents a model of the Critical Success Factors of a tourist destination located in the Caribbean area, where there are other destinations, some already well-established and others emerging. In this context, this research will be of value not only for understanding the attractions of the destination as a tourism offer from the customers' perspective, but also for the design of the most effective strategies and policies to ensure the sustainability of the destination, the competitiveness of the companies providing tourist services, etc., and for encouraging management chains to invest and participate in the area, and particularly in the destination studied. The methodology employed is that of Fuzzy Cognitive Maps based on a content analysis as a tool of strategic diagnosis; this is the principal contribution of the study. This approach has enabled us to reach important conclusions on the central elements to be strengthened in formulating Innovation Strategies that would allow an integrated outcome for the destination, through those factors that should ensure its success.

Keywords: Fuzzy maps, Critical success factors, Innovation, Tourism.

1 Introduction

In the 1960's Ronald Daniel argued that "in most industries there are usually three to six factors that determine success; these key jobs must be done exceedingly well for a company to be successful". In 1979, J. Rockart defined these factors, critical success factors (CSFs, from now on), as "the limited number of areas in which results, if they are satisfactory, will ensure successful competitive performance for the organization. They are the few key areas where -things must go right- for the business to flourish", and, given their importance for the company, these factors must be central to three key management tasks or functions: the management information systems, the formulation of the company's strategy, and its implementation. With these antecedents, the strategic management of the company will integrate with the CSFs: without this, the company's competitiveness in its sector of activity is inconceivable (Kotler 2000).

A.M. Gil-Lafuente et al. (Eds.): Soft Comput. in Manag. and Bus. Econ., STUDFUZZ 287, pp. 361–373.
springerlink.com

In the work described here, a more in-depth study is made of Cognitive Maps (CMs) given that this tool is of particular interest for determining the CSFs of a competitive enterprise, and also its innovative potential trough Modeling by Fuzzy CMs .The application of the Fuzzy CMs methodology is considered particularly useful and it allows us to demonstrate, as result of the qualitative study, the key factors for the success of a tourism destination.

These studies illustrate the significance of knowledge of those factors that determine operating efficiency in the hotel activity sector, and of the factors that ensure effectiveness in the marketing of a tourist destination, particularly when this sector has a strategic role in the national economy, as is the case of the research presented in the paper on Cuba as a tourist destination, in all its aspects. More efficient and efficacious policies and Innovative strategies will be designed in function of these factors, not only in order to ensure the investment and growth of the sector but also to improve its attractiveness and promotion.

2 Theoretical Background

Modeling by means of maps can be utilized for diverse ends: for planning, prospection, structuring of data or, as in the case presented here, as a tool for determining the factors of greatest strategic importance. Representation in the form of maps allows information to be structured and thus described more clearly and precisely, and understood more easily and rapidly, than with other types of model (Ortigueira 2007). At the practical level, the capacities offered by mapping could be described in terms of the generation of ideas, the description of phenomena, and the rational presentation of structure.

Modeling can be considered to fall into three main types: conceptual (Novak 1984); cognitive (Eden 1988); and mental mapping (Buzan 1993). Although the visual representation of each type shows a certain degree of similarity, both the objective and the use of the model show significant differences.

The difference between mental, conceptual and cognitive maps is that, in the mental map, the point of departure is one single principal or central idea from which a branching structure is created; branches in which the relationships are passive only represent an association, not a cause-and-effect.

In the conceptual map, the nodes are labeled by using descriptive texts and, unlike cognitive maps, the links are labeled expressing the relationship existing between the nodes (Novak 1984) and subsequently what William, Trochim & Rhoda (1986) call Concept Mapping: a pictorial representation that is the result of combining a multidimensional scaling and a cluster analysis, from a matrix of data.

A cognitive map is a representation of how people think about one situation in particular, by means of the organization and structuring of the elements or problems that comprise it; then the concepts that form the connections between those elements are modeled graphically. Thus causes and effects are identified, and the causal links are explained (Ackermann, Eden and Cropper 1992). According to these authors, the cognitive map is based on the Personal Construct Theory, an approach from previous research that was conceived as an aid to obtaining the system of constructs that a person uses to make sense of a repertoire of elements in a particular situation; in this case, short phrases are used that express an idea or opinion.

In conceptual and cognitive maps, there may also be various different approaches in the construction of complex networks. On this point, in the more recent literature, extensive use is made of what are termed Fuzzy Cognitive Maps (Axelrod 1976). An augmented perspective has even been described, that consists of the creation of maps from the aggregation of several adjacency matrices (Salmeron 2009), referred to as augmented Fuzzy Cognitive Maps (FCMs, in the rest of this paper).

The idea of FCMs was introduced by Axelrod in 1976; these maps examine the causal relationships existing between the nodes of a cognitive map, taking as their bases the relationships displayed in an adjacency matrix. Among the elements that will form part of the matrix, and of the resulting map, are the nodes (variables, attributes, concepts, key terms, etc.) and the relationships of causality. According to Kosko (1986), FCMs constitute "fuzzy-graph structures for representing causal reasoning". The type of causality is considered fuzzy to the extent that it admits degrees forming a vector, which indicates the state of the system at a particular moment in time. This state may undergo changes, which are identified as feedback in the map, and so the state of the system is interrelated with the rest of the elements in the map (Ortigueira 2007).

There are, moreover, various different kinds of analysis can be performed in the map, as is the case of indirect effects and centrality. The indirect effect analyzes the effect observed between two nodes that do not show a direct causal relationship; centrality represents the importance of the node for the causal flow in the cognitive map.

The starting point for producing a FCM is the creation of a non-symmetrical squared adjacency matrix, whose rows and columns are formed by the nodes selected. The relationship between the nodes of the map is represented in this matrix. The FCM represents relationships of causality between nodes of a network of relationships that are considered relevant in a particular field; in its augmented model it also enables various different maps with nodes in common to be integrated. The relationships are represented in the model by means of arrows that indicate the direction of the causal relationship.

Salmeron (2009) identifies two characteristics possessed by a cognitive map model. The first characteristic is depiction of the intensity of the relationship existing between the nodes of the network of causal relationships, represented by a number, either 0 or 1. The causal value can take a positive or negative sign; the positive sign is substituted by the value +1, the negative by the value -1; the value 0 indicates the absence of any causal relationship. The second characteristic is that feedback is included, whereby the effect of change in one node can affect one or more other nodes. The adjacency matrix (A) contains the individual result of the relationships between nodes (Cij) established by each expert (aij), giving as the result:

$$A = \begin{matrix} & \begin{matrix} C_1 & C_2 & \cdots & C_j \end{matrix} \\ \begin{matrix} C_1 \\ C_2 \\ \vdots \\ C_i \end{matrix} & \begin{pmatrix} a_{11} & a_{12} & \ldots & a_{1j} \\ a_{21} & a_{22} & \ldots & a_{2j} \\ \ldots & \ldots & \ldots & \ldots \\ a_{i1} & a_{i2} & \ldots & a_{ij} \end{pmatrix} \end{matrix} \tag{1}$$

where all the aij elements can take the value: -1 or 0 or +1, representing the relationship between each node (C1, C2,….,Cn).

From the individual matrices generated by the analyzed opinion of each expert, the next step is to carry out a process of aggregation of these matrices, giving rise to the augmented adjacency matrix (B).

$$B = \begin{pmatrix} \ldots & \ldots \\ \ldots & W_{ij} \end{pmatrix} \tag{2}$$

In this aggregated matrix, the row and column vectors would include the nodes of all the individual matrices, and the Wij values of the resulting augmented adjacency matrix B would be:

$$W_{ij} = \frac{\sum_{k=1}^{n} W_{ij}^{k}}{n} \tag{3}$$

where n is the number of adjacency matrices aggregated, k is the identifier of each expert, and i and j are the identifiers of the relationship between nodes.

The resulting map with the augmented model combines the opinion of each and every one of the experts considered, so there is no need for any expert to change their opinions in order to reach a consensus view.

3 Augmented Fuzzy Cognitive Maps, as Tool for Strategic Management

In the services sector, to cite an example, knowledge of the CSFs has constituted a competitive advantage for the leading companies. For this reason, the search for techniques by which the CSFs can be identified has been the topic of studies such as those of Geller (1985) and Brotherton and Shaw (1996), who did this for the case of tourism, a sector that which was showing marked dynamism in the 1990's.

Another relevant empirical study, presented by Haven-tang, Jones and Webb (2007), is a case study on the CSFs of the tourism business in the United Kingdom. In Baker and Cameron (2008) an analysis was made of the CSFs associated with market research and the design of strategies for the promotion of a tourist destination; this latter is understood as "a geographically-delimited place to which visitors are temporarily attracted and includes continents, regions, countries, states, cities and towns" (Bull, 1995; Pike 2004).

Table 1 gives the results of an analysis of the bibliography.

Table 1 Factors studied by other authors

Type of Factor	*Cognitive*			*Operational*	
Authors	Receptivity of the residents	Culture	Natural resources	Access routes	Security
Sternquist (1985)	x	x	-	-	-
Haahti (1986)	x	x	x	-	-
Gartner & Hunt (1987)	x	-	x	-	-
Calantone et al. (1989)	x	x	-	x	x
Gartner (1989)	x	x	x	-	-
Ahmed (1991)	x	x	x	-	-
Chon (1991)	x	x	x	x	x
Fakeye & Crompton (1991)	x	x	x	x	-
Crompton et al. (1992)	x	-	x	-	-
Chon (1992)	x	x			
Echtner & Ritchie (1993)	x	x	x	x	x
Driscoll et al. (1994)	x	x	-	-	x
Dadgostar & Isotalo (1995)	-	x	x	-	-
Muller (1995)	x	x	-	-	x
Ahmed (1996)	x	x	x	-	-
Opperman (1996)	-	x	-	x	x
Schroeder (1996)	x	x	x	-	-
Baloglu (1997)	x	x	x	x	x
Crouch & Ritchie (1999)	-	-	-	-	x
Baloglu &McCleary (1999)	x	x	-	-	x

This table 1 shows the growing importance attributed to knowledge of those items that contribute to visitors' preferences, attraction and competitiveness, in the management of the tourism offer of any particular destination.

The tourism sector underwent a structural change in the 1990's, by which it became a strategic sector, highly attractive for investment. At that time, several Caribbean destinations came to prominence, apart from those already well-established such as Riviera Maya (Mexico) and others. This stage was characterized by the entry in the Caribbean area of multinational and family hotel chains with international experience and recognized brands, which secured presence in the principal destinations, as shown in Figure 1, up to the year 2008; it is also demonstrated in this graph that the Spanish chains are those with the strongest presence.

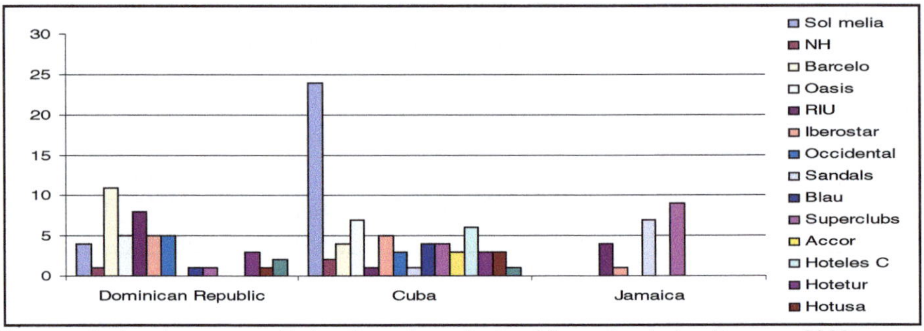

Source: Authors' own compilation from data obtained from the chains (January 2009).

Fig. 1 Presence of Multinational Hotel Chains in the Caribbean (Dominican R., Jamaica and Cuba)

With an average annual growth rate of 14 %, Cuba is situated in eighth place in the ranking of preference among international tourist destinations, and in third place among destinations in the Caribbean (Hosteltur, 2008). Surprisingly, the increase per period in the preference for the island of Cuba is superior to that for the destinations that flourished at the end of the 1980's: Jamaica and Punta Cana (Dominican Republic).

The eight Spanish chains present in the Island administer 6,519 rooms in 23 hotels; next in size are two French companies with 1,410 rooms in six hotels; Italian, German and Jamaican companies are also present; these latter companies have innovated in the concept of marketing and provision of service, introducing the "all-inclusive" product in the destination.

Figure 2 and 3 presenting a comparative analysis between the three emerging destinations, in which a turning point is demonstrated in 1998 in the hotel capacity as the result of the expansion in investment. The consequences of this are seen in terms of the attraction of tourists, through the increase in the numbers of arrivals, which are significant in the case of Cuba and Dominican R., whereas Jamaica shows a less pronounced rise.

It can be appreciated in the previous graphics that Cuba has shown a favorable behavior as a tourist destination. This has been the result of implementing the tourism strategy conceived in the 1990's to strengthen the tourism sector and hotel activity in the Island by opening the country up to investment and cooperation with foreign chains (Romero and Gómez, 2009).

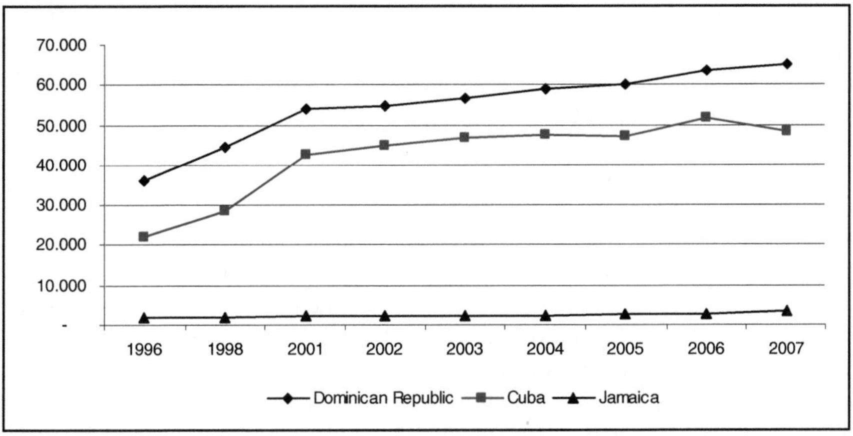

Source: Authors' own compilation, from official statistical data of Dominican R., Jamaica and Cuba.

Fig. 2 Comparison of Hospitality Capacity (number of rooms/country) 1996-2007

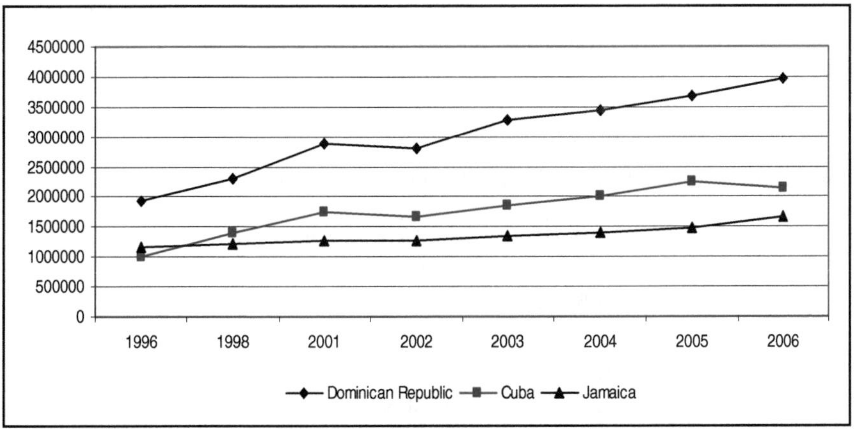

Source: Authors' own compilation, from the World Development Indicators database (2008).

Fig. 3 International Tourism, number of arrivals 1996-2006

4 Modeling the Critical Success Factors of Cuba as a Tourist Destination

In the study presented here on the CSFs, the nodes are represented by the CSFs analyzed; thus the cognitive map is capable of representing the relationship existing between cognitive and operational factors, and the degree of influence exerted by those factors, obtained by integrating the opinions of all the experts considered.

To arrive at the final model, we start from an analysis of the information content (Diefenbach, 2001) of interviews conducted by Cuban communications officials with management experts, knowledgeable about the destination; a total of seven such interviews were conducted in the period from October 2007 to May 2008.

Those interviewed were considered to be experts as a consequence of having participated to a significant degree in the decision-making processes and the conception of the tourism strategy of the Island. They include persons holding senior positions in relevant government ministries and official associations.

Specifically, the interviewees are: two senior members of staff of the Ministry of Tourism of Cuba, two senior members of staff of the Ministry of Culture, the official Historian of the City of La Habana, the president of the Union of Writers and Artists of Cuba, and an external tourism adviser from an international consultancy of recognized prestige. This methodology also took into account the criteria proposed by Okoli and Pawlowski (2004), who stated that a range of between 5 and 18 experts is considered valid for such interviews.

From the content analysis the key concepts were extracted that are representative of the meaning of the texts (i.e. the transcripts of the interviews) and that determine both the nodes of the map and the relationships existing between the nodes. Essentially, by producing the augmented FCM, it was possible to 'discover' the expert knowledge incorporated in the unstructured texts. The concepts translate into the 17 CSFs presented in table 2.

Table 2 Critical success factors of Cuba as a tourist destination

LABEL	*CRITICAL SUCCESS FACTOR*
A	Systematic promotion of cultural events
B	Advance knowledge of agents and tour operators
C	Tourist motivation
D	Quality leisure and recreation
E	Receptivity of the residents
F	Richness of culture
G	Richness of heritage
H	Richness of history
I	Protection of the heritage
J	Traditions
K	Political, social and economic stability
L	Communication
M	Conservation of cultural values
N	Protection of biodiversity
O	Landscape, nature and climate
P	Cost of air travel
Q	Security

The next step is to model the relationships and their intensities, giving rise to the augmented adjacency matrix (figure 4). Then the FCM is constructed from this matrix.

	A	*B*	*C*	*D*	*E*	*F*	*G*	*H*	*I*	*J*	*K*	*L*	*M*	*N*	*O*	*P*	*Q*
A	0	0.14	0.14	0	0	0	0	0	0	0	0	0	0	0	0	0	0
B	0	0	0.29	0	0	0	0	0	0	0	0	0	0	0	0	0	0
C	0	0	0	0	0	0	0	0	0	0	0	0	0	0	0	0	0
D	0	0	-0.14	0	0	0	0	0	0	0	0	0	0	0	0	0	0
E	0	0	0.29	0	0	0	0	0	0	0	0	0	0	0	0	0	0
F	0	0	0.71	0	0	0	0	0	0	0	0	0	0	0	0	0	0
G	0	0	0.43	0	0	0	0	0	0	0	0	0	0	0	0	0	0
H	0	0	0.14	0	0	0	0	0	0	0	0	0	0	0	0	0	0
I	0	0	0	0	0	0	0.14	0	0	0	0	0	0	0	0	0	0
J	0	0	0	0	0	0.14	0	0	0	0	0	0	0	0	0	0	0
K	0	0	0.14	0	0	0	0	0	0	0	0	0	0	0	0	0	0
L	0	0.14	0	0	0	0	0	0	0	0	0	0	0	0	0	0	0
M	0	0	0	0	0	0.29	0	0	0	0	0	0	0	0	0	0	0
N	0	0	0	0	0	0	0.14	0	0	0	0	0	0	0	0	0	0
O	0	0	0	0	0	0	0.14	0	0	0	0	0	0	0	0	0	0
P	0	0	-0.14	0	0	0	0	0	0	0	0	0	0	0	0	0	0
Q	0	0	0.14	0	0	0	0	0	0	0	0	0	0	0	0	0	0

Fig. 4 Augmented Adjacency Matrix

The resulting model (figure 5) shows the causal relationships and intensities of the CSFs of Cuba as a tourist destination, and allows the hierarchical positions or rankings of these factors to be established. It can be observed from the model that, according to the criteria of the experts, the principal cause motivating both the tourist and the foreign investor is the cultural richness of the Island (0.71), followed by the richness of its heritage (0.43).

In comparison with Jamaica and Dominican Republic this perception is understandable, considering the artistic recognition and prestige of Cuba, the numerous World Heritage cities, the places of historical interest, etc.

The advance knowledge of agents and tour operators (0.29) and the receptivity of the residents (0.29) also appear as relevant factors. With respect to the critical factors that have a negative influence on Cuba as a tourist destination, the experts identify the cost of air travel to the Island (-0.14), which is high in comparison with other emerging destinations of the area, and is significantly higher than the market average. Despite the artistic, cultural and musical potential of the Island, which differentiates it from the rest of the countries of the Caribbean, deficient management of the leisure and recreation facilities (-0.14) is identified as a negative factor.

The results demonstrated by the augmented FCM (figure 5) are consistent with the model proposed by Crouch and Ritchie (1999) for measuring the competitiveness of a destination. That model serves to identify the factors of success in the tourism industry that will determine the competitive success of a particular tourist

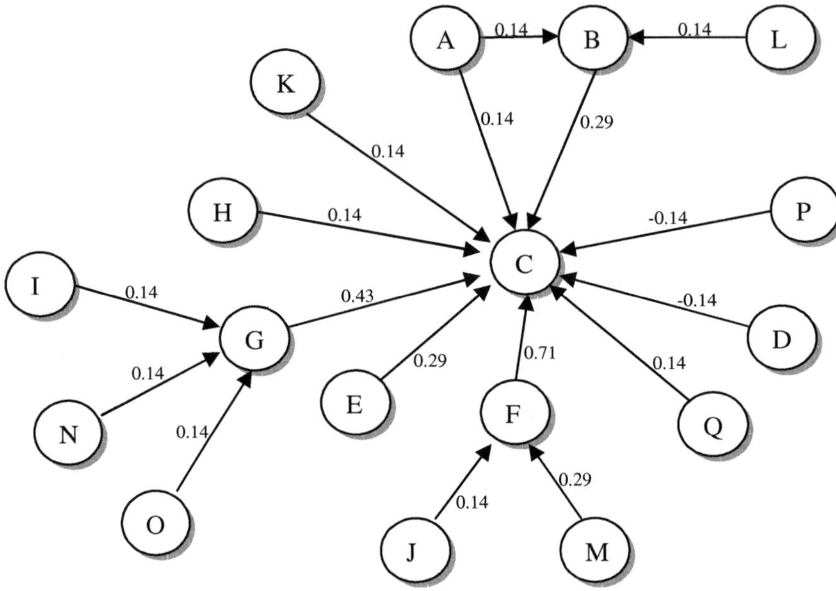

Fig. 5 Augmented fuzzy cognitive map

destination, and includes among these factors elements that coincide with the results obtained in our research, such as: political and social stability, security and the available means of transport or access routes. The high value attributed to the causal relationship between richness of culture and motivation of the tourist reveals the importance of this relationship in the formulation of the tourism strategy for Cuba.

These results are consistent and in agreement with other similar studies, such as that of Bailey (2008), who argues that growing importance is attributed in the literature regarding "the symbolic and cultural meanings of tourism and the ways in which cities are increasingly competing for tourists through the promotion of cultural assets and different forms of spectacle" offered.

More efficient and efficacious policies and Innovative Strategies will be designed in function of these factors, not only in order to ensure the investment and growth of the sector but also to improve its attractiveness and promotion. This strategy should be focused on increasing the number of tourists and the foreign investment in the sector, through strategies that are not based solely on Beach and Sunshine tourism, given the weight in the FCM accounted for by factors such as the richness of the culture and heritage. Furthermore, given the geographic location of the Island, the cultural and geographic proximity to other tourist destinations in the Caribbean, together with the presence of international management chains that operate in these destinations, the creation of multi destination tourism packages is also considered appropriate; this would be a way of exploiting the CSFs of other destinations in combination with Cuba's own.

5 Final Remarks

Modeling using augmented Fuzzy Cognitive Maps, and taking the content analysis as the methodology for this, constitute an appropriate procedure for the analysis of strategies, in the context of identifying and exploiting the critical success factors of a tourism sector, as has been demonstrated in this study of developing Cuba as a tourist destination.

From the experts consulted, it has been possible to specify 17 critical success factors; these have then been integrated in a model that shows the direction and intensity of the causal relationships between the factors. Models like this should serve as an important tool for those the involved in the process of decision-making and the conception of the tourism strategy of national and foreign chains present in the Island, particularly the Spanish chains.

The article provides useful information for understanding the Cuban tourism industry and the particular characteristics of the destination, especially in the context of the beach and sunshine holiday destinations in the Caribbean. The great value of this information rests on its usefulness not only as an aid for understanding the expectations of the customers but also for the market studies of investors and managers. Appropriate strategies can then be formulated and implemented. Those factors that are perceived as positive can be strengthened; action can be taken to create awareness of those factors that are positive and already present there but that are not visible and not considered critical; and lastly, action can be taken to modify factors that are evidently negative, or are perceived negatively without justification.

The methodological proposal presented in this paper is appropriate for application to research on the attractions of other destinations. The authors believe it would be of great interest to conduct similar studies in the other destinations mentioned: the Dominican Republic and Jamaica.

References

Ahmed, Z.U.: The Influence of the Components of a State's Tourist Image on Product Positioning Strategy. Tourism Management 12, 331–340 (1991)

Ahmed, Z.U.: The Need for the Identification of the Constituents of a Destination's Tourist Image a Promotional Segmentation Perspective. Revue du Tourisme 51(2), 44–57 (1996)

Axelrod, R.: Structure of Decision: The Cognitive Maps of Political Elites. Princeton University Press, Princeton (1976)

Bailey, N.: The Challenge and Response to Global Tourism in the Post-modern Era: The Commodification. Reconfiguration and Mutual Transformation of Habana Vieja, Cuba. Urban Studies 45(5-6), 1079–1096 (2008)

Baker, M.J., Cameron, E.: Critical success factors in destination marketing. Tourism and Hospitality Research 8(2), 79–97 (2008)

Baloglu, S.: The Relationship between Destination Images and Sociodemographic and Trip Characteristics of International Travellers. Journal of Vacation Marketing 3, 221–233 (1997)

Baloglu, S., McCleary, K.W.: A Model of Destination Image Formation. Annals of Tourism Research 26, 808–889 (1999)

Brotherton, B., Shaw, J.: Towards an identification and classification of Critical Success Factors in UK Hotels Plc. International Journal of Hospitality Management 15(2), 113–135 (1996)

Bull, A.: The Economics of Travel and Tourism. Pitman Longman Cheshire, London (1995)

Buzan, T.: The Mind Map Book. BBC Books, London (1993)

Calantone, R.J., Di Benedetto, C.A., Hakam, A., Bojanic, D.C.: Multiple Multinational Tourism Positioning Using Correspondence analysis. Journal of Travel Research 28(2), 25–32 (1989)

Chon, K.-S.: The Role of Destination Image in Tourism: An Extension. Revue du Tourisme 1, 2–8 (1992)

Crouch, I., Ritchie, B.: Tourism, competitiveness and societal prosperity. Journal of Business Research 44(3), 137–152 (1999)

Dadgostar, B., Isotalo, R.M.: Content of City Destination Image for Near-Home Tourists. Journal of Hospitality and Leisure Marketing 3(2), 25–34 (1995)

Daniel, D.D.: Management Information Crisis. Harvard Business Review 39, 111–121 (1961)

Diefenbach, L.: Historical foundations of computer-assisted content analysis. In: West, D. (ed.) Theory, Method, and Practice in Computer Content Analysis, vol. 16, pp. 13–41. Ablex Publishing, Westport (2001)

Driscoll, A., Lawson, R., Niven, B.: Measuring Tourists Destination Perceptions. Annals of Tourism Research 21, 499–511 (1994)

Echtner, C.M., Ritchie, J.R.: The Measurement of Destination Image: An Empirical Assessment. Journal of Travel Research 31(4), 3–13 (1993)

Eden, C.: Cognitive mapping. European Journal of Operational Research 36, 1–13 (1988)

Gartner, W.C., Hunt, J.D.: An Analysis of State Image Change over a Twelve-Year Period (1971-1983). Journal of Travel Research 26(2), 15–19 (1987)

Gartner, W.C.: Tourism Image: Attribute Measurement of State Tourism Products using Multi-dimensional Scaling Techniques. Journal of Travel Research 28(2), 16–20 (1989)

Geller, N.: Tracking the critical success factors for hotel companies. Cornell Hotel and Restaurant Administration Quarterly 25(2), 76–81 (1985)

Haahti, A.J.: Finland's Competitive Position as a Destination. Annals of Tourism Research 13, 11–35 (1986)

Haven-Tang, C., Jones, E., Webb, C.: Critical Success Factors for Business Tourism Desti-na-tions: Exploiting Cardiff's National Capital City Status and Shaping Its Business Tourism Offer. Journal of Travel & Tourism Marketing 22, 109–120 (2007)

Kosko, B.: Fuzzy cognitive maps. International Journal of Man-Machine Studies 24(1), 65–75 (1986)

Kotler, P.: Marketing Management, The Millennium Edition. Prentice Hall, Upper Saddle River (2000)

Muller, T.E.: How Personal Values Govern the Post-Visit Attitudes of International Tourists. Journal of Hospitality and Leisure Marketing 3(2), 3–24 (1995)

Novak, J.D., Gowin, D.B.: Learning How to Learn. Cambridge University Press, Cambridge (1984)

Okoli, C., Pawlowski, S.: The Delphi method as a research tool: an example, design considerations and applications. Information and Management 42(1), 15–29 (2004)

Oppermann, M.: Convention Destination Images Analysis of Association Meeting Planners' Perceptions. Tourism Management 17, 175–182 (1996)

Ortigueira, L.C.: A Systems Approach to Public Policies, Working Paper Series WP BSAD 07.01, Research Papers in Economics, Department of Business Administration, Pablo de Olavide University (2007)

Pike, S.: Destination Marketing Organisations. Elsevier, Oxford (2004)

Rockart, J.F.: Chief executives define their own data needs. Harvard Business Review 57(2), 81–93 (1979)

Romero, I., Gomez-Selemeneva, D.: La Cooperación en el Sector Hotelero Cubano; valoración de su impacto en la implantación de la estrategia turística. In: XIX Congreso Nacional ACEDE, Toledo, España (2009)

Salmeron, J.L.: Augmented Fuzzy Cognitive Maps for modelling LMS Critical Success Factors. Knowledge-Based Systems 22(4), 275–278 (2009)

Schroeder, T.: The Relationship of Residents' Image of their State as a Tourist Destination and their Support for Tourism. Journal of Travel Research 34(4), 71–73 (1996)

Sternquist, W.B.: Attitudes about Resort Area a Comparison of Tourists and Local Retailers. Journal of Travel Research 24(1), 14–19 (1985)

William, M., Trochim, K., Rhoda, L.: Conceptualization for Planning and Evaluation. Evaluation and Program Planning 9, 289–308 (1986)

[illegible]

The Forgotten Effects of Sport

Anna M. Gil-Lafuente, Fabio Raúl Blanco-Mesa, and César Castillo-López

Department of Business Administration, University of Barcelona,
Av. Diagonal 690, 08034, Barcelona, Spain
amgil@ub.edu, frblamco@yahoo.com, cclopez@economistes.com

Abstract. The concern for the increase in chronic diseases, urban violence and armed conflicts, promoting the development of nations and education have become increasingly. Sports and physical activity are taking more importance as social and economic aims of governments and corporations, such are economic resources provided are significant. However, in these times of economic uncertainty and austerity major cuts were made on budgets related to the welfare of people, therefore, are necessary tools for the decision making of available resources to be optimized, to generated multiplier effect and greatest benefit to populations. The theory of Forgotten Effects is used in the analysis of overall causes in sports and physical activity, the effects they have on society and which can produce a significant social benefit. The conclusion of the paper shows those incidents that can generate greater social benefit.

Keywords: Forgotten effects, Sport management, decision making, sport and physical activity.

1 Introduction

Sports and physical activity as a social phenomenon is a source for promoting education, health, development and peace are efforts to achieve the Millennium Development Goals (ONU, 2003). These involve international institutions, governments, associations of professional and amateur athletes, industries producing goods and sports services, transportation, communication, sponsors, and others. Where all decision making and its effects may become to have a considerable impact on the problems related to the planet's population as violence, health and environment which represents an improvement of the welfare state.

It's essential to find new models that will allow Sports' leaders and managers to take a better decision making, especially in environments of uncertainty to maximize the resources available to obtain a greater benefit for social welfare. Although a decision-making process should be as assertive as possible and adjusted to reality. Occasionally does not take into account other relations of cause and effect incidents that are hidden, as the effects on the accumulation of effects and underlying causes (Gil Lafuente and Barcellos de Paula, 2010).

A.M. Gil-Lafuente et al. (Eds.): Soft Comput. in Manag. and Bus. Econ., STUDFUZZ 287, pp. 375–391.
springerlink.com

The aim of this paper is to establish the causality relationships among the causes of sports; physical activity and scope (societal effects) in a general way, showing how, a cause can generate a general social welfare and take advantage it. In order, we will use the theory of the forgotten effects, which is a sequential processing technique, that allows cause and effect relationships be develop from incidence matrices to obtain or recover items, which were not considered by experts because they are hidden or indirect incidents (Gil Lafuente and Barcellos de Paula, 2010)

2 Antecedent

In 1988, Kaufmann and Gil Aluja established from previous studies on the incidence or causal relationship "Theory forgotten effects." This model allows obtaining all direct and indirect relationships, without the possibility of error or omission and retrieving. In this approach, phenomena and events overall surround us are part of a system or subsystem. We can ensure that virtually all activity is subject to some type of incidence of cause-effect.

2.1 Methodological Approach

Despite a good control system, there is the possibility of omitting or forgetting voluntarily or involuntarily some causal relationships that are not always explicit, obvious, visible or directly perceived. Usually those relationships are hidden by effects on outcomes, and there is therefore a backlog of causes that generate them. Our thinking needs to rely on tools and models to create a technical basis on which to work with all the information. Comparing them with those obtained from the environment and brings out all the relations of direct and indirect causality can be inferred.

The concept of incidenc (Gil-Lafuente, 2001) which carries the idea of function is present in all the actions of living beings. Precisely in all processes of nature sequential, it is usual to omit any voluntary or involuntary link.

Each oblivion, as a result has side effects ranging repercussions throughout the web of influence on a kind of combinatorial process. The incidence is eminently subjective concept, often difficult to measure, but their analysis can improve reasoned action and decision-making. To proceed, roughly, to show the operation of the theory of the Forgotten effects, we begin by briefly delve into their methodological foundations. If we have two sets of elements:

$$A - \{a_i \mid i - 1,2,\ldots,n\} \tag{1}$$

$$B = \{b_j \mid j = 1,2,\ldots,m\} \tag{2}$$

There is an incidence of ai over bj if the values of the characteristic function of pair's relevance (a_i, b_j) valuation in [0.1], therefore:

$$\forall(a_i, b_j) \Rightarrow \mu(a_i, b_j) \in [0,1] \tag{3}$$

The set of pairs of items valued will define what we call "direct incidence matrix," which shows the relationship of cause and effect that occur with different degrees of the elements of set A (causes) and the elements of set B (effects)

$$\tilde{M} \quad \begin{array}{c|cccc} & b_1 & b_2 & b_3 & b_j \\ \hline a_1 & \mu_{a_1b_1} & \mu_{a_1b_2} & \mu_{a_1b_3} & \mu_{a_1b_j} \\ a_2 & \mu_{a_2b_1} & \mu_{a_2b_2} & \mu_{a_2b_3} & \mu_{a_2b_j} \\ a_3 & \mu_{a_3b_1} & \mu_{a_3b_2} & \mu_{a_3b_3} & \mu_{a_3b_j} \\ a_4 & \mu_{a_4b_1} & \mu_{a_4b_2} & \mu_{a_4b_3} & \mu_{a_4b_j} \\ a_5 & \mu_{a_4b_1} & \mu_{a_5b_2} & \mu_{a_5b_3} & \mu_{a_5b_j} \\ a_i & \mu_{a_ib_1} & \mu_{a_ib_2} & \mu_{a_ib_3} & \mu_{a_ib_j} \end{array}$$

One set of incidences show three ways of presenting the cause-effect relationships, which take place into two sets of elements represented on the direct incident matrix (also called first-order).

These have been considered at the moment of establishing the impact of some elements over others. In fact, it is the first step in the approach of the model, which allows recovering different levels of incidence among items that have not been detected or simply, initially have been forgotten. Assume, for example, which appears a third set of elements

$$C = \{c_k \mid k = 1, 2, \ldots, z\} \tag{4}$$

Which is conformed of elements that act as effects of set B, therefore

$$\tilde{N} \quad \begin{array}{c|cccc} & c_1 & c_2 & & c_z \\ \hline b_2 & \mu_{b_1b_1} & \mu_{b_1c_2} & & \mu_{b_1c_z} \\ b_2 & \mu_{b_2b_1} & \mu_{b_2c_2} & & \mu_{b_2c_z} \\ \vdots & & & & \\ b_m & \mu_{b_mb_1} & \mu_{b_{m2}b_1} & & \mu_{b_mb_z} \end{array}$$

Thereafter be derived two incidence matrices, which will have elements common of set B:

$$\tilde{M} \quad \begin{array}{c|cccc} & b_1 & b_2 & & b_m \\ \hline a_1 & \mu_{a_1b_1} & \mu_{a_1b_2} & & \mu_{a_1b_m} \\ a_2 & \mu_{a_2b_1} & \mu_{a_2b_2} & & \mu_{a_2b_m} \\ \vdots & & & & \\ a_n & \mu_{a_nb_1} & \mu_{a_nb_2} & & \mu_{a_nb_m} \end{array}$$

$$\tilde{N} \quad \begin{array}{c|cccc} & c_1 & c_2 & & c_z \\ \hline b_2 & \mu_{b_1b_1} & \mu_{b_1c_2} & & \mu_{b_1c_z} \\ b_2 & \mu_{b_2b_1} & \mu_{b_2c_2} & & \mu_{b_2c_z} \\ \vdots & & & & \\ b_m & \mu_{b_mb_1} & \mu_{b_{m2}b_1} & & \mu_{b_mb_z} \end{array}$$

Therefore, there are two incidence relations:

$$M \subset A \times B \;\; y \;\; N \subset B \times C \tag{5}$$

The mathematical operator, which allows establish the incidences of A on C is the max-min composition. In fact, when raised the three uncertain incidence relations:

$$M \subset A \times B, N \subset B \times C, P \subset A \times C \tag{6}$$

The result of the composition is:

$$M \circ N = P \tag{7}$$

The symbol $\circ$ represents precisely the composition max-min. The composition of two relations uncertain is such that:

$$\forall (a_i, c_z) \in A \times C: \tag{8}$$

$$\mu(a_i, c_z)_{M \circ N} = V_{bj}(\mu_M(a_i, b_j) \wedge \mu_N(b_j, c_z)) \tag{9}$$

Can say that the incidence matrix P define the causality relationships between elements of the first set A and elements of the third set C, in the intensity which involves to consider of the elements belonging to set B. Relationship direct and indirect causality

2.2 Relationship Direct and Indirect Causality

After an analysis of the methodology used to understand the relationships of incidence having considered three sets of elements, now, we aim to posing a methodology for know the cause-effect relationships which are hidden, when is execute a study of causality between different elements. The approach (Gil-Lafuente, 2005) begins with the existence of a direct effect relationship, therefore, a cause-effect uncertain matrix defined by two sets of elements

$$A = \{a_i \mid i = 1, 2, \ldots, n\} \tag{1}$$

$$B = \{b_j \mid j = 1, 2, \ldots, m\} \tag{2}$$

And a causality relationship M defined by the matrix:

$$[\tilde{M}] = \{\mu_{a_i b_j} \in [0,1]/i = 1,2,\ldots,n; j = 1,2,\ldots,m\} \tag{10}$$

Being $\mu_{a_i b_j}$ the characteristic functions belonging of each of the elements of the matrix $[\tilde{M}]$ (formed by the rows corresponding to elements of the set A-causes- and the columns corresponding to elements of set B-effects-)

Is correct to say, that the matrix $[\tilde{M}]$ is composed by estimates made around all the effects which elements of set A exert on the elements of the set B. The more significant is this relationship of incidence, the higher be valuation assigned to each of the elements of the matrix, therefore, it is assumed that the characteristic function of belonging is in to the interval [0, 1], it is understood which the higher the ratio of incidence, closer to 1 will prove assigned valuation, instead, the weaker is considered a causality relationship between two elements, closer to 0 will prove the assigned valuation.

It should be emphasized which matrix $[\tilde{M}]$ is made from the direct causal relationships. The aim is based on obtaining a new matrix of incidents, which reflect not only the direct causality link, but rather also those are not obvious, exist and would be able to be fundamental for assessing phenomena. To achieve this goal it is necessary to introduce devices that will allow at the different causes to able have effects on themselves and at the same time, keep in mind that certain effects can also lead to incidents on themselves. Therefore, it is necessary to build two additional incidents relationships which will gather the possible effects arising from causes related to each other and effects each other. These two auxiliary matrices are defined as follows:

$$[\tilde{A}] = \{\mu_{a_i a_j} \subset [0,1]/i,j = 1,2,\ldots,n\} \tag{11}$$

$$[\tilde{B}] = \{\mu_{b_i b_j} \in [0,1]/i,j = 1,2,\ldots,m\} \tag{12}$$

The matrices $[\tilde{A}]$ and $[\tilde{B}]$ contains the relations of incidences that can occur between each of their elements (causes or effects) and being both reflexives matrices, therefore:

$$\mu_{a_i a_j} = [0,1]/i,j = 1,2,\ldots,n \tag{13}$$

$$\mu_{b_i b_j} = [0,1]/i,j = 1,2,\ldots,m \tag{14}$$

And it means that an element is cause or effect, incide with maximum presumption over itself.

In contrast neither $[\tilde{A}]$ nor $[\tilde{B}]$ are symmetric matrices, therefore:

$$\mu_{a_i a_j} \neq [0,1]/i,j = 1,2,\ldots,n \tag{15}$$

$$\mu_{b_i b_j} \neq [0,1]/i,j = 1,2,\ldots,m \tag{16}$$

Once built the matrices $[\widetilde{M}]$, $[\widetilde{A}]$ and $[\widetilde{B}]$must proceed to the establishment of direct and indirect effects, therefore, incidences involved simultaneously any cause or effect inserted. To do this we proceed to the max-min composition of the three matrices:

$$[\tilde{A}]\circ[\tilde{M}]\circ[\tilde{B}] = [\tilde{M}^*] \tag{17}$$

The order of composition must allow always match the number of elements in the first row of the matrix with the number of elements in the second column matrix. The result will be a matrix $[\tilde{M}^*]$ which gathers incidences between causes and effects of second generation, therefore, the initial causal relationships affected by the possible interposed incidence in some cause or effect. In this sense we have:

$$\underset{[\tilde{A}]}{\begin{array}{c|cccc} & a_1 & a_2 & & a_n \\ \hline a_1 & 1 & \mu_{a_1 a_2} & & \mu_{a_1 a_n} \\ a_2 & \mu_{a_2 a_1} & 1 & & \mu_{a_2 a_n} \\ \vdots & & & & \\ a_n & \mu_{a_n a_1} & \mu_{a_n a_2} & & 1 \end{array}} \circ \underset{[\tilde{M}]}{\begin{array}{c|cccc} & b_1 & b_2 & & b_m \\ \hline a_1 & 1 & \mu_{a_1 b_2} & & \mu_{a_1 b_m} \\ a_2 & \mu_{a_2 b_1} & 1 & & \mu_{a_2 b_m} \\ \vdots & & & & \\ a_n & \mu_{a_n b_1} & \mu_{a_n b_2} & & 1 \end{array}} \circ \underset{[\tilde{B}]}{\begin{array}{c|cccc} & b_1 & b_2 & & b_m \\ \hline b_1 & 1 & \mu_{b_1 b_2} & & \mu_{b_1 b_m} \\ b_2 & \mu_{b_2 b_1} & 1 & & \mu_{b_2 b_m} \\ \vdots & & & & \\ a_n & \mu_{b_m b_1} & \mu_{b_m b_2} & & 1 \end{array}}$$

$$\underset{\tilde{M}^*}{=} \begin{array}{c|ccc} & b_1 & b_2 & b_m \\ \hline a_1 & \mu^*_{a_1 b_1} & \mu^*_{a_1 b_2} & \mu^*_{a_1 b_m} \\ a_2 & \mu^*_{a_2 b_1} & \mu^*_{a_2 b_2} & \mu^*_{a_2 b_m} \\ \vdots & & & \\ a_n & \mu_{a_n b_1} & \mu_{a_n b_2} & \mu^*_{a_n b_{mz}} \end{array}$$

Therefore, the difference between the matrix of the effects of second-generation and direct incidences matrix will allow knowing the degree of casuals' relationships that have been forgotten or obviates:

$$[\tilde{O}] = [\tilde{M}^*](-)[\tilde{M}] \tag{18}$$

It is also possible to know from a degree forgetting of some incidence, the element (cause-effect) that is the link. In order, just must follow the steps made from max-min composition of matrices outlined above:

$$
\widetilde{[O]} \quad
\begin{array}{cccc}
 & b_1 & b_2 & b_m \\
a_2 & \mu^*_{a_1b_1} - \mu_{a_1b_1} & \mu^*_{a_1b_2} - \mu_{a_1b_2} & \mu^*_{a_1b_m} - \mu_{a_1b_m} \\
a_2 & \mu^*_{a_2b_1} - \mu_{a_2b_1} & \mu^*_{a_2b_2} - \mu_{a_2b_2} & \mu^*_{a_1b_m} - \mu_{a_1b_m} \\
\vdots & & & \\
a_n & \mu^*_{a_nb_1} - \mu_{a_nb_1} & \mu^*_{a_nb_2} - \mu_{a_nb_2} & \mu^*_{a_nb_m} - \mu_{a_nb_m}
\end{array}
$$

Finally, between higher the value of the characteristic function of membership of the matrix $\widetilde{[O]}$, the higher the degree of forgetting in the relationship of initial incidence. This means that the implications of some incidents that are not considered neither taken into account in their right intensity, can lead to some erroneous actions or at least, poorly estimated.

3 Applications

Now proceed with the application of this model to see generating the beneficial effects of sport relationship, in its several strands, levels and segments across society (age, location, economic, cultural level, welfare and others). We will observe how the practice of sport has a positive influence in all the social fields, even when apparently not be aware of this cause-effect relationship. The approach following will provide results that will enable know not only the real implications that the sport poses social, but it will show the multiplier effect on related variables and the level at which happen various incidences.

The ultimate aim of this study must allow better decision making at social agents involved about investments which must become more beneficial and with greater at the social welfare multiplier effect.

It proceeds in the first phase explain the application of the model used:

3.1 Study Variables

Table 1 Taxonomy of the causes of physical activity and sport

C1	Organized Physical Activity	C7	Sports Marketing
C2	Unorganized Physical Activity	C8	Adapted Sports
C3	Competitive Sports	C9	Educational Sport
C4	Professional Sports	C10	Sports of Initiation
C5	Sport and Business	C11	Sports Management
C6	Sports Spectacle	C12	Sports Laws and Policy

Table 2 Taxonomy Areas for Action and its Social Effects

Areas for Action		Social Effects
Physical Fitness	E1	Physical Fitness Health
	E2	Performance Fitness
	E3	Physiological Fitness

Table 2 *(continued)*

Welfare	E33	The Physical Context
	E4	The social
	E5	Intellectual
	E6	Emotional
	E7	Spiritual
	E8	Professional
	E9	Environment
	E10	Social Equality
	E11	Conflict Resolution and Peace Search
Health	E12	Fitness and health Prevention
	E13	Maintenance
	E14	Recuperation- rehabilitation
Education	E15	Improve school performance
	E16	Reduction of School Failure
	E17	Development of new areas of study
Economic and Sport (goods and services)	E18	Production
	E19	Trading
	E20	Distribution
	E21	Consumption
Public Budget	E22	Positive impacts on health budgets
	E23	Overcoming poverty
Technological Development	E24	New materials
	E25	New methods
	E34	Better process
	E26	New technologic (computer sciences, mechanics, electronics, chemistry, etc)
	E27	Innovation
Mass Media	E28	Radio
	E29	Television
	E30	News
	E31	Internet: Web 2.0
	E32	Reviews, journals

– Source: Owner Elaboration

3.1.1 Definition of Variables: Taxonomy of the Causes of Physical Activity and Sport

C1. Organized Physical Activity: is done within the framework of an organization and under the supervision of a person in charge of conducting the activity (technical, sports, coach, monitor, to among others) outside school hours at least once a week (Institut Barcelona Esports, 2011).

C2. Unorganized Physical Activity also called spontaneous: is characterized by independently execution, is outside the framework of an organization and without the supervision of a person in charge of conducting the activity outside school hours and at least once a week (Institut Barcelona Esports, 2011).

C3. Competitive Sports: It is which is practiced with the intention of defeating on opponent or to surpass himself (Blázquez, 1999).

C4. Professional Sports: Accepts natural's person like competitors with remuneration, in accordance with the standards of the International Federation (Blázquez, 1999).

C5. Sport and Business: it is the involvement of companies in associations or in collaboration of any organization for to work together in development of aims through sport and harness the power of sport to promote its reputation, risk management in a context of internationalization or marketing or social investment (May and Phelan, 2005).

C6. Sport as entertainment: Beginning the base of competitive is: search great sporting results with competitive demands high, where athletes are considered professionals and perceived pressure and influence of socioeconomic and sociopolitical demands (Cagigal, 1981) likewise can include mega sporting events in this definition.

C7. Sports Marketing: Sports marketing is comprised of various activities are designed to analyze the wishes and needs of consumers through exchange processes. It has developed two main objectives: marketing of sports products and services aimed at consumers of sport and marketing to other consumers of industrial products and service through sports promotions (Mullin et al., 2007).

C8. Adapted Sports: Sports specialty make use of alternatives others to the usual to be able to be practiced by athletes different than usual (Garcia, 2004)

C9. Sport Educational: Is one whose fundamental intention is to contribute to the harmonious development and strengthen the values of the individual. This activity is important in school-age individuals (Blázquez, 1999).

C10. Sports of Initiation: The teaching-learning process, followed by an individual, for the acquisition to the capacity for knowledge and practice of a sport (Blázquez, 1999).

C11. Sports Management: Can be defined as the sum of operations, business and marketing techniques which are developed to achieve a maximum level of functioning and quality of sports entity (Gutiérrez, 1996).

C12. Sports Law and Policy: it is regarding legislation applied by governments and government agencies to promote regulate and control the sport in economic, competitive, organizational and equality (European Commission, 2007).

3.1.2 Definition Variables: Taxonomy Areas for Action and Its Social Effects

E1-E3: Physical Fitness: In relation to physical fitness is the ability that the person has to perform tasks that demand their daily lives in order to improve quality of life, which has three classifications: **Physical Fitness health:** basic attributes such as cardio respiratory endurance, muscular strength, muscular endurance, body composition and flexibility like components that allow to promote health and wellness. **Performance fitness**: have aim high performance sports with motor skills specific competitive activity or sport. **Physiological fitness:** Reference the operation of biological systems such as metabolism, morphology and integrity of bones, and with minimum increments of physical activity, these systems can have a significant improvement (Sánchez, 2006).

E4-E11 y E33: Welfare: It is a positive component of health, a subcategory, which can be referred to as a state of being and reflects the individual's ability to enjoy life successfully, therefore, feeling good: the physical, social, intellectual, emotional, spiritual, professional, environmental (Sánchez, 2006) as well as, we can link those reference to social equity and conflict resolution and peace search (UN 2003)

E12-E14 Health: it is self-regulatory processes of the organism against environmental requirements and adapts for enjoy life as we grow, mature, get older, injure and wait for death (Sánchez, 2006). The sports movement as a physical activity beneficial to health (AFBS, HEPA) has a greater influence than any other social movement in: prevention, maintenance, recovery and rehabilitation of non-communicable diseases (NCDs) (European Commission, 2007; O.M.S, 2004)

E15-E17 Education: The multidirectional process which transmit knowledge, values, customs and ways of acting in combination with the sport improves a child's ability to learn, increases concentration, attendance, development of knowledge, motivation, skills and readiness for personal effort, factors that contribute to improved school performance, reduced school failure and the development of new areas (ONU, 2003; UNESCO, 1979).

E18-E21 Economy and Sport: it is an economic process, production, trade, distribution and consumption of goods and services of professional sports. Production factors, like workforce (athletes, coaches and managers) combined with the capital (of sport, equipment, etc.) produce a product which is sold to consumers (spectators and followers) usually in a stadium or media (Downward et al., 2009) dynamic sector is growing rapidly, with macroeconomic impacts which contribute to the growth aims, job creation and tool for local and regional development, urban regeneration or rural development(European Commission, 2007).

E22-E23 Public Budget: The ordered list of revenue and expenditures publics, where their forecasting is embodied in a law passed by the legislative body authorizing the execution by the government (Gimeno et al., 2008). The lack of physical activity increases the incidence in: overweight, obesity and chronic diseases like cardiovascular disease or diabetes, which reduce the quality of life, threatening the lives of people and represent a burden on health budgets and the economy (European Commission, 2007).The sports industry and hosting major sporting events, generate employment opportunities will help fight poverty (ONU, 2003).

E24-E27 E34. Technological Development: The way to find and solve problems or generate an idea on new materials, methods, new technologies or improvement product, production and management processes and become an innovation, which develop from contact with science and socio-economic environment that surrounds it (Escorsa and Valls, 2003).

E28-E32 Mass Media: Refers to the instrument or form of content which makes the communication process or communication. The sport has been a driving force in the traditional media (radio, press, review, etc.), the arrival of new media (Internet: Web2.0) and interactive TV services (European Commission, 2007).

4 Results

Proceed to show the results obtained of the combinatorial process and detailed each of the intermediate steps performed according to the model described above[1]:

Table 3 Estimated incidence between causes and effects $\left[\underset{\sim}{M}\right]$

	E_1	E_2	E_3	E_4	E_5	E_6	E_7	E_8	E_9	E_{10}	E_{11}	E_{12}	E_{13}	E_{14}	E_{15}	E_{16}	E_{17}	E_{18}	E_{19}	E_{20}	E_{21}	E_{22}	E_{23}	E_{24}	E_{25}	E_{26}	E_{27}	E_{28}	E_{29}	E_{30}	E_{31}	E_{32}	E_{33}	E_{34}
C_1	1	0,8	0,8	0,7	0,9	0,7	0,8	0,7	1	1	0,9	1	1	1	1	1	0,7	0,9	0,6	0,7	1	1	0,8	0,6	0,6	0,6	0,8	0,9	0,8	0,7	0,7	0,9	0,8	1
C_2	0,8	0,9	1	1	1	1	0,9	0,7	1	1	1	0,7	0,7	0,4	0,7	0,7	0,9	0,6	0,6	0,7	0,7	1	0,9	0,6	0,7	0,6	0,7	0,9	0,4	0,4	0,4	1	0,9	0,6
C_3	1	0,9	1	0,8	1	0,7	0,8	0,8	1	1	1	0,9	0,9	1	1	0,8	1	0,8	0,8	0,8	1	1	0,8	1	1	0,7	0,8	1	0,6	0,6	0,6	0,9	1	1
C_4	1	1	1	1	1	1	1	0,7	1	1	1	1	1	0,6	1	1	0,7	0,6	0,6	0,6	0,7	1	0,8	0,4	0,7	0,4	1	0,4	0,3	0,3	0,3	1	1	1
C_5	1	1	1	0,7	1	1	1	0,7	1	1	1	1	1	0,9	0,9	0,9	0,7	0,6	0,6	0,6	0,7	1	0,8	0,6	0,9	0,6	1	0,4	0,4	0,4	0,4	1	1	1
C_6	1	0,7	0,7	0,7	0,7	0,7	0,7	0,7	0,7	0,6	0,8	0,6	0,6	1	0,9	1	1	0,9	0,9	0,9	1	1	0,7	0,6	0,9	0,7	1	0,7	0,9	0,8	0,8	1	0,7	1
C_7	1	0,3	0,3	1	1	1	0,4	1	0,7	0,3	1	0,2	0,2	1	0,2	0,2	1	1	1	1	1	0,7	0,2	1	1	1	1	1	1	1	1	1	0,3	1
C_8	0,1	0,1	0,1	0,1	0,1	0,1	0,1	1	0,7	0	0	0	0	0	0	0	1	1	1	1	1	0,8	0	1	1	1	1	1	1	1	1	1	0,1	0
C_9	0	0	0	0	0,8	0	0	1	0,7	0,8	0,9	0,9	0,9	0,7	0,3	0,3	0,9	1	1	1	1	0,7	0	0,6	0,6	0,6	1	1	1	1	1	1	0	0
C_{10}	0,9	0,9	0,9	0,9	0,9	0,9	0,9	1	0,9	1	1	0,8	0,8	0,8	0,2	0,2	1	1	1	1	1	0,6	0,9	0,7	0,7	0,6	1	0,7	0,4	0,4	0,4	1	0,9	0,8
C_{11}	0,2	0,2	0,2	0,9	0,9	0,9	0,9	1	1	1	1	1	0,8	1	0,1	0,1	0,1	0,8	0,8	0,8	0,8	1	1	0,1	0,1	0,1	1	0,1	0,1	0,8	0,8	0,8	1	0,9
C_{12}	1	0,2	0,2	0,9	1	1	0,3	1	1	0,2	1	0,2	0,2	1	0,2	0,2	0,9	1	1	1	1	0,7	0,2	1	1	1	1	1	0,8	0,9	1	1	0,2	1

Table 4 Convulacion among matrices max-min: $\left[\underset{\sim}{A}\right]\circ\left[\underset{\sim}{M}\right]$

	E_1	E_2	E_3	E_4	E_5	E_6	E_7	E_8	E_9	E_{10}	E_{11}	E_{12}	E_{13}	E_{14}	E_{15}	E_{16}	E_{17}	E_{18}	E_{19}	E_{20}	E_{21}	E_{22}	E_{23}	E_{24}	E_{25}	E_{26}	E_{27}	E_{28}	E_{29}	E_{30}	E_{31}	E_{32}	E_{33}	E_{34}
C_1	1	1	1	1	1	1	1	1	1	1	1	1	1	1	1	1	1	1	1	1	1	1	0,9	1	1	0,8	1	1	0,8	0,8	0,8	1	1	1
C_2	0,9	0,9	1	1	1	1	0,9	1	1	1	1	0,9	0,9	0,9	0,9	0,8	1	1	1	1	1	1	0,9	0,9	0,9	0,7	1	0,9	0,6	0,7	0,7	1	0,9	0,9
C_3	1	0,9	1	0,9	1	0,9	0,9	1	1	1	1	0,9	0,9	1	1	0,9	1	1	1	1	1	1	0,9	1	1	0,8	1	1	0,9	0,8	0,8	1	1	1
C_4	1	1	1	1	1	1	1	1	1	1	1	1	1	1	1	1	1	1	1	1	1	1	0,9	0,9	0,9	0,9	1	0,9	0,9	0,9	0,9	1	1	1
C_5	1	1	1	1	1	1	1	1	1	1	1	1	1	1	1	1	1	0,9	0,9	0,9	1	1	1	1	1	0,9	1	1	0,9	0,9	0,9	1	1	1
C_6	1	1	1	1	1	1	1	1	1	1	1	1	1	1	1	1	1	1	1	1	1	1	1	1	1	1	1	1	0,9	0,9	1	1	1	1
C_7	1	1	1	1	1	1	1	1	1	1	1	1	1	1	1	1	1	1	1	1	1	1	1	1	1	1	1	1	1	1	1	1	1	1
C_8	1	0,9	0,9	1	1	1	0,9	1	1	1	1	0,9	0,9	1	0,7	0,7	1	1	1	1	1	0,8	0,9	1	1	1	1	1	1	1	1	1	0,9	1
C_9	1	1	1	1	1	1	1	1	1	1	1	1	1	1	1	1	1	1	1	1	1	1	0,9	1	1	1	1	1	1	1	1	1	1	1
C_{10}	1	1	1	1	1	1	1	1	1	1	1	1	1	1	1	1	1	1	1	1	1	1	0,9	1	1	1	1	1	1	1	1	1	1	1
C_{11}	1	1	1	1	1	1	1	1	1	1	1	1	1	1	1	1	1	1	1	1	1	1	1	1	1	1	1	1	1	1	1	1	1	1
C_{12}	1	0,9	0,9	1	1	1	0,9	1	1	1	1	1	0,9	1	0,9	1	1	1	1	1	1	1	1	1	1	1	1	1	1	1	1	1	1	1

[1] On this occasion we have proceeded to use the model to perform FUZZYLOG calculations as the amount of variables incorporated in the process.

Table 5 convaluacion among matrices max-min: $[\underset{\sim}{A}] \circ [\underset{\sim}{M}] \circ [\underset{\sim}{B}] = [\underset{\sim}{M}^{*}]$

↱	E1	E2	E3	E4	E5	E6	E7	E8	E9	E10	E11	E12	E13	E14	E15	E16	E17	E18	E19	E20	E21	E22	E23	E24	E25	E26	E27	E28	E29	E30	E31	E32	E33	E34
C1	1	1	1	1	1	1	1	1	1	1	1	1	1	1	1	1	1	1	1	1	1	1	1	1	1	1	1	1	1	1	1	1	1	1
C2	1	1	1	1	1	1	1	1	1	1	1	1	1	1	1	1	1	1	1	1	1	1	1	1	1	1	1	1	1	1	1	1	1	1
C3	1	1	1	1	1	1	1	1	1	1	1	1	1	1	1	1	1	1	1	1	1	1	1	1	1	1	1	1	1	1	1	1	1	1
C4	1	1	1	1	1	1	1	1	1	1	1	1	1	1	1	1	1	1	1	1	1	1	1	1	1	1	1	1	1	1	1	1	1	1
C5	1	1	1	1	1	1	1	1	1	1	1	1	1	1	1	1	1	1	1	1	1	1	1	1	1	1	1	1	1	1	1	1	1	1
C6	1	1	1	1	1	1	1	1	1	1	1	1	1	1	1	1	1	1	1	1	1	1	1	1	1	1	1	1	1	1	1	1	1	1
C7	1	1	1	1	1	1	1	1	1	1	1	1	1	1	1	1	1	1	1	1	1	1	1	1	1	1	1	1	1	1	1	1	1	1
C8	1	1	1	1	1	1	1	1	1	1	1	1	1	1	1	1	1	1	1	1	1	1	1	1	1	1	1	1	1	1	1	1	1	1
C9	1	1	1	1	1	1	1	1	1	1	1	1	1	1	1	1	1	1	1	1	1	1	1	1	1	1	1	1	1	1	1	1	1	1
C10	1	1	1	1	1	1	1	1	1	1	1	1	1	1	1	1	1	1	1	1	1	1	1	1	1	1	1	1	1	1	1	1	1	1
C11	1	1	1	1	1	1	1	1	1	1	1	1	1	1	1	1	1	1	1	1	1	1	1	1	1	1	1	1	1	1	1	1	1	1
C12	1	1	1	1	1	1	1	1	1	1	1	1	1	1	1	1	1	1	1	1	1	1	1	1	1	1	1	1	1	1	1	1	1	1

Table 6 Forgotten effects: $[\underset{\sim}{O}] = [\underset{\sim}{M}^{*}] (-) [\underset{\sim}{M}]$

↱	E1	E2	E3	E4	E5	E6	E7	E8	E9	E10	E11	E12	E13	E14	E15	E16	E17	E18	E19	E20	E21	E22	E23	E24	E25	E26	E27	E28	E29	E30	E31	E32	E33	E34
C1	0	0,2	0,2	0,3	0,1	0,3	0,2	0,3	0	0	0,1	0	0	0	0	0	0,3	0,1	0,4	0,3	0	0	0,2	0,4	0,4	0,4	0,2	0,1	0,2	0,3	0,3	0,1	0,2	0
C2	0,2	0,1	0	0	0	0	0,1	0,3	0	0	0	0,3	0,3	0,6	0,3	0,3	0,1	0,4	0,4	0,3	0,3	0	0,1	0,4	0,3	0,4	0,3	0,1	0,6	0,6	0,6	0	0,1	0,4
C3	0	0,1	0	0,2	0	0,3	0,2	0,2	0	0	0	0,1	0,1	0	0	0,2	0	0,2	0,2	0,2	0	0	0,2	0	0	0,3	0,2	0	0,4	0,4	0,4	0,1	0	0
C4	0	0	0	0	0	0	0	0,3	0	0	0	0	0	0,4	0	0	0,3	0,4	0,4	0,4	0,3	0	0,2	0,6	0,3	0,6	0	0,6	0,7	0,7	0,7	0	0	0
C5	0	0	0	0,3	0	0	0	0,3	0	0	0	0	0	0,1	0,1	0,1	0,3	0,4	0,4	0,4	0,3	0	0,2	0,4	0,1	0,4	0	0,6	0,6	0,6	0,6	0	0	0
C6	0	0,3	0,3	0,3	0,3	0,3	0,3	0,3	0,3	0,4	0,2	0,4	0,4	0	0,1	0	0	0,1	0,1	0,1	0	0	0,3	0,4	0,1	0,3	0	0,3	0,1	0,2	0,2	0	0,3	0
C7	0	0,7	0,7	0	0	0	0,6	0	0,3	0,7	0	0,8	0,8	0	0,8	0,8	0	0	0	0	0	0,3	0,8	0	0	0	0	0	0	0	0	0	0,7	0
C8	0	0	0	0	0	0	0	0	0,3	1	1	1	1	1	1	1	0	0	0	0	0	0,2	1	0	0	0	0	0	0	0	0	0	0	1
C9	1	1	1	1	0,2	1	1	0	0,3	0,2	0,1	0,1	0,1	0,3	0,7	0,7	0,1	0	0	0	0	0,3	1	0,4	0,4	0,4	0	0	0	0	0	0	1	1
C10	0,1	0,1	0,1	0,1	0,1	0,1	0,1	0	0,1	0	0	0,2	0,2	0,2	0,8	0,8	0	0	0	0	0	0,4	0,1	0,3	0,3	0,4	0	0,3	0,6	0,6	0,6	0	0,1	0,2
C11	0,8	0,8	0,8	0,1	0,1	0,1	0,1	0	0	0	0	0	0,2	0	0	0	0	0,2	0,2	0,2	0,2	0	0	0	0	0	0	0	0	0,2	0,2	0,2	0	0,1
C12	0	0,8	0,8	0,1	0	0	0,7	0	0	0,8	0	0,8	0,8	0	0,8	0,8	0,1	0	0	0	0	0,3	0,8	0	0	0	0	0	0,2	0,1	0	0	0,8	0

4.1 Analysis

Proceed to explain some relevant causal relationships and meaningful in the context of the aims pursued in this study. Starting of incidence relationships which were not detected by previous approaches and are importance for research, several highlights:

The causality relationship between the Sports and Business and the positive impact on healthcare budgets is established, through, of the relevant key relationship

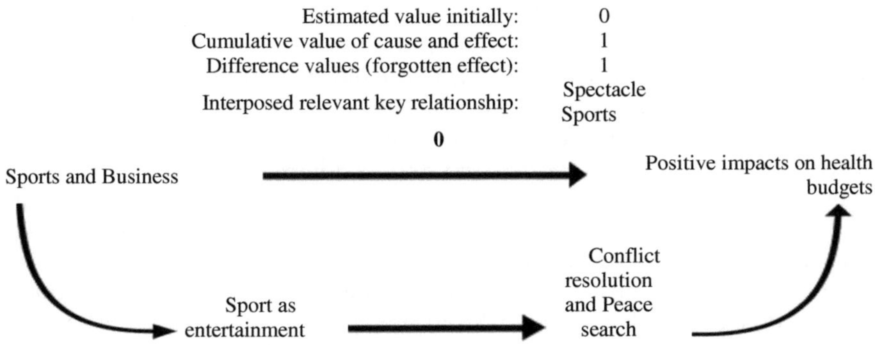

Fig. Graphic 1

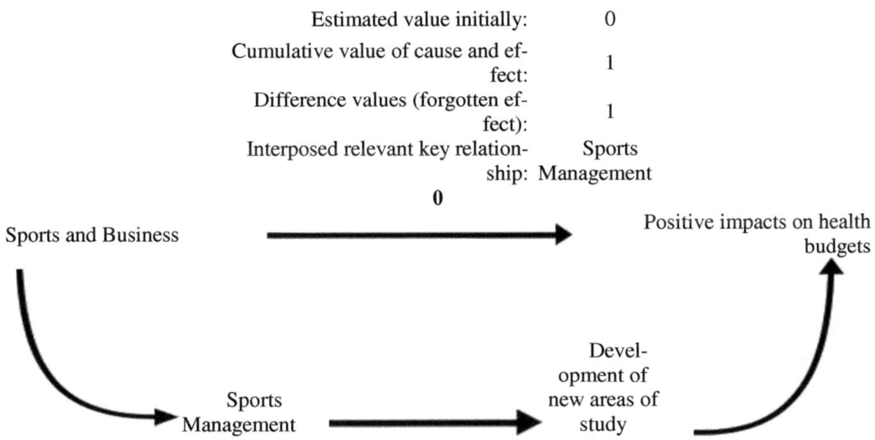

Fig. Graphic 2

with the Sport as entertainment which have direct incidence the resolution of conflicts and the search for peace. The Sport as entertainment and mega-sports events, its effects are classed usually across four balanced perspectives: economic impact, sports development, media and sponsor evaluation, and place marketing effects (Dolles and Soderman, 2008), however, this classification does not explain every effects possible. The effect of media in sport can be used as an effective mechanism for knowledge transfer and capacity in areas such as peace, tolerance, understanding and respect for opponents, despite the ethnic, cultural , religious or other nature(ONU, 2003), preventing youth crime, participation in armed militias and gangs, with a effect directly proportional in the health budgets of a nation, such as lower costs of policing and surveillance, a lower cost to victims of crime through to decrease crime rates and reduction of street gangs or militia activities(Right to Play, 2008).

The causality relationship between the Sports and Business and the positive impact on healthcare budgets is established, through, of the relevant key relationship with the sports management which have direct incidence the development of new areas of study. The sports management how career is developed in the organization, marketing, implementation and evaluation concerning of a sport-related activity, such as, secondary or college sports, charities non-profit that are linked in sports, corporate events, sports organizations (national and international), sports and mega-sports events or in the trade(Ratten, 2010) although, it is recent study area in last years has been rapid growth in various fields, circumstance that influences to the development of new areas of study and research, which influences highest Productivity Economic and a lowest social cost, thanks to an more educated youth and better capacity to obtain employment(Right to Play, 2008).

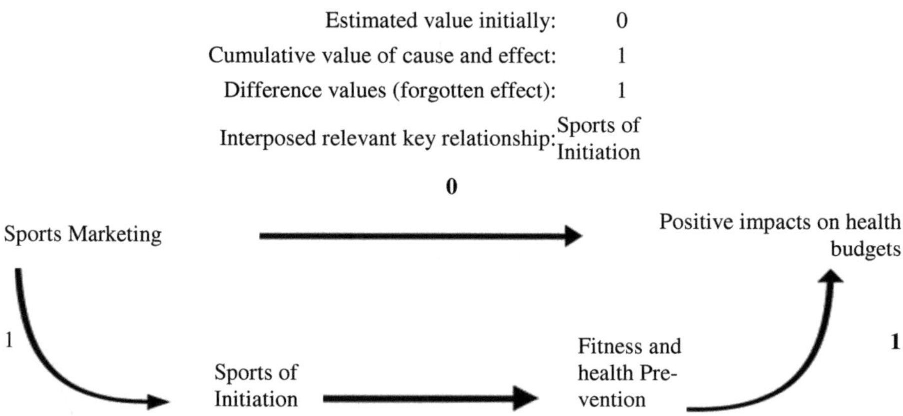

Fig. Graphic 3

The causality relationship between the Sports marketing and the positive impact on healthcare budgets is established, through, of the relevant key relationship with the sports of initiation which have direct incidence fitness and health prevention. To explain this relationship, sports marketing classify it in three dimensions: emotional, symbolic, environmental with active participation of consumers(Ratten, 2010) dimensions, that positively influence the initiation of individual sports activities, which generate feelings of belonging and membership of a specific sport as well as the benefits of a healthy and active life, establishing a relationship to the prevention of physical and emotional health, particularly of non-communicable diseases that are affecting more young people(O.M.S, 2004), this are reflected in lower costs related Health Care with Chronic Diseases and Mental Disorders, lower payroll costs for the employer due to reduced absenteeism, reduced charge for people caused by the loss of its job and Sickness Medical costs, and Economic Benefits from increased Productivity(Right to Play, 2008).

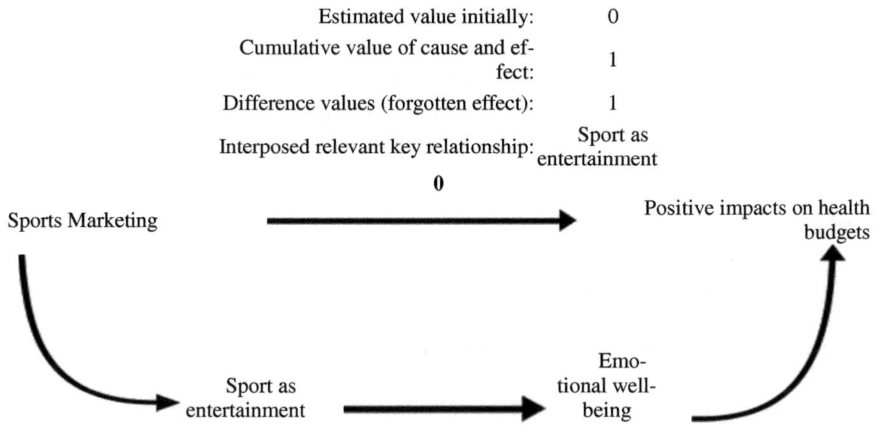

Fig. Graphic 4

The causality relationship between the Sports marketing and the positive impact on healthcare budgets is established, through, of the relevant key relationship with the Sport as entertainment which has direct incidence emotional well-being. sport as entertainment and sports marketing have an important emotional component that is used for different purposes, particularly the commercial, however, it also has great impact on the mood of people to generate senses of belonging and affiliation, issues related social integration, social inclusion, social capital and social cohesion refered to "feel-good" processes of states of being, as opposed to the 'dark sides' of segregation, exclusion, isolation and fragmentation(Bottenburg and Sterkenburg, 2005), concepts that are linked to an individual's emotional wellbeing and influence the productivity individual, therefore, an emotionally stable person has a higher throughput, reduce the possibility of absence from work due to incapacity, which is positive for both the company and for public health care, which will have fewer cases of emotional instability or mental illness(Right to Play, 2008).

5 Conclusions

The study developed as the development and promotion of Sports Marketing and Sports and Business have influence positively in social expenditure relating to public health budgets, without having to affect the state's welfare population.

Second, it shows as the sports marketing and sports and business might influence social expenditure significantly in a nation, considering the synergies created by other factors in society and the relationship between them.

The public administration budgeting and assignment of resources is based on the anticipation of potential and future needs of the population, considering was done in preceding period. Nowadays, these budgets have been affected by cuts to cope with major economic needs of the state, mainly affecting welfare state of the population and causing a sequential effect on the population and in the economy where sports is directly affected.

The results arising from the application of the model will allow decision making about the investment selection process or budgetary aims. Thereby, assign minimum resources in activities which produce the maximum benefit.

References

Blázquez, D., Ramírez, F.: La iniciación deportiva y el deporte escolar. Inde, Barcelona (1999)

Bottenburg, M., Sterkenburg, J.: Consultation conference with the european sport movement on the funtion social of sport, volunteering in sport and the fight against doping. European Commission Web (2005), http://ec.europa.eu/sport/library/documents/b1/doc322_en.pdf (accessed December 12, 2011)

Cagigal, J.M.: Deporte: espectáculo y acción. Salvat, Madrid (1981)

Dolles, H., Soderman, S.: Mega-sporting events in Asia - impacts on society, business and management: An introduction. Asian Business and Management 7, 147–162 (2008)

Downward, P., Alistair, D., Trudo, D.: Sports economics theory, evidence and policy. Butter-worth-Heinemann Publications, Great Britain (2009)

Escorsa, P., Valls, J.: Tecnológia e innovación en la empresa. Ediciones UPC, Barcelona (2003)

European Commission: White papper on sport. European Commission Web (2007), http://ec.europa.eu/sport/documents/white-paper/whitepaper-full_en.pdf (accessed November 29, 2011)

García de Mingo, J.M.: El deporte adaptado en el ámbito escolar. Educación y Futuro Revista de investigación aplicada y experiencias educativas 10, 81–90 (2004)

Gil-Lafuente, A.M.: Fuzzy logic in financial analysis. Springer, Berlin (2005)

Gil-Lafuente, A.M.: Nuevas Estrategias para el análisis financiero en la empresa. Arial, Barcelona (2001)

Gil-Lafuente, A.M., Barcellos de Paula, L.: Una aplicación de la metodología de los Efectos Olvidados: Los factores que contribuyen al crecimiento sostenible de la Empresa. Cuadernos del Centro de Investigación en Metodología Borrosa Aplicada a la Gestión y Economía; Cuadernos CIMBAGE 12, 23–52 (2010)

Gimeno, J., Guirola, J., De la Concepción, M., Ruiz, J.: Principios de Economía. McGraw Hill, Madrid (2008)

Gutiérrez, J.F.: Administración Deportiva. Physical Education and Sport Journal 18, 101–107 (1996)

Institut Barcelona Esports: Estudi del hàbits esportius escolars a Barcelona, Ajuntamiento de Barcelona, Barcelona (2011)

May, G., Phelan, J.: Shared Goals: Sport and business in partnerships for development. Tool kit sport development Web (2005), http://www.toolkitsportdevelopment.org/html/resources/9A/9A141603-39AE-4360-8749-FB2BA9B54676/shared_goals_1.pdf (accessed December 01, 2011)

Mullin, B., Hardy, S., Sutton, W.: Marketing Deportivo. Paidotribo, Barcelona (2007)

ONU.: El año como el Año Internacional del Deporte y la Educación Física: "El deporte como medio para fomentar la Educación, la Salud, el Desarrollo y la Paz". Unit Nation Web (2003), http://www.un.org/spanish/sport2005/concepto.html (accessed November 06, 2011)

O.M.S: Estrategia mundial sobre régimen alimentario, actividad física y salud. Estrategia mun-dial sobre régimen alimentario, actividad física y salud. WHO Web (2004), http://www.who.int/dietphysicalactivity/strategy/eb11344/strategy_spanish_web.pdf (accessed November 6, 2011)

Ratten, V.: The future of sports management: A social responsibility, philanthropy and entrepreneurship perspective. Journal of Management and Organization 16, 488–494 (2010)

Right to Play: Aprovechamiento del poder del deporte para el desarrollo y la paz: Recomendaciones a los gobiernos. Informe final grupo internacional de trabajo sobre el deporte para el desarrollo y la paz. Unit Nation Web (2008), http://www.un.org/wcm/webdav/site/sport/shared/sport/pdfs/Reports/Final_Report_Spanish.pdf (accessed December 1, 2011)

Sánchez, J.C.: Definición y Clasificación de Actividad y Salud. Actividad fisca Web (2006), http://www.actividadfisica.net/actividad-fisica-definicion-clasificacion-actividad-fisica.html (accessed November 6, 2011)

UNESCO: Carta Internacional de la Educación física y el deporte. UNESCO Web (1979), http://unesdoc.unesco.org/images/0011/001140/114032s.pdf#page=30 (accessed November 06, 2011)

Author Index

MIX
Papier aus verantwortungsvollen Quellen
Paper from responsible sources
FSC® C105338

If you have any concerns about our products,
you can contact us on
ProductSafety@springernature.com

In case Publisher is established outside the EU,
the EU authorized representative is:
Springer Nature Customer Service Center GmbH
Europaplatz 3, 69115 Heidelberg, Germany

Printed by Libri Plureos GmbH
in Hamburg, Germany